Contractual Procedures in the Construction Industry

Contractual Procedures in the Construction Industry, Seventh Edition aims to provide students with a comprehensive understanding of the subject, and reinforces the changes that are taking place within the construction industry. The book looks at contract law within the context of construction contracts, it examines the different procurement routes that have evolved over time and the particular aspects relating to design and construction, lean methods of construction and the advantages and disadvantages of PFI/PPP and its variants. It covers the development of partnering, supply chain management, design and build and the way that the clients and professions have adapted to change in the procurement of buildings and engineering projects.

This book is an indispensable companion for students taking undergraduate courses in Building and Surveying, Quantity Surveying, Construction Management and Project Management. It is also suitable for students on HND/C courses in Building and Construction Management as well as foundation degree courses in Building and Construction Management.

Key features of the new edition include:

- A revised chapter covering the concept of value for money in line with the greater emphasis on added value throughout the industry today.
- A new chapter covering developments in information technology applications (building information modelling, blockchains, data analytics, smart contracts and others) and construction procurement.
- Deeper coverage of the strategies that need to be considered in respect of contract selection.
- Improved discussion of sustainability and the increasing importance of resilience in the built environment.
- Concise descriptions of some of the more important construction case laws.

Allan Ashworth was, until recently, a visiting professor at the University of Malaya, Malaysia and also at the University of Salford, UK, where he was involved with the Centre for Education in the Built Environment.

Srinath Perera is the Chair and Professor of Built Environment and Construction Management at Western Sydney University, Sydney, Australia. He is also the Director of Centre for Smart Modern Construction (c4SMC). He is a Chartered Quantity Surveyor and Member of the Royal Institution of Chartered Surveyors (RICS) with over 25 years' experience in both industry and academia. He is a Fellow of both the Royal Society of NSW and the Australian Institute of Building and is also a member of the Australian Institute of Quantity Surveyors.

Contractual Procedures in the Construction Industry

Seventh Edition

Allan Ashworth and Srinath Perera

LONDON AND NEW YORK

Seventh edition published 2018
by Routledge
2 Park Square, Milton Park, Abingdon, Oxon, OX14 4RN

and by Routledge
711 Third Avenue, New York, NY 10017

Routledge is an imprint of the Taylor & Francis Group, an informa business

First edition published by Pearson Education 1985
Sixth edition published by Routledge 2013

British Library Cataloguing-in-Publication Data
A catalogue record for this book is available from the British Library

Library of Congress Cataloging-in-Publication Data
Names: Ashworth, Allan, 1944- author. | Perera, Srinath, author.
Title: Contractual procedures in the construction industry/Allan Ashworth and Srinath Perera.
Description: Seventh edition. | Milton Park, Abingdon, Oxon; New York, NY: Routledge, 2018. | Includes bibliographical references and index.
Identifiers: LCCN 2017042239 | ISBN 9781138693920 (hardback : alk. paper) | ISBN 9781138693937 (pbk. : alk. paper) | ISBN 9781315529059 (ebook : alk. paper)
Subjects: LCSH: Construction contracts–England. | Construction industry–Law and legislation–England.
Classification: LCC KD1641 .A985 2018 | DDC 343.4207/8624–dc23
LC record available at https://lccn.loc.gov/2017042239

ISBN: 978-1-138-69392-0 (hbk)
ISBN: 978-1-138-69393-7 (pbk)
ISBN: 978-1-315-52905-9 (ebk)

Typeset in Times New Roman
by Sunrise Setting Ltd., Brixham, UK

Contents

Figures

Tables

Boxes

Preface

'Nothing endures but change', Heraclitus *c* 540–480 BC.

This book, which is now in its seventh edition, was first written in 1985, now over thirty years ago. There is hardly any resemblance between that edition and this newly revised edition. The above quotation captures not only this book, but the world in which we live. In any lifetime, there is a sense that change appears to increase more rapidly as time goes by. The Russian economist, N. D. Kondratiev, developed a concept of long-wave developments, each of less than fifty years' duration. This is a simplified picture of the world in which we live, but illustrates the great inventions which include steam power, railways, electricity, automobiles, petrochemicals and electronics. There are of course many other items that can be added to this list. According to his thinking, we are at the crest of yet further innovation: intelligent machines. When we look back at history we can see the impact that the above new ideas had on their generations. Historians in the future will look back at the impact of information communication technologies in the world in which we now live. This has an impact on so many areas of our lives today and it is difficult to keep abreast of the rapid changes that this is making. Its impact is so great that a new chapter has been added to this book dealing with how this is affecting the way in which we procure buildings.

The construction industry is sometimes thought of as lagging behind, for example, the manufacturing industry. Certainly, the latter got to grips with computer-controlled machines long before the construction industry got involved. But let us not assume that this industry is dormant. Far from it! One has to look at the manner in which new construction projects are designed and built to see advances in ideas and processes that were hitherto unknown. So, it is with the procurement of construction projects. The internet can be identified as one of the single sources of most social and lifestyle change in the current time. It is true of business processes and construction, too. The future will see the propagation of technologies having a greater impact on procurement itself. This book attempts to explore some of the possibilities in its new added chapter.

There are lots of vested interests in the best way to procure construction projects successfully today. There remains a great deal of variation in approaches, not just in the UK but throughout the world. This suggests that we have not found a panacea solution even for projects in the broad ranges of small, medium and large. One might have expected that by now the best of the different ideas from around the world would have been condensed into a relatively few safe and secure procurement systems, but that is not so. It is also a fallacy to assume that one day we will, by a process of evolution, arrive at the perfect system. The different vested interests will see to that, but for very good reasons of experience, practice and familiarity.

Hence the need for this book that explores the principles involved in the procurement of construction projects and what needs to be properly considered to ensure that a construction project is built with a minimum of fuss and effort. It recognises that the process of constructing is changing and will continue to do so as new ideas, skills, materials and manufacturing processes are employed. Time, cost and quality will remain the three important targets to satisfy. I am sometimes bemused at some of the ideas from the past that have now fallen into disuse. These still get a mention in this book, but I never imagined at the time that they would work properly and catch on. So it is with practices today. Without being arrogant, we need to remember that the customer, the client, is always right and it is important to develop construction procurement with this in mind. We also need to remember that we are human beings, with all of our frailties, and these need to be accounted for within any system recommended by those in positions of authority.

Allan Ashworth, York, England
Srinath Perera, Sydney, Australia
2017

Acknowledgements

We are grateful to the following for permission to reproduce copyright material:

Figures

Figures 19.1 to 19.3 from *Building on Success: The Future Strategy for Achieving Excellence in Construction* (Cabinet Office and Office of Government Commerce (2003)), available at http://webarchive.nationalarchives.gov.uk/20100503135839/http://www.ogc.gov.uk/index.

Tables

Tables 7.1, 7.2, 7.3, 7.4 and 7.5 adapted from *Contracts in Use Survey* (RICS 2010); Table 12.1 from *Added Value in Design and Construction* (Ashworth and Hogg 2000); Tables 17.2, 17.3 and 17.4 from *Construction Skills Foresight Report 2003* (CITB 2003); Tables 19.2 and 19.2 from *Rethinking Construction: The Report of the Construction Task Force* (Egan 1998), Crown copyright material is reproduced with the permission of the Controller of Her Majesty's Stationery Office and the Queen's Printer for Scotland; Table 19.1 from 'The Reporting of Injuries, Diseases and Dangerous Occurrences Regulations 1995', which contains public sector information published by the Health and Safety Executive and licensed under the Open Government Licence v1.0.

Text

Chapter 13 case studies, 'Pacific Contracting of San Francisco' and 'Neenan Company, Colorado' from *Rethinking Construction: The Report of the Construction Task Force* (Egan 1998), Crown copyright material is reproduced with the permission of the Controller of Her Majesty's Stationery Office and the Queen's Printer for Scotland; Boxes 17.1, 17.2, 17.3 and 17.4 from *The Standard Form of Building Contract* (The Joint Contracts Tribunal 2016).

In some instances, we have been unable to trace the owners of copyright material and we would appreciate any information that would enable us to do so.

Table of relevant statutes (Acts of Parliament)

A complete list of statutes can be found in Halsbury's *Statutes of England and Wales* (1991)

Administration of Justice Act 1985
Arbitration Acts 1934, 1950, 1975, 1979, 1996
Architects (Registration) Acts 1931, 1938
Architects Registration (Amendment) Act 1969
Banking Act 2009
Bankruptcy Acts 1869, 1914, 1966
Bills of Exchange Act 1882
Bills of Sale Act 1891
Building Act 1984
Building Societies Acts 1986, 1997
Capital Allowances Act 2001
Charities Acts 1992, 1993, 2006
Civil Jurisdiction and Judgments Act 1982
Civil Liability Acts 1978, 2003
Companies Act 2006
Compulsory Purchase Act 1965
Construction (Design and Management) Regulations 1994, 2007, 2015
Construction Acts 1996, 2009
Consumer Credit Act 2006
Consumer Protection Act 1987
Contempt of Court Act 1981
Contracts (Applicable Law) Act 1990
Contracts (Rights of Third Parties) Act 1999
Control of Pollution Act 1974
Copyright, Design and Patents Act 1988
Corporation Acts 1661, 2001
Corporation Taxes Act 2010
County Court Act 1984
Courts Act 2003
Courts and Legal Services Act 1990
Defective Premises Act 1972
Employer's Liability (Compulsory Insurance) Act 1969
Employment Acts 1946, 2008
Employment Protection Acts 1975, 1978

Employment Relation Acts 1999, 2004
Employment Rights Act 1996
Enterprise Act 2002
Environmental Protection Acts 1990, 1995
European Communities Act 1972
Factories Acts 1937, 1961
Fair Trading Act 1973
Family Law Reform Act 1969
Finance Acts 1972, 2010
Financial Services Act 2010
Government Trading Act 1990
Health Act 2006
Health and Safety at Work Act 1974
Highways Acts 1959, 1980
Housing Acts 1988, 2004
Housing Grants, Construction and Regeneration Act 1996
Income and Corporation Taxes Acts 1988, 2010
Income Tax (Subcontractors in the Construction Industry) Regulations 1993
Income Tax (Subcontractors in the Construction Industry) (Amendment) Regulations 1998
Industrial Development Act 1982
Industrial Development (Financial Assistance) Act 2003
Industrial Training Acts 1964, 1982
Insolvency Acts 1986, 2000
Land Drainage Act 1994
Landlord and Tenant Acts 1927, 1985
Land Registration Act 2003
Late Payment of Commercial Debts (Interest) Act 1998
Latent Damage Act 1986
Law of Property Acts 1925, 1994
Limitation Acts 1939, 1980
Limited Liability Partnerships Acts 2000, 2009
Local Democracy, Economic Development and Construction Act 2009
Local Government Acts 1972, 2003
Local Government and Housing Act 1989
London Building Acts 1667, 1894, 1930, 1939
Magistrates' Court Act 1980
Management of Health and Safety at Work Regulations 1999
Mental Health Act 1959
Misrepresentation Act 1967
National Heritage Acts 1983, 2005
National Insurance Acts 1911, 1965
New Roads and Street Works Act 1991
Noise and Statutory Nuisance Act 1993
Official Secrets Act 1989
Ontario Construction Lien Act 1993
Partnership Act 1980
Patents Act 2004
Planning Act 2008

Planning (Listed Buildings and Conservation) Act 1990
Planning and Compensation Act 1991
Planning and Compulsory Purchase Act 2004
Prevention of Corruption Acts 1889, 1916
Property Misdescriptions Act 1991
Public Health Act 1961
Public Utilities Street Works Act 1950
Rating (Valuation) Act 1999
Restrictive Practices Act 1976
Restrictive Trade Practices Act 1956
Rights of Light Act 1959
Sale and Supply of Goods Act 1994
Sale of Goods Act 1979
Scheme for Construction Contracts (England and Wales) Regulations 1998
Social Security Pensions Act 1975
Solicitors Act 1974
Supply of Goods and Services Act 1982
Sustainable Energy Act 2003
Theft Act 1968
Town and Country Planning Acts 1971, 1990, 2015
Town and Country Planning (Assessment of Environmental Effects) Regulations 1988
Town and Country Planning (Making of Applications) Regulations 2012
Trade Descriptions Act 1968
Unfair Contract Terms Act 1977
Utilities Act 2000
Value Added Tax Act 1994
War Damage Act 1965
Warm Homes and Energy Conservation Act 2000
Water Act 2003
Water Industry Act 1999
Water Resources Act 1991

Part 1

Contract law

1 The English legal system

There are many individuals within the construction industry who will, at some time in their careers, become professionally involved in either litigation, arbitration or adjudication. The laws which are applied in the construction industry are both of a general and a specialist nature. They are general in the sense that they embrace the tenets of law appropriate to all legal decisions, and are special since the interpretation of construction contracts and documents requires a particular knowledge and understanding of the construction industry. Note, however, that the interpretation and application of law will not be contrary to or in opposition to the established legal principles and precedents found elsewhere. It is appropriate at this stage to consider briefly the framework of the English legal system.

The legal system of England and Wales is separate from those of Scotland and Northern Ireland. It differs from them in law, judicial procedure and court structure. However, there is a common distinction between civil law (disputes between individuals) and criminal law (acts harmful to the community). The supreme judicial authority for England and Wales is the Supreme Court (see 1.4.4). This is the ultimate court of appeal from all courts in Great Britain and Northern Ireland (except for criminal courts in Scotland) for all cases except those concerning the interpretation and application of European Community law.

1.1 The nature of law

Law, in its legal sense, may be distinguished from scientific law, or the law of nature, and from the rules of morality. In the first case, scientific laws are not man-made and are not therefore subject to change. In the case of morality, it is less easy to draw a distinction between legal rules and moral precepts. It may be argued, for example, that the legal rules follow naturally from a correct moral concept. The difference between the two is, perhaps, that obedience to law is enforced by the state whereas morals are largely a matter of conscience and conduct. The laws of a country are, however, to some extent an expression of its current morality, since laws can generally only be enforced by a common consensus. Law, therefore, may be appropriately defined as a body of rules for the guidance of human conduct but which may be enforced by the authorities concerned.

1.2 Classification of law

Law is an enormous subject and some specialisation is therefore essential. A complete classification system would require a very detailed chart. Essentially, the basic division in the English legal system is the distinction between criminal and civil law. Usually, the distinction will be obvious. It is the difference between being prosecuted for a criminal offence and being sued for a civil wrong. If the aim of the person bringing the case is to punish the defendant, then it will probably be a criminal case. However, if the aim is to obtain some form of compensation or other benefit, then it will generally be a civil case.

Alternative methods of classification are to subdivide the offences that are committed against persons, property or the state under these headings. Laws may also be classified as either public or private. Public law is primarily concerned with the state itself. Private law is that part of the English legal system which is concerned with the rights and obligations of the individuals.

1.3 Sources of English law

Every legal system has its roots, the original sources from which authority is drawn. The sources of English law can be categorised in the following ways.

1.3.1 Custom

In the development of the English legal system, the common law was derived from the different laws associated with the different parts of the country. These were adapted to form a national law common to the whole country. Since the difference between the regions stemmed from their different customary laws, it is no exaggeration to say that custom was the principal original source of the common law. The term 'custom' has three generally accepted meanings:

- *General custom*: accepted by the country at large.
- *Mercantile custom*: principles established on an international basis.
- *Local custom*: applicable only to certain areas within a country.

The following conditions must be complied with before a local custom will be recognised as law:

- The custom must have existed from 'time immemorial'. The date for this has been fixed as 1189.
- The custom must be limited to a particular locality.
- The custom must have existed continuously.
- The custom must be a reasonable condition in the eyes of the law.
- The custom must have been exercised openly.
- The custom must be consistent with, and not in conflict with, existing laws.

In some countries, the writings of legal authors can form an important source of law. In England, however, because of tradition, such writings have in the past been treated with comparatively little respect. They are therefore rarely cited in the courts. This general rule has always been subject to certain exceptions and there are 'books of authority' which are almost treated as equal to precedents. Many of these books are very old, and in some cases date back to the twelfth century.

1.3.2 Legislation

The majority of new laws are made in a documentary form by way of an Act of Parliament. Statute has always been a source of English law and by the nineteenth century it rivalled decided cases as a source. If statute and common law clash, then the former will always prevail, since the courts cannot question the validity of any Act. The acceptance by the courts of Parliament's supremacy is entirely a matter for history. Today, it is the most important new source of law because:

- The complex nature of commercial and industrial life has necessitated legislation to create the appropriate organisations and legal framework.
- Modern developments such as drugs and the motor car have necessitated legislation to prevent their abuse.
- There are frequent changes in the attitudes of modern society, such as that relating to females, and the law must thus keep in step with society.

Before a legislative measure can become law, it must undergo an extensive process:

- The measure is first drafted by civil servants who present it to the House of Commons or the House of Lords as a bill.
- The various clauses of the bill will already have been accepted and agreed by the appropriate government department prior to its presentation.

Before the bill can become an Act of Parliament, it must undergo five stages in each house:

- *First reading*: the bill is introduced to the house.
- *Second reading*: a general debate takes place upon the general principles of the measure.
- *Committee stage*: each clause of the bill is examined in detail.
- *Report stage*: the house is brought up to date with the changes that have been made.
- *Third reading*: only matters of detail are allowed to be altered at this stage.

The length of time which is necessary for the bill to pass through these various stages depends upon the nature and length of the bill and how politically controversial it is. Once the bill has been approved and accepted by each house, it then needs the royal assent for it to become law.

A public bill is legislation which affects the public at large and applies throughout England and Wales. Scottish law is similar to English law, but it is not exactly the same. A private bill is legislation affecting only a limited section of the population, for example, in a particular locality. A private member's bill is a public bill introduced by a back-bench member of Parliament, as distinct from a public bill, which is introduced by the government in power.

Delegated legislation arises when a subordinate body makes laws under specific powers from Parliament. These can take the form of:

- orders in council
- statutory instruments
- by-laws.

Whilst these are essential to the smooth running of the nation, the growth of delegated legislation can be criticised, because law-making is transferred from the elected representatives to the minister, effectively the civil servants. The validity of delegated legislation can be challenged in the courts as being *ultra vires*, i.e. beyond the powers of the party making it, and thus making it void. The judicial safeguard depends on the parent legislation, i.e. the Act giving the powers. Often this is extremely wide and such a restraint may therefore be almost ineffectual.

All legislation requires interpreting. The object of interpretation is to ascertain Parliament's will as expressed in the Act. The courts are thus, at least in theory, concerned with what is stated and not with what it believes Parliament intended. A large proportion of cases reported to the House of Lords and the High Court involve questions of statutory interpretation and in many of these the legislature's intention is impossible to ascertain because it never considered the question before the court. The judge must then do what he or she thinks Parliament would have done had they considered the question.

Since Britain's entry into the European Community on 1 January 1973, it has been bound by Community law. All existing and future Community law which is self-executing is immediately incorporated into English law. A self-executing law therefore takes immediate effect and does not require action by the UK legislature.

1.3.3 Case law

Case law is often referred to as judicial precedent. It is the result of the decisions made by judges who have laid down legal principles derived from circumstances of the particular disputes coming before them. Importance is attached to this form of law in order that some form of consistency in application in practice can be achieved. The doctrine of judicial precedent is known as *stare decisis*, which literally means 'to stand upon decisions'. In practice, therefore, a judge trying a case must always look back to see how previous judges dealt with similar cases. In looking back, the judge will expect to discover those principles of law which are relevant to the case now being decided. The decision made will therefore seek to be in accordance with the already established principles of law and may in turn develop those principles further. Note that the importance of case law is governed by the status of the court which decided the case. The cases decided in a higher court will take precedence over the judgments in a lower court.

Here are the main advantages claimed for judicial precedent:

- *Certainty*: because judges must follow previous decisions, a barrister can usually advise a client on the outcome of a case.
- *Flexibility*: it is claimed that case law can be extended to meet new situations, thereby allowing the law to adjust to new social conditions.

A direct result of the application of case law is that these matters must be properly reported and published and should be readily available for all future users. Consequently, there is now available within the English legal system an enormous collection of law reports stretching back over many centuries. Within the construction professions, a number of different firms and organisations now collate and publish law reports which are relevant to this industry. Computerised systems are also available to allow for rapid access and retrieval from such reports.

It is not the entire decision of a judge that creates a binding precedent. When a judgment is delivered, the judge will give the reason for the decision. This is known as *ratio decidendi*, and is a vital part of case law. It is the principle which is binding on subsequent cases that have similar facts in the same branch of law. The second aspect of judgments, *obiter dicta*, are things said 'by the way', and these do not have to be followed. Although the facts of a case appear similar to a binding precedent, a judge may consider that there is some aspect or fact which is not covered by the *ratio decidendi* of the earlier case. The judge will therefore 'distinguish' the present case from the earlier one which created the precedent.

Examples

Here are some examples of how sources of English law are appropriate to the construction industry:

- Custom
    - Right to light
    - Right of way
- Legislation
    - Highways Act 1980
    - Town and Country Planning Act 1990
    - Local Government Act 2003
    - Control of Pollution Act 1974
- Cases
    - *Hadley* v. *Baxendale* (1854)
    - *Sutcliffe* v. *Thackrah and Others* (1974)
    - *Dawber Williamson Roofing* v. *Humberside County Council* (1979)

A higher court may also consider that the *ratio decidendi* set in a lower court is not the correct law which should be followed. When another case is argued on similar facts, the higher court will overrule the previous precedent and set a new precedent to be followed in future cases. Such a decision does not affect the parties in the earlier case, unlike a decision that is reversed on appeal.

Finally, a superior court may consider that there is some doubt as to the standing of a previous principle, and it may disapprove, but not expressly overrule, the earlier precedent.

1.3.4 EU law

A completely new source of English law was created when Parliament passed the European Communities Act 1972. Section 2(1) of the Act provides that:

> All such rights, powers, liabilities, obligations and restrictions from time to time created or arising by or under the Treaties, and all such remedies and procedures from time to time provided for by or under the Treaties, as in accordance with the treaties are without further enactment to be given legal effect or used in the United Kingdom shall be recognised and available in law, and be enforced, allowed and followed accordingly.

The effect of this section is that all UK courts have to recognise European Union (EU) law, whether it comes directly from treaties or other Community legislation. As soon as the 1972 Act became law, some aspects of English law were changed to bring them into line with EU law.

There are several institutions to implement the work of the EU. These include the European Parliament, the Council of Ministers, the European Commission and the European Court of Justice.

1.4 The courts

The structure of the courts in all three jurisdictions in the UK tends to be arranged with regard to the subject matter of cases brought before the courts. The following particularly appertain to England and Wales. The legal system differentiates between civil and criminal actions. The court system is illustrated in Figure 1.1, which identifies the different courts for different purposes. Contractual procedures are largely concerned with civil actions that are first brought by the plaintiff – private person, company or civil authority – against the defendant. The plaintiff must try to prove the case on the balance of probabilities. The sorts of cases resulting in civil actions are typically about contracts and torts. The choice of court often depends on the amount of the claim that is being pursued. The more substantial claims, in excess of about £30,000, are tried in the High Court and other claims in the county courts. Relatively small claims can be handled by a small claims procedure. This involves a quick hearing without lawyers and before a district judge.

Cases may be moved up to the higher courts, such as the Court of Appeal. When the matter is still unresolved, then it can be brought to the Supreme Court. In the event that justice is still felt not to have taken place, then further action can be instigated before the European Court of Justice.

Alternatively, the parties involved in a dispute can choose arbitration or one of the forms of alternative dispute resolution (see Chapter 5).

1.4.1 County courts

There are 250 county courts around the country, which deal with civil matters. The main advantages claimed for county courts are their lower costs and shorter delays before coming to trial. These are also the reasons given for using arbitration or alternative dispute resolution. Cases that cannot be resolved in the county court, or which are appealed, will usually be held in the High Court. The cases are dealt with by a district judge or circuit judge. The district judge deals with a great majority of matters of procedure and directions. They are usually solicitors.

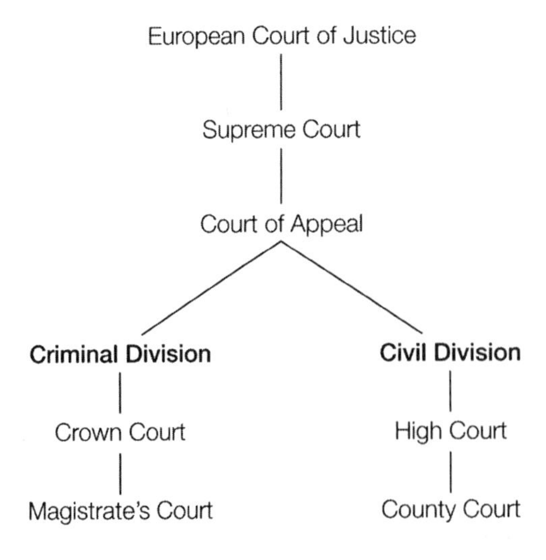

Figure 1.1 The court system.

Deputy district judges sit part-time and continue to practise. The circuit judges are the senior judges in a locality. These tend to deal with cases that require a lengthy hearing.

1.4.2 The High Court

The High Court hears all the more important civil cases. It is the lower half of the Supreme Court of Judicature and was brought into being under the Judicature Acts 1873–1875. It comprises three divisions, which all have equal competence to try any actions, according to the pressure of work, although certain specific matters are reserved for each of them:

- *The Queen's Bench Division* (QBD) deals with all types of common law work, such as contract and tort. This is the busiest division. Matters concerning the construction industry usually come to this High Court. The division is headed by the Lord Chief Justice and there are about 40–50 lesser judges. These are known as puisne (pronounced 'puny') judges. There are two specialist courts within QBD. The Commercial Court hears major commercial disputes, usually in private, with the judge hearing the case in the more informal role of an arbitrator. The Admiralty Court hears maritime disputes.
- *The Chancery Division* deals with such matters as trusts, mortgages, deeds, land, taxation and partnership disputes. The division, whilst nominally headed by the Lord Chancellor and the Master of the Rolls, is actually run by a vice-chancellor, with the help of about ten to twelve lesser judges.
- *The Family Division* deals with matters of family disputes such as probate and divorce. This division is headed by a president and three lesser judges.

The High Court normally sits at the Strand in London but there are fifteen other towns to which judges of the High Court travel to hear common law claims.

1.4.3 The Court of Appeal

Once a case has been heard, either party may consider an appeal. This means the case is transferred to the Court of Appeal, where three judges usually sit to form a court. The High Court has the right to refuse an appeal. In civil appeals, the appellant has six weeks from the date of judgment in which to give the Court of Appeal formal notice of appeal. The appellant must specify the exact grounds on which the appeal is based and on which the lower court reached an 'incorrect' decision.

The Civil Division hears appeals on questions of law and of fact, rehearsing the whole of the evidence presented to the court below relying on the notes made at the trial. If the appeal is allowed, the court may reverse the decision of the lower court, or amend it or order a retrial. It can hear appeals about the exercise of discretion, for example, discretion as to costs.

Most appeals are heard by three judges, although some (e.g. appeals from county court decisions) can be heard by only two judges. Decisions need not be unanimous. The head of the court is the Master of the Rolls, perhaps the most influential appointment in our legal system.

1.4.4 The Supreme Court

Appeals from the decisions of the Court of Appeal were formerly made to the House of Lords. Permission to do this was sparingly given and the appeal allowed only in matters of general legal importance. The court traditionally sat in the House of Lords, but since 1948 it has usually sat as an appellate committee room in the Palace of Westminster. Cases must go through the different courts and progress slowly up the judicial hierarchy. In 2009 the Supreme Court replaced the House of Lords as the highest court in the United Kingdom. The Supreme Court is now explicitly separate from government and Parliament. The Supreme Court continues in the tradition of the House of Lords as the final court of appeal for all civil law cases in the United Kingdom. It hears appeals on arguable points of law of general importance and concentrates on those cases which are of the greatest public and constitutional importance. The Supreme Court maintains and develops a role as a court of importance in the common law.

All of the courts must apply statute law in reaching their decisions and, in general, the lower courts are bound by the decisions of the higher courts. In practice, the law resulting from a case to the Supreme Court can only be changed by an Act of Parliament. For more information on the Supreme Court and the changes in the legal system since 2009 see *The English Legal System* (Elliott and Quinn 2012).

1.4.5 Court of Justices of the European Communities

This court was set up under the Treaty of Rome in 1957 at the time that the European Community was established. The court sits in Luxembourg and consists of judges from all of the countries who form a part of the European Community. It has been made a part of the English legal system by virtue of the European Communities Act 1972.

Its intervention can arise in two ways under article 177 of the Treaty of Rome. Firstly, the UK courts can ask for a preliminary judgment on a point of Community law. Secondly, a case can be brought to this court where a House of Lords judgment is questioned. The Court of Justice can overrule all other courts on matters of Community law.

1.4.6 The Technology and Construction Court

Over the past few years the practices and procedures used for civil justice have undergone major changes. Cases that in the past would have come before the courts have sometimes been transferred to one of the alternative dispute procedures, such as mediation or adjudication under the Housing Grants, Construction and Regeneration Act 1996. In the context of such reform, the existing Official Referees Court could not be left unchanged, even though it had provided a much-needed service. In 1998, their work was transferred to the new Technology and Construction Court (TCC). A new judiciary was introduced, since, prior to this, judges did not sit in the Official Referees Court. In 2005 a second edition of the TCC Guide came in to force. The TCC deals with issues that are principally concerned with technology and construction disputes. TCC cases are perhaps even more complex than other legal cases and often raise technical issues and issues of procedure, which, for those in practice, are accepted and taken at face value. The experts involved in such cases often come from architecture, engineering and quantity surveying. These courts often deal with matters that challenge previous adjudication decisions.

The TCC is based in London and deals initially with all High Court TCC claims. There are three High Court judges and senior circuit judges who, in addition to sitting in London, also sit in regional courts. TCC judgments which are of relevance to practitioners can be accessed on the court's website (available at www.justice.gov.uk). The following are some examples of the types of claims which may be brought to the TCC for judgment:

- Building, construction or engineering disputes, including the enforcement of the decisions of adjudicators under the Housing Grants, Construction and Regeneration Act 1996.

- Professional negligence in services provided, for example, by engineers, architects, surveyors, accountants.
- Matters relating to the statutory duties of local authorities concerning the development of land or the construction of buildings.
- Breaches of repairing covenants between landlords and tenants.
- Matters of tort, such as nuisance, trespass, etc. between neighbours, owners and occupiers of land.
- Environmental matters such as pollution.
- Losses caused by fires.
- Challenges to decisions of arbitrators in construction and engineering disputes.

1.4.7 Adjudication and arbitration

Adjudication and arbitration are alternatives to litigation in the courts, and are widely used for the settlement of disputes which involve technical or commercial elements. The tribunal is chosen by the parties concerned, and the powers of the arbitrator largely depend upon agreement between these parties. Adjudication and arbitration are more fully explained in Chapter 5.

1.4.8 Courts and Legal Services Act 1990

The Courts and Legal Services Act 1990 was an Act of Parliament that reformed the legal profession. The Act was the culmination of a series of reports and reforms that started with the Benson Commission in the 1970s. This significantly changed the way that the legal profession and court system operated. The changes introduced in this Act covered a wide variety of areas and important changes were made to the judiciary, particularly in terms of appointments, the introduction of district judges, alternative dispute resolution and the procedure in the courts. Changes were made in terms of the distribution of civil business between the Court of Justice and the county courts.

The most significant changes were made in the way the legal profession was organised and regulated. The Act broke the monopoly held on conveyancing work, creating an Authorised Conveyancing Board, which could certify any individual, corporation or employee of a corporation as an authorised conveyancer subject to certain requirements. The Act also broke the monopoly of barristers on advocacy and litigation in the higher courts by granting solicitors in the Crown Court.

1.5 The lawyers

1.5.1 Solicitors

There were around 140,000 solicitors and 10,500 solicitor firms in England and Wales by mid-2017. They are the lawyers that the public most frequently meets and as such are the general practitioners of the legal profession. A solicitor's work falls into two main categories of court work and non-court work. The latter accounts for about three-quarters of their business. A solicitor operates in many ways like a business person, with an office to run, clients to see and correspondence to be answered. Traditionally, property (conveyance and probate) has been one of the main fee earners.

The Solicitors Act 1974 gave solicitors three monopolies: conveyancing; probate and suing; and starting court proceedings. The conveyancing monopoly was, however, significantly eroded by the Administration of Justice Act 1985. This allowed for licenced conveyancers to do this work.

The Law Society is the regulatory and representative body for solicitors in England. It was founded in 1825. It has important public responsibilities for:

- regulating and setting standards for solicitors to make sure that they deliver good advice to consumers
- representing solicitors
- supporting solicitors, to help them achieve the standards expected of them
- influencing law reform to achieve a better system of justice.

It acts as a regulatory body through admission, discipline and continuing professional development. The society maintains its validating and monitoring role for undergraduate and postgraduate training contracts and compulsory professional development.

The role of solicitors includes:

- helping with everyday problems, which include drawing up wills and dealing with relationship breakdown
- promoting business by providing the legal basis for commercial transactions
- protecting the rights of individuals to ensure that they are treated fairly
- supporting the community, for example through legal aid programmes.

1.5.2 Barristers

There were more than 16,000 barristers in England and Wales in 2016, of which about 10,000 were male and 6,000 were female. They are specialist advocates and the specialist advisers of the legal profession. About 1,600 (10 per cent) of the Bar is made up of Queen's Counsel. Barristers have been providing expert advice and advocacy since the thirteenth century. For many years they had a monopoly on the right to represent people in higher courts. Although that monopoly has gone, the Bar remains a thriving profession offering high-quality advice and advocacy. Some of their work is non-court work, such as advising on difficult points of law or on how a particular case should be conducted. Barristers are specialist legal advisers and courtroom advocates. They are independent and objective and are trained to advise clients on the strengths as well as the weaknesses of their case.

The Bar Council is the regulatory and representative body for barristers in England and Wales. It deals with qualification and conduct rules governing barristers and those wishing to become barristers. It deals with complaints against barristers. It also puts the Bar's view on matters of concern about the legal system and acts as a source of information about the Bar. There are four Inns of Court, which provide support for barristers through a wide range of activities. Anyone wishing to train for the Bar must join one of the Inns.

Barristers practise as self-employed, referral professionals. Until recently, it was not normally possible for members of the public to go to a barrister direct. They needed to use some other recourse professional or licenced access client. However, the Bar Council decided to relax these rules relating to direct access. There are now three main routes of access to a barrister: Professional Client Access, Licensed Access and Public Access.

The country is divided into regions or *circuits* for the purposes of administration of justice. As the law became more complex, barristers increasingly chose to specialise in areas of work. Barristers are individual practitioners who work in groups of offices known as chambers, which are situated in cities and towns throughout England and Wales.

1.5.3 Judges

In contrast with many other European countries, the judiciary in England and Wales is not a separate career. Judges are appointed from both branches of the legal profession. They serve in the Supreme Court (the final appellate court), the Court of Appeal, the High Court, Crown Court or as circuit or district judges.

The circuit judges sit either in a Crown Court to try criminal cases or in county courts to try civil cases. District judges sit in county courts. There also part-time judges appointed from both branches of the practising legal profession, who serve in Crown Court, county courts or on various tribunals; for example, those dealing with unfair dismissal from employment.

However, the majority of cases are not dealt with by judges, but by laypersons who are appointed to various tribunals because of their special knowledge, experience and good standing. For example, the majority of minor criminal cases are judged by justices of the peace in magistrates' courts. They are not legally qualified and do not receive a remuneration, but are respected members of the community who sit part-time.

All members of the judiciary are appointed by the Lord Chancellor – a member of the government and Speaker of the House of Lords. The Lord Chancellor holds a function similar to that of a minister of justice, although some matters concerning the administration of justice are the responsibility of the Home Secretary.

Once appointed, judges are completely independent of both the legislature and the executive, and so are free to administer justice without fear of any political interference.

1.5.4 Expert witnesses

An expert witness is someone who has extensive knowledge and experience of the issue in a dispute and who is able to provide expert evidence either in writing or orally to a court or other tribunal. It used to be common for both sides in a dispute to have their own experts. However, the Civil Procedure Rules, which govern civil litigation in the county court and High Court, encourage the use of single experts. This is an attempt by the government to limit the cost of expert evidence to the parties and thus keep the costs of litigation down. In practice, judges may direct that both sides in a dispute agree the identity of a single expert who will give an expert opinion to the court. If the parties cannot agree, then the judge may direct that the parties seek a nomination from a relevant professional body.

Independent expert determination is a procedure whereby the parties to a dispute agree to be bound by the decision of a third party who has expert knowledge of the subject matter in dispute. It is used as a form of alternative dispute resolution (see 5.7).

1.6 Modernisation

Law is sometimes seen as a question of how far you can afford to go rather than how good your case is. A commission helps to remedy the huge problems of cost, delay, complexity and inequality in the civil justice system. The key is to recognise that justice is not an abstract quality. It has to be proportionate, it has to be within the means of the parties and it has to be expeditious. Many people are denied access to the courts because the costs involved are disproportionate to their claims. Judges should become trial managers able to dictate the pace of legal cases. Also, attitudes must change, but the powers to encourage settlement or to strike out unworthy cases must also be provided. Judges should also be able to encourage litigants to look at other ways of settling disputes such as mediation or alternative dispute resolution (see Chapter 5). These procedures encourage early settlement. They should also be able to give summary judgment, leaving only the core of the dispute to go to trial. The following are some recommendations:

- Small claims court expanded.
- Fast-track cases with capped costs and fixed hearings.
- New multi-track providing hands-on management teams by judges for heaviest cases.
- New post to run all civil courts as a single system; post to be filled by a senior judge.
- Better use of technology, with laptop computers for all judges and videoconferencing facilities.
- Incentives for early settlement, including 'plaintiff's offer' and referral to alternative dispute procedures.
- Solicitors to inform clients of charges as the bill for legal services mounts.
- Elimination of courtroom Latin.

The present consequences of the current system are that it is often excessive, disproportionate and unpredictable. The delay is frequently unreasonable. For example, in London, High Court cases take on average 160 weeks just to reach trial. Outside London they take even longer, average 190 weeks. In the county courts, the typical figure is eighty weeks.

A report chaired by Lord Justice Wolfe recommended a three-track civil system with a single entry point, headed by a senior judge in a new post called head of civil justice. A key point of the proposed system is to encourage settlement. Both parties will be able to make offers to settle at any stage, relating to the whole case or just one of the issues involved.

Annually, almost 300,000 writs are lodged with the Lord Chancellor's Office, with less than 3 per cent ever coming to trial.

1.7 Some legal jargon

abate	To reduce or make less.
acquittal	Discharge of defendant following verdict or direction of not guilty.
Act	Law, as in an Act of Parliament.
actus reus	The guilty act.
ad infinitum	Without limit.
adjudication	Judgment or decision of a court or tribunal.
ad valorem	According to the value. For example, stamp duty on sale of land is charged according to the price paid.
advocate	A barrister or solicitor representing a party in a hearing before a court.
affidavit	A written statement to be used as evidence in court proceedings.
ancient lights	Windows which have had an uninterrupted access of light for at least twenty years. Buildings cannot be erected which interfere with this right of light.
appeal	Application to a higher court or authority for a review of a decision of a lower court or authority.
appellant	A person who appeals.
attestation	The signature of a witness to the signing of a document by another person.
bona fide	In good faith.
brief	Written instructions to counsel to appear at a hearing on behalf of a party prepared by a solicitor and setting out the facts of a case and any case law relied upon.

burden of proof	The obligation of proving the case.
caveat emptor	Let the buyer beware.
certiorari	An order of the High Court to review and quash the decision of the lower court which was based on an irregular procedure.
chattels	All property other than freehold real estate.
common law	The law, established by judicial precedent, from judicial decisions and accepted within a community.
consideration	Where a person promises to do something for another, it can only be enforced if the other person gave or promised to give something of value in return. Every contract requires consideration.
contempt of court	Disobedience or wilful disregard to the judicial process.
corroboration	Evidence by one person confirming that of another or supporting evidence.
costs	The expenses relating to legal services.
counterclaim	When a defendant is sued, any claim against the plaintiff may be included, even if it arises from a different matter.
custom	An unwritten law dating back to time immemorial.
damages	An amount of money claimed as compensation for physical or material loss.
de facto	As a matter of fact.
defendant	A person who is sued or prosecuted, or who has any court proceedings brought against them.
enactment	An Act of Parliament, or part of an Act.
erratum	An error.
estoppel	A rule which prevents a person denying the truth of a statement or the existence of facts which another person has been led to believe.
ex gratia payment	A payment awarded without acceptance of any liability or blame.
ex parte	An application to the court by one party to the proceedings without the other party being present.
expert witness	One who is able to give an opinion on a subject. This is an exception to the rule that a witness must only tell the facts.
fieri facias	A court order to the sheriff requiring the seizure of a debtor's goods to pay off a creditor's judgment.
force majeure	The meaning of *force majeure* is imprecise, but it is generally accepted as 'exceptional circumstances beyond the control of either of the parties of the contract'.
frustration	A contract is frustrated if it becomes impossible to perform because of a reason that is beyond the control of the parties. The contract is then cancelled.
good faith	Honesty.
goodwill	The whole advantage, wherever it may be, of the reputation and connection of the firm.
High Court	A civil court divided into the three divisions of Queen's Bench, Family and Chancery (see 1.4.2).
in camera	When evidence is not heard in open court.
injunction	A court order requiring someone to do, or to refrain from doing, something.
in situ	In its original situation.
ipso facto	The reliance of facts that together prove a point.
judicial review	An application made to the divisional court when a lower court or tribunal has behaved incorrectly.

legislation seal	This used to be an impression of a piece of wax to a document. Now a small red sticky label is used instead. The absence of the seal will not invalidate the document, since to constitute a sealing requires neither wax, wafer, piece of paper nor even an impression.
lien	A legal right to withhold goods or materials of another until payment is made.
limitation	Court proceedings must begin within a limitation period. Different periods exist for different types of claim.
liquidated sum	A specific sum, or a sum that can be worked out as a matter of arithmetic.
liquidator	A person who winds up a company.
litigation	Legal proceedings.
Lord Chancellor	The cabinet minister who acts as Speaker in the House of Lords and oversees the hearings of the Law Lords.
Lord Chief Justice	Senior judge of the Court of Appeal.
mitigation	Reasons submitted on behalf of a guilty party in order to excuse or partly excuse the offence that has been committed.
moiety	One half.
nota bene	Note well (NB).
obiter dictum	A statement of opinion by a judge which is not relevant to the case being tried. It is not of such authority as if it had been relevant to the case being tried.
official referee	A layperson appointed by the High Court to try complex matters in which he or she is a specialist.
order	A direction by a court.
particulars	Details relevant to a claim.
plaintiff	Person who sues.
plc	A public limited company. Most used to call themselves Ltd but changed this when UK company law was brought into line with EC law in 1981.
pleadings	Formal written documents in a civil action. The plaintiff submits a statement of claim and the defendant a defence.
precedent	The decision of a case which established principles of law that act as an authority for future cases of a similar nature.
pre-trial review	Preliminary meeting of parties in a county court action to consider administrative matters and what agreement can be reached prior to the trial.
prima facia	Evidence sufficient to prove a case unless disproved.
quantum meruit	As much as has been earned.
ratio decidendi	The reason for a judicial decision. A statement of legal principle in a *ratio decidendi* is more authoritative than if in an *obiter dictum.*
res inter alios acta	A thing done between others which does not harm or benefit others.
res ipsa loquitor	The matter speaks for itself.
retrospective	An Act that applies to a period before the Act was passed.
sine die	Indefinitely.
special damage	Financial loss that can be proved.
specific performance	When a party to a contract is ordered to carry out their part of the bargain. Only ordered where monetary damages would be an inadequate remedy.
stare decisis	To stand by decided matters. Alternative name for the doctrine of precedent.
statute	An Act of Parliament.

statutory instrument	Subordinate legislation made by the Queen in Council or a minister, in exercise of a power granted by statute.
stay of proceedings	When a court action is stopped by the court.
sub judice	Whilst a court case is under consideration details cannot be disclosed.
subpoena	A court order that a person attends court, either to give evidence or to produce documents.
summary judgment	Judgment obtained by a plaintiff where there is no defence to the case or the defence contains no valid grounds.
summons	Order to appear or to produce evidence to a court. Also the old name for a claim form.
tort	A civil wrong committed against a person for which compensation may be sought through a civil court.
uberrimae fidei	Of utmost good faith.
ultra vires	An act that falls outside or beyond the jurisdiction of a court.
unenforceable	A contract or other right that cannot be enforced because of a technical defect.
vicarious liability	When one person is responsible for the actions of another because of their relationship.
void	Of no legal effect.
voidable	Capable of being set aside.
with costs	The winner's cost will be paid by the loser.
writ	The document that commences many High Court actions.

2 Legal aspects of contracts

A number of Acts of Parliament affect construction contracts, although it is only during this present century that they have begun to play any significant part. Historically, the law of contract has evolved by judicial decisions, so that there now exists a body of principles which apply generally to all types of construction contract. These principles have been accepted on the basis of proven cases that have been brought before the courts.

Construction contracts are usually made in writing, using one of the standard forms available. The use of a standard form provides many advantages, and although standard forms are not mandatory, their use should be encouraged in all possible circumstances. It is important to remember, however, that the making of a contract does not require any special formality. A binding contract could be made by an exchange of letters between the parties, rather than signing an elaborate printed document. On some occasions, a binding contract could be made by a gentlemen's agreement, i.e. by word of mouth. There are, however, many practical reasons why construction contracts for all but the simplest projects should be made using an approved and accepted form of contract.

2.1 Definition of a contract

A contract has been defined by Sir William Anson as 'a legally binding agreement made between two or more parties, by which rights are acquired by one or more to acts or forbearances on the part of the other or others'. The essential elements of this definition are as follows:

- *Legally binding*: not all agreements are legally binding; in particular, there are social or domestic arrangements which are made without any intention of creating legal arrangements.
- *Two or more parties*: in order to have an agreement, there must be at least two parties; in law, one cannot make bargains with oneself.
- *Rights are acquired*: an essential feature of a contract is that legal rights are acquired; one person agrees to complete part of a deal and the other person agrees to do something else in return.
- *Forbearances*: to forbear is to refrain from doing something; there may thus be a benefit to one party to have the other party promise not to do something.

2.2 Agreement

The whole basis of the law of contract is agreement. Specifically, a contract is an agreement bringing with it obligations which are able to be enforced in the courts if this becomes necessary.

Most of the principles of modern contract date from the eighteenth and nineteenth centuries. The concept of a contract at that time was of equals coming together to bargain and reach agreement, which they would wish to be upheld by the courts. Whilst it is still true that individuals come together to form agreements, note that many contracts are formed between parties who are not equals in any way, even where the law may pretend that they are. A major criticism of contract law in recent years has been that the wealthy, experienced and legally advised corporations have been able to make bargains with many people who are themselves of limited resources and poorly legally represented. Because of this, the law of contract has gradually moved away from a total commitment to enforce, without qualification, any agreement which has the basic elements of a contract. In particular, Parliament has introduced statutes which are often designed to protect relatively weak consumers from business persons having greater bargaining power. Despite this, the courts are still reluctant to set aside an agreement having all of the elements of a contract and, in this respect, follow their nineteenth-century predecessors.

2.3 The elements of a contract

2.3.1 Capacity

In general, every person has full legal powers to enter into whatever contracts they might choose. There are, however, some broad exceptions to this general rule. Infants and minors, that is anyone under the age of eighteen years as set out in the Family Law Reform Act 1969, cannot contract other than in certain circumstances, such as for necessaries or benefits. Persons of an unsound mind, as defined in the Mental Health Act 1959, can never make a valid contract. Other persons of an unsound mind and those unbalanced by intoxication are treated alike. Their contracts are divided into two types, those for necessary goods (the situation with minors above) and other contracts where the presumption is one of validity. Corporations are legal entities created by a process of law. A company can only contract on matters falling within its objects clause, and since the records of companies are matters of public record available for inspection at Companies House, it used to be the case that a company could not have a contract enforced against it if it lay outside its objects clause. The presumption was that those entering into contracts with a company knew or ought to have known the contents of the objects clause. Consequently, anyone making an *ultra vires* 'outside the powers' contract with a company only had themselves to blame. On entry into the European Community in 1973 this *ultra vires* doctrine had to be revised, since it was not followed in the other EU countries (Figure 2.1).

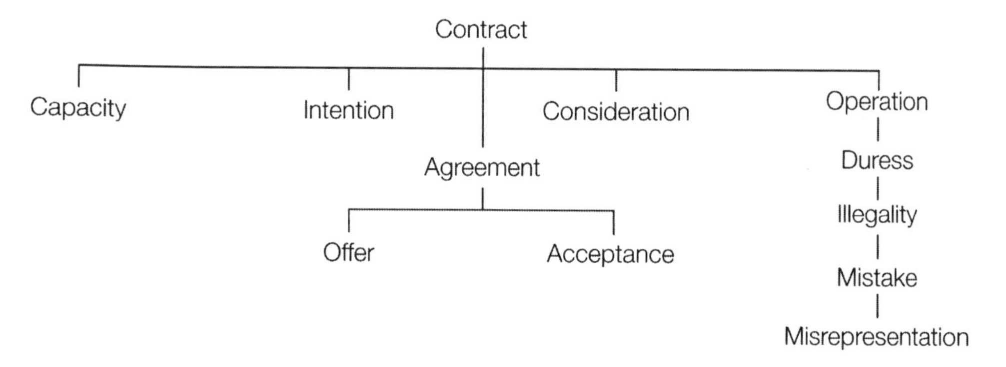

Figure 2.1 The elements of a contract.

2.3.2 Intention to create legal relations

Merely because there is an agreement, it cannot be assumed that an enforceable contract exists. English law requires that the parties to a contract actually intended to enter into legal relations. These are relations actionable and enforceable in the courts. If it can be demonstrated that no such intention existed, then the courts will not intervene, despite the presence of both agreement and consideration. In commercial agreements, the courts presume that the parties do intend to enter into legal relations. (This is different from social, family and other domestic agreements, where the general rule is that the courts presume there is no intention to enter into legal relations.) These are the general rules and it is also possible to demonstrate the opposite intention.

In contract law, we need to know what we have agreed. It is possible for two parties to use words which are susceptible to interpretation in different ways, so they do not have the same idea in mind when they agree. It is important that both parties have agreement to the same idea (*consensus ad idem*). The classic case to which students should refer is *Raffles* v. *Wichelhaus* (1864).

2.4 Offer and acceptance

The basis of the contract is agreement and this is composed of two parts: offer and acceptance. In addition, conditions are generally required by law (in all but the simplest of contracts) to make the offer and acceptance legally binding.

2.4.1 The offer

An offer must be distinguished from a mere attempt to negotiate. An offer, if it is accepted, will become a binding contract. An invitation for contractors to submit tenders is inviting firms to submit offers for doing the work. The invitation often states that the employer is not bound to accept the lowest or any tender or to be responsible for the costs incurred.

An offer may be revoked by the person who made it at any time before it is accepted. Thus, a contractor may submit a successful tender in terms of winning the contract. However, the contractor may choose to revoke this offer prior to formal acceptance.

Tenders for building works do not remain on offer indefinitely. If they are not accepted within a reasonable time then the offer may lapse, or be subject to some monetary adjustment should it later be accepted. The building owner may stipulate in the invitation to the tenderers that the offer should remain open for a prescribed period of time.

Offers concerned with building projects are generally made on the basis of detailed terms and conditions. The parties to the contract will be bound by these conditions, as long as they know that such conditions were incorporated in the offer, even though they may never have read them or acquainted themselves with the details.

In many instances with building contracts, the offer must follow a stipulated procedure. Such procedures often incorporate delivery of the offer by a certain date and time, in writing, on a special form and in a particular envelope, and stipulate that the offer must not be disclosed to a third party. Failure to comply with these procedures will result in the offer being rejected.

2.4.2 The acceptance

Once an agreeable offer has been made, there must be an acceptance of it before a contract can be established. The acceptance of the offer must be unconditional and it must be communicated to the person who made the offer. The unconditional terms of acceptance must correspond precisely with the terms of the offer. In practice, the parties may choose to negotiate on the basis of the offer. For example, the building owner may require the project to be completed one month earlier and this may result in the tenderer revising the tender sum. A fresh offer made in this way is known as a counter-offer, and is subject to the conditions now applied.

An offer or acceptance is sometimes made on the basis of 'subject to contract'. In practice, the courts tend to view such an expression as of no legal effect. No binding contract will come into effect until the formal contract has been agreed.

In a leading contract law case (*Hyde* v. *Wrench* (1840)) in relation to the issue of counter-offers and their relation to initial offers, it was ruled that any counter-offer cancels the original offer. In this case, Wrench offered to sell his farm to Hyde for the sum of £1,200. This offer was declined by Hyde. Sometime later, Wrench made a final offer to sell the farm for £1,000. Hyde subsequently offered £950. After examining this offer, Wrench refused to accept, and informed Hyde accordingly. Hyde then offered to buy the farm for £1,000, but Wrench now refused to sell the farm to him. Hyde then attempted to sue for a breach of contract.

The judgment read as follows:

> Under the circumstances stated in this bill ... there exists no valid binding contract between the parties for the purchase of the property. The defendant offered to sell it for £1,000, and if that had been at once unconditionally accepted there would ... have been a perfectly binding contract; instead of that, the plaintiff made an offer of his own, to purchase the property for £950, and he thereby rejected the offer previously made by the defendant ... it was not afterwards competent for him to revive the proposal of the defendant, by tendering an acceptance of it ... therefore, there exists no obligation of any sort between the parties.

2.5 Form

The word *form* means some peculiar solemnity or procedure accompanying the expression of agreement. It is this formality which gives to the agreement its binding character. The formal contract in English law is the contract under seal – that is, one made by deed. The contract is executed – that is, it is made effective – by being signed, sealed and delivered:

- *Signature*: doubts have been expressed regarding the necessity for a signature; some statutes make a signature a necessity.
- *Sealing*: today this consists of affixing an adhesive wafer, or in the case of a corporate body an impression in the paper, and the party signs against this and acknowledges it as his or her seal.
- *Delivery*: this is not now necessary for legal effectiveness; as soon as the party acknowledges the document as the deed, it is immediately effective.
- *Witnesses*: these attest by signing the document; they are not usually a legal necessity.

2.6 Consideration

Another essential feature of a binding contract, other than a contract made under seal, is that the agreement must be supported by consideration. The most common forms of consideration are payment of money, provision of goods and the performance of work. Consideration has been judicially defined as 'some right, forbearance, detriment, loss or responsibility given, suffered or undertaken by the other in respect of the promise'. In building contracts, the consideration of the contractor to carry out the works in accordance with the contract documents is matched by that of the building owner to pay the price. The following rules concerning consideration should be adhered to:

- Every simple contract requires consideration to make it valid.
- The consideration must be worth something in the eyes of the law. The courts are not concerned whether the bargain is a good one, but simply that there is a bargain.
- Each party must get something in return for the promise, other than something that they are already entitled to, otherwise there is no consideration.
- The consideration must not be such that it conflicts with the established law.
- The consideration must not relate to some event in the past.

2.7 Duress and undue influence

Duress is actual or threatened violence to, or restraint of, the person of a contracting party. If a contract is made under duress, it is at once suspect because consent has not been freely given to the bargain supposedly made. The contract is voidable at the option of the party concerned. Duress is a common law doctrine which relates entirely to the person and has no relation to that person's goods. As such it is a very limited doctrine and is one where cases are rare.

The dictionary defines duress as: restraint of liberty or illegal compulsion. Whilst the concept of economic duress is easily understood, it is often difficult to prove in practice. The difficulty for those making judgments is to make a distinction between commercial practices that may be both legal and tough and those practices that occur because of coercion. Two recent cases where economic duress was raised are: *DSND Subsea* v. *Petroleum Geo-Services* (2000) and *Carillion Construction Limited* v. *Felix (UK) Limited* (2001).

Economic duress might, for example, be applied in the first place to the signing of a contract or then subsequently in dealing with variations or change orders. The pressure applied by one party must be both illegal and significant in inducing the other party to agree to a course of action that otherwise would not be followed. The victim's conduct must be affected in a significant way by the duress, and a reasonable alternative must not be available at the time to the victim of the duress. The Privy Council has identified four factors to consider in assessing whether economic duress is present:

- Did the person claiming to be coerced protest?
- Did that person have any other available course of action?
- Were they independently advised?
- After entering into the contract, did they take steps to avoid it?

2.8 Unenforceable contracts

Contracts may be described as void, or voidable or unenforceable. A void contract creates no legal rights and cannot therefore be sued upon. It may occur because of a mistake as to the nature of the contract, or because it involves the performance of something illegal that is prohibited by a statute. A void contract will also result because of the incapacity of the parties, as in the case of infants. Corporations cannot make contracts beyond their stated powers which are said to be *ultra vires* 'beyond one's powers'.

A contract is said to be voidable when only one of the parties may take advantage. In cases involving misrepresentation, only the party who has been misled has the right to a void in one of the ways previously described.

Unenforceable contracts are those that are valid, but owing to the neglect of the formalities involved, a party seeking to enforce it will be denied a remedy.

2.9 Mistake

The law recognises that, in some circumstances, although a contract has been formed, one or both of the parties are unable to enforce the agreement. The parties are at variance with one another and this precludes the possibility of any agreement. Mistake may be classified as follows:

- *Identity of subject matter*: this occurs where one party intends to contract with regard to one thing and the other party with regard to another; the parties in this situation cannot be of the same mind and no contract is formed.
- *Identity of party*: if the identity of either party enters into consideration, this will negate the contract.
- *Basis of contract*: if two parties enter into a contract on the basis that certain facts exist, and they do not, then the contract is void.
- *Expressing the contract*: if a written contract fails to express the agreed intentions of the two parties, then it is not enforceable; courts may, however, express the true intention of the parties and enforce it as amended.

2.10 Misrepresentation

Misrepresentation consists in the making of an untrue statement which induces the other party to enter into a contract. The statement must relate to fact rather than opinion. Furthermore, the injured party must have relied on the statement and it must have been a material cause of their entering into the contract. Where such a contract is voidable it may be renounced by the injured party but, until such time, it is valid. Misrepresentation may be classified as:

- *Innocent misrepresentation*: an untrue statement is made in the belief that it is true.
- *Fraudulent misrepresentation*: an untrue statement is made with the knowledge that it is untrue or made recklessly without attempting to assess its validity.
- *Negligent misrepresentation*: a statement is made honestly, but without reasonable grounds for belief that it is true. It is really a special case of innocent misrepresentation because, although the statement is made in the belief that it is true, insufficient care has been taken to check it.

Where misrepresentation occurs, the injured party has several options:

- The injured party can affirm the contract, when it will then continue for both parties.
- The injured party can repudiate the contract and set up misrepresentation as a defence.
- An action can be brought for rescission and restitution:
    - Rescission involves cancelling the contract and the restoration of the parties to the state that they were in before the contract was made.
    - Restitution is the return of any money paid or transferred under the terms of the contract.

- The injured party can bring an action for damages. The claim for damages is only possible in circumstances of fraudulent misrepresentation.

2.11 Disclosure of information

When entering into a contract, it is not always necessary to disclose all the facts that are available. A party may observe silence in regard to certain facts, even though it may know that such facts would influence the other party. This is summed up in the maxim *caveat emptor* 'let the buyer beware'.

There are, however, circumstances where the non-disclosure of relevant information may affect the validity of the contract. This can occur where the relevant facts surrounding the contract are almost entirely within the knowledge of one of the parties, and the other has no means of discerning the facts. These contracts are said to be *uberrimae fidei* 'of the utmost good faith'.

2.12 Privity of contract

A contract creates something special for the parties who enter into it. The common law rule of privity is that only the persons who are party to the contract can be affected by it. A contract can neither impose obligations nor confer rights upon others who are not privy to it. For example, the clause in the standard form of building contract which allows the employer to pay money direct to the subcontractor may be used by the employer. It cannot be enforced by the subcontractor, who is not a party to the main contract. The subcontractor may seek to persuade the employer to adopt this course of action, but cannot enforce the employer to do so in a court of law.

2.13 Express and implied terms

The terms of contract can be classified as either express or implied. Those terms which are written into the contract documents or expressed orally by the parties are described as express terms. Those terms which were not mentioned by the parties at the time that they made the contract are implied terms, so long as they were in the minds of both parties. The courts will, where it becomes necessary, imply into building and engineering contracts a number of implied terms. Although the implied term is one which the parties probably never contemplated when making the contract, the courts justify this by saying that the implication is necessary in order to give business efficacy to the contract. This does not mean, however, that the courts will make a contract more workable or sensible. The courts will generally imply into a building contract the following terms:

- The contractor will be given possession of the site within a reasonable time, should nothing be stated in the contract documentation.
- The employer will not unreasonably prevent the contractor from completing the work.
- The contractor will carry out the work in a workmanlike manner.

Implied terms which have evolved from decided cases often act as precedents for future events. In the majority of the standard forms of building contract, all of these matters are normally express terms, since the contracts themselves are very comprehensive and hope to cover every eventuality.

There are some notable terms which are not normally to be implied into building and engineering contracts, such as the practicability of the design. However, there will be express or implied terms that the work will comply with the appropriate statutes.

2.13.1 Express terms

An express term is a clear stipulation in the contract which the parties intend should be binding upon them. Traditionally, the common law has divided terms into two categories, conditions and warranties:

- *Conditions* are terms which go to the root of the contract, and for breach of which the remedies of repudiation or rescission of the contract and damages are allowed.
- *Warranties* are minor terms of the contract, for breach of which the only remedy is damages.

2.13.2 Implied terms

Implied terms are those which, although not expressly stated by the parties by words or conduct, are by law deemed to be part of the contract. Terms may be implied into contracts by custom, statute or the courts:

- *By custom*: in law, this means an established practice or usage in a trade, locality, type of transaction or between parties. If two or more people enter into a contract against a common background of business, it is considered that they intend the trade usage of that business to prevail unless they expressly exclude it.
- *By statute*: there are many areas in the civil law where Parliament has interfered with the right of parties to regulate their own affairs. This interference mainly occurs where one party has used a dominant bargaining position to abuse this freedom. Thus, in the sale of goods, the general principle *caveat emptor* 'let the buyer beware' has been greatly modified, particularly in favour of the consumer by the provisions in the Sale of Goods Act 1979. In addition, there have been important changes in relation to exemption clauses brought about by the Unfair Contract Terms Act 1977.
- *By the courts*: the court will imply a term into a contract, under the doctrine of the implied term, if it was the presumed intention of the parties that there should have been a particular term, but they have omitted to expressly state it.

2.14 Limitations of actions

Generally speaking, litigation is a costly and time-consuming process which becomes more difficult as the time between the disputed events and the litigation increases. Also, rights of action cannot be allowed to endure for ever. Parties to a contract must be made to prosecute their causes within a reasonable time. For this reason, Parliament has enacted limitation acts which set a time limit on the commencement of litigation. The rules and procedures in respect of limitation of actions are contained in the Limitation Act 1980. The right to bring an action can be discharged in three ways:

- The parties to a contract might decide to discharge their rights.
- Through the judgment of a court.
- Through lapse of time.

If an action is not commenced within a certain time, then the right to sue is extinguished. Actions in a simple contract (not under seal) and tort become statute barred after six years. In the case of contracts under seal, this period is extended to twelve years. The Act does permit certain extensions to these time limits in very special circumstances; for example, where a person may be unconscious as the result of an accident. If damage is suffered at a later date, then this will not affect the limitation period, although an action for negligence may be pursued. The period of limitation can be renewed if the debtor acknowledges the claim or makes a part payment at some time during this period. A number of international conventions, particularly with respect to the law of carriage, lay down shorter limitation periods for action than those specified in the Act. For example, in carriage by air under Warsaw Rules the period is two years.

2.15 The Unfair Contract Terms Act 1977

Contractual clauses designed to exonerate a party, wholly or partly, from liability for breaches of express or implied terms first appeared in the nineteenth century. The common law did not interfere, but took the view that parties forming a contractual relationship were free to make a bargain within the limits of the law. This is still largely the rule today, although the growth of large trading organisations has led to an increase in both excluding and restricting clauses to the severe detriment of other parties.

The efforts of the courts to mitigate against the worst effects of objectionable exclusion clauses have been reinforced by the Unfair Contract Terms Act 1977. This Act restricts the extent to which liability can be avoided for breach of contract and negligence. The Act relates only to business liability, so transactions between private individuals are not covered. The reasonableness is further extended to situations where parties attempt to exclude liability for a fundamental breach, i.e. where performance is substantially different from that reasonably expected or there is no performance at all of the whole or any part of the contractual obligations. The following are subject to the Act's provisions:

- Making liability or its enforcement subject to restrictive or onerous conditions.
- Excluding or restricting any right or remedy.
- Excluding or restricting rules of evidence or procedure.
- Evasion of the provisions of the Act by a secondary contract is prohibited.

The Society of Construction Law has published guidance on this Act (see www.scl.org.uk).

2.16 *Contra proferentem*

Any ambiguity in a clause in a contract will be interpreted against the party who put it forward. It is a general rule of construction of any document that it will be interpreted *contra proferentem*, that is, against the person who prepared the document. As an exclusion clause is invariably drafted by the imposer of it, this is an extremely useful weapon against exclusion clauses. The effect of the rule is to give the party who proposed the ambiguous clause only the lesser of the protections possible. The one who draws up the contract has the choice of words and must choose them to show clearly the intention.

2.17 Collateral warranties

A collateral warranty is a contract which runs alongside another contract and is subsidiary to it. Warranties need not be in writing and it is unusual for them to be in a form of specially drafted document. A warranty is a term of a contract, the breach of which may give rise to a claim for damages, but not the right to treat the contract as repudiated. It is therefore a less important term in a contract. It is one which is collateral to the main purpose of the contract, the breach of which by one party does not entitle the other party to treat the obligations as discharged.

Collateral warranty is one of those terms that thrives in the modern-day construction industry. The actual collateral warranties themselves can often be a source of seemingly endless correspondence amongst the professional team, their legal advisers and lenders.

A collateral security is a separate obligation attached to another contract to guarantee performance.

The House of Lords decision in *Murphy* v. *Brentwood District Council* (1991) has sounded the death knell for tortious claims from third parties acquiring an interest in a building who suffer economic loss as a result of latent defects in the building. As a result of this and other cases, the use of collateral warranties in the construction industry has become widespread in order to give tenant, purchasers and funders contractual remedies against contractors and consultants for latent defects in buildings in which they acquire an interest. Without such a contractual link, those parties would have no means of recovery against the contractor and the consultants for economic loss, due to their not being party to the relevant construction contracts and having no tortious remedy.

If latent defects do arise in buildings, parties who have received collateral warranties from the construction team will be able to seek to recover damages from the contractor and the consultant under the terms of those warranties. Where the contractor and consultant have acted independently of each other so as to cause the latent defect, the contractor and consultants will be jointly and severally liable for that defect.

2.18 Agency

Agency is a special relationship whereby one person (the agent) agrees on behalf of another (the principal) to conclude a contract between the principal and a third party. Providing that the agent acts only within the scope of the authority conferred upon them, those acts become the acts of the principal, and the principal must therefore accept the responsibility for them. The majority of contractual relationships involve some form of agency. For example, the architect, in ordering extra work, is acting in the capacity of the employer's agent. The contractor may presume, unless anything is known to the contrary, that the carrying out of these extras will result in future payment by the client.

A contract of agency may be established by:

- *Express authority*: authority that has been directly given to an agent by his or her principal.
- *Implied authority*: because a person is engaged in a particular capacity, others dealing with that person are entitled, perhaps because of trade custom, to infer that that person has the necessary authority to contract within the limits usually associated with that capacity.
- *Ratification*: a principal subsequently accepts an act done by the principal's agent, even where this exceeds the agent's authority. This becomes as effective as if the principal originally authorised it.

In contracts of agency, a principal cannot delegate to an agent powers which the principal does not already possess. The capacity of the agent is therefore determined by the capacity of the agent's principal.

2.19 Subcontracting, assignment and novation

There are very few construction contracts where it is intended that all of the construction work will be undertaken by only one of the parties, the principal contractor. It is recognised that such firms no longer, if they ever did, possess all of the knowledge and skills to carry out all of the trades. Traditionally, some of the work would be subcontracted to a third party. In some cases, this work was purported to be of a specialist nature. Historically, these firms might have been separated between nominated or named subcontractors and direct or domestic subcontracts. Most of the forms of contract often required the principal contractor to tell the client who these firms were. Approval would not normally be withheld without good reason. However, the obligations under the contract would remain the responsibility of the principal contractor.

Assignment is a transfer that is recognised by the law of a right or obligation of one person to another. Whilst assignment can transfer the rights and benefits to a new party, the obligations continue to remain with the first party.

Novation occurs in those circumstances where one party transfers all its obligations and benefits under a contract to a third party. This third party then effectively replaces the original party as a party to the contract. When a contract is novated, the other party remains in the same position as that prior to the novation taking place. Novation requires the agreement of all three parties involved in the contract. With assignment, the parties do not change, but with novation, a new party is introduced that takes on the rights and responsibilities of one of the original parties to the contract.

Contracts used in the construction industry for building and civil engineering works normally contain terms restricting or prohibiting assignments or subletting without approval of the employer. Such terms have the effect of making any purported assignment invalid as against the other party to the contract, without approval of the other party.

3 Discharge of contracts

A contract is said to be discharged when the parties become released from their general contractual obligations. The discharge of a contract may be brought about in several ways.

3.1 Discharge by performance

In these circumstances, the party has undertaken to do a certain task and nothing further remains to be done. In general, only the complete and exact performance of the contractual obligations can discharge the contract. In practice, where a contract has been substantially performed, payment can be made with an adjustment for the work that is incomplete.

A building or civil engineering contract is discharged by performance once the contractor has completed all the work, including the making good of defects under the terms of the contract. The architect or engineer must have issued all the appropriate certificates and the owner paid the requisite sums. If, however, undisclosed defects occur beyond this period, the owner can still sue for damages under the statute of limitations. The contract has not been properly performed if there are hidden defects.

Thus, A undertakes to sell to B 1,000 roof tiles. A will be discharged from the contract when they have delivered the tiles, and B when they have paid the price. A question sometimes arises whether performance by another party will discharge the contract. The general rule is that where personal qualifications are a factor of consideration, then that person must perform the contract. Where, however, personal considerations are unimportant it would not matter who supplied the goods.

An exception exists where a court is satisfied that substantial performance has been achieved. The court may then award the contractually agreed price and deduct sums to reflect the amount not performed. If, however, the performance is not held to amount to substantial performance, then the claimant may be entitled to nothing. Difficulty arises as to what amounts to substantial performance. There are no precise rules agreed and each case is determined on the facts of individual circumstances. Where one party freely agrees to accept partial performance, then an amount can be calculated to reflect this situation.

3.2 Discharge under condition

Contracts consist of a large number of stipulations or terms. In many types of contract there are conditions which, although not expressed, are intended because the parties must have contracted with these conditions in mind. It will, however, be obvious that the terms and conditions of a construction contract are not of equal importance. Some of the terms are fundamental to the contract, and are so essential that if they are broken the whole purpose of the contract is defeated.

Thus, if a contractor agreed to design and build a building for a specific use, and it is incapable of such a use, the building owner would be able to reject the project and recover the costs from the contractor.

Contracts frequently contain a number of terms forming a specification. The constructor agrees to construct the project in accordance with this specification. If the constructor deviates from this in some small way, then this will give rise to an action for damages and the contract will not be discharged.

It is, however, well established that, subject to an express or implied agreement to the contrary, a party who has received substantial benefit under a contract cannot repudiate it for breach of condition.

3.3 Discharge by renunciation

This is effected when one of the parties refuses to perform obligations. Thus, A employs an architect to design and supervise a proposed building project. On completion of the design, A decides not to continue with the project. This is renunciation of the contract and the architect can sue for fees on a *quantum meruit* basis – for as much as has been earned.

3.4 Discharge by fresh agreement

A contract can be discharged by a fresh agreement being made between the parties, which is both subsequent to and independent of the original contract. Such a contract may, however, discharge the parties altogether. This is known as a rescission of the original contract. Where one party to a contract is released by a third party from undertaking obligations, then the original contract discharged is termed 'novation'. Any alteration to a contract made with the consent of the parties concerned has the effect of making a new contract.

3.5 Frustration

A contract formed between two parties will expressly or impliedly be subject to the condition that it will be capable of performance. If the contract becomes incapable of performance then the parties will be discharged from it. Impossibility of performance is usually called frustration of contract. It occurs whenever the law recognises that, without default of either party, a contractual obligation has become incapable of execution.

For example, the event causing the impossibility may be due to a natural catastrophe. A agrees to carry out a contract for the repair of a road surface, but prior to starting the work the road subsides to such an extent that it disappears completely down the side of a hill. Impossibility may also occur because a government may introduce a law that makes the contract illegal. For example, A contracts to carry out some insulation work using asbestos. The government subsequently introduces legislation forbidding the use of this material in buildings, thus making the contract impossible to carry out. A contractor cannot, however, claim that a contract is frustrated if by deliberate actions the contractor causes a delay to avoid completion. For example, a contract for the building of a sea wall must be completed prior to winter weather setting in, otherwise practical performance will become impossible. The knowledge that this predictable event will occur, coupled with a deliberate delay, does not result in a frustration of the contract.

Examples of construction contracts being frustrated are extremely rare. Note that hardship, inconvenience or loss do not decide whether an occurrence frustrates the contract, because these are accepted risks. The legal effect of frustration is that the contract is discharged and money prepaid in anticipation of performance should be returned. A party receiving benefit from a partly executed contract should reimburse the other party to the value of that benefit.

3.6 Determination of contract

There are provisions in all the common forms of contract that allow either party to terminate the contract. There must, of course, be good reasons to support any party who decides to determine, otherwise a breach of contract can occur. In one sense, they are fairly exceptional happenings, and so they should be, since they provide a final option to the employer. The decision to determine must be taken very carefully and reasonably. It generally occurs because the contractor is failing to take notice of instructions from the architect to remedy an already existing breach of contract. Determination may arise because of either a refusal or interference with payments due to the contractor. It may also occur where the works are suspended for an unreasonable length of time. Again, the action to determine must be carefully taken, and is a last-resort decision on the part of the contractor. The contractor takes this decision when it is felt it is no longer possible to continue, perhaps after a number of delays in payment, to work with the employer. Determination of contract by either party results in two losers and no winners. Although there are provisions in the contract for financial recompense to the aggrieved party, these rarely suffice.

3.7 Assignment

It sometimes happens that one party to a contract wishes to dispose of its obligations under it, but the extent to which this is permitted is limited. The other party may have valid reasons why it prefers the obligations to be performed by the original contractor. The rule is that liabilities can only be assigned by novation. This is the formation of a new contract between the party who wishes performance and the new contractor, who is accepted as adequately qualified to perform as the original contractor. Liabilities can only be assigned by consent. By contrast, rights under a contract can usually be assigned without consent of the other party, except where the subject matter involves a personal service. However, even in those circumstances where the contract does not involve personal service, but specifically restricts the right to assign the contract or any interest in it to a third party, then the assignment of rights will not be permitted.

4 Remedies for breach of contract

A breach of contract occurs when one party fails to perform an obligation under the terms of the contract. For example, if:

- the contractor refuses to obey an architect's or engineer's instruction
- the employer fails to honour a certificate
- the contractor refuses to hand over certain antiquities found on the site
- the contractor fails to proceed regularly and diligently with the works.

Defective work is not necessarily in breach of contract, as long as the contractor rectifies it in accordance with instruction. A breach will occur where the contractor either refuses to remove the defective work or ignores the remedial work required.

A breach of contract may have two principal consequences. The first is damages. Then, if the breach is sufficiently serious, there may be determination by the aggrieved party under the terms of the contract. The aggrieved party may decide to terminate the contract with the other, and also sue for damages, or alternatively take only one of these courses of action. Damages may include the loss resulting from the breach, the loss flowing from the termination and the additional costs of completing the contract.

There is, with few exceptions, the rule that no one who is not a party to the contract can sue or be sued in respect of it. There may be other remedies, for instance in tort, but these are beyond the scope of this book.

The legal remedy for breach of contract is damages. This consists of the award of a sum of money to the injured party, designed to compensate for the loss sustained. The basis of the award of damages is by way of compensation. The damages awarded are made in an attempt to recompense the actual loss sustained by reason of the breach of contract. The injured party is to be placed, as far as money can do it, in the same situation as if the contract had been performed. It should be clearly understood that not every breach automatically results in damages. In order for damages to be paid, the injured party must be able to prove that a loss resulted from the breach. Furthermore, the innocent party must take all reasonable steps to mitigate any loss. Damages may be classified as follows.

4.1 Nominal damages

Where a party can show a breach of contract, but cannot prove any sustained loss as a result, then nominal damages may be awarded. These comprise merely a small sum in recognition that a contractual right has been infringed.

4.2 Substantial damages

Substantial damages represent the measure of loss sustained by the injured party. Despite their name, they might be quite small.

4.3 Remoteness of damage

A breach of contract can, in some circumstances, create a chain of events resulting in considerable damage, and the question may arise whether the injured party or parties can claim for the whole of the damage sustained. In ordinary circumstances, the only damages that can be claimed are those which arise immediately because of the breach. Damage is not considered too remote if the parties at the time the contract was entered into contemplated that it could occur. If the contract specifies the extent of the liability, then no question of consequential damages arises. Defects liability clauses, which require the builder to rectify defects, do not limit the extent of the remedial work, but may also require other work to be put right that has resulted from these defects.

4.4 Special damage

Damages resulting from special circumstances are recoverable if they flow from a breach of contract, and if the special circumstances were known to both parties at the time of making the agreement.

4.5 Liquidated damages

Each of the standard forms of contract provide for payment of agreed damages by the contractor when completion of work is not within the stipulated time. These payments are known as liquidated damages, and their amount should be recorded on the appendix to the form of a contract. The sum stated should be a genuine estimate of the damage that the building owner may suffer. If, however, the sum stated is excessive and bears no relation to the actual damage, then it may be regarded as a penalty. In these circumstances, the courts will not enforce the amount stated in the contract but will assess the damage incurred on an unliquidated basis. Liquidated damages may be estimated on the basis of loss of profit in the case of commercial projects, but they are much more difficult to determine in the case of public works projects such as roads.

In some circumstances, a building owner may seek the occupation of a project even though it remains incomplete. This would normally deprive the owner of the right to enforce a claim for liquidated damages.

4.6 Unliquidated damages

When no liquidated damages have been detailed in the contract, the employer can still recover damages should the contractor fail to complete on time. The sum awarded in this case is that regarded as compensation for the loss actually sustained by the breach.

4.7 Specific performance

Specific performance was introduced by the courts for use in those cases where damages would not be an adequate remedy. In building contracts, this remedy is only really available in exceptional circumstances. The courts will enforce a party to do what it has contracted to do, in preference to awarding damages to the aggrieved party. A decree of specific performance will not be granted where the court cannot effectively supervise or enforce the performance.

4.8 Injunction

Injunction is a remedy for the enforcement of a negative undertaking. A party is prohibited from carrying out a certain action. It is often awarded by the courts as an effective remedy against nuisance.

4.9 Rescission

This is an equitable remedy, which endeavours to place the parties in the pre-contractual position by returning goods or money to the original owners. This means that the parties are no longer bound by the contract. This is granted at the discretion of the court, but will not be awarded where:

- the injured party was aware of the misrepresentation and carried on with the contract
- the parties cannot be returned to their original position
- another party has acquired an interest in the goods
- the injured party waited too long before claiming this remedy.

4.10 *Quantum meruit*

A claim for damages is a claim for compensation for loss. Where, under the terms of the contract, one party undertakes its duty for the other, and the other party breaks the contract, the former can sue upon a *quantum meruit* basis; that is, to claim a reasonable price for the work carried out. It is payment earned for work carried out. If the two parties cannot agree, the question of what sum is reasonable is decided by the court. A claim on a *quantum meruit* basis is appropriate where there is an express agreement to pay a reasonable sum upon the completion of some work. In assessing a *quantum meruit* claim, the parties may choose to use the various means that are available. For example, it may be based upon the costs of labour and materials plus profit, or the measurement of the work using reasonable rates.

Payment may be made on this basis in a number of different situations:

- Where work is carried out under a contract, but there are no express provisions on how the work will be paid.
- Where there is an express agreement to pay a reasonable sum for the work that has been carried out.
- If work has been carried out on a project assuming a valid contract, but where the contract turns out to be void.
- Where work is carried out by one party at the request of another following a letter of intent.

5 Settlement of disputes

> And God said unto Noah, the Ark shall be finished within seven days. And Noah saith, it shall be so. But it was not so. And the Lord saith, what seemeth to be the trouble this time? And Noah saith, mine subcontractor hath gone bankrupt. The pitch which thou commandest me to use has not arrived. The plumber hath gone on strike. And the glazier departeth on holiday to Majorca, even tho I did offer him double time.
>
> Christopher Taylor, Engineering Manager, Shell UK (quoting from 'The Lord said unto Noah', by Keith Waterhouse)

Construction contracts in the distant past consisted of a document of about five pages long. They generally concluded with a handshake, but underlying such agreements were an essential set of values of competence, fairness and honesty. Today things are different. We have developed a complex and onerous set of conditions that attempt to cover every eventuality and in so doing create loopholes that the legal professions can feast upon. It has often been suggested that the only individuals to make any money in the construction industry are the lawyers! Precious time and resources are thus drawn away from the main purpose of the industry: getting the project built on time, to the right design and use of construction technology at an agreed price and quality.

Everyone in the industry agrees that construction contracts need to become less adversarial and more simply constructed, emphasising the positive needs of the project. Consultants and contractors should be allowed and encouraged to use their best endeavours and to work together as a team (Chapter 11) rather than watching their individual backs all of the time. When a project ends up in a protracted dispute, the project will fail to meet its original goals and expectations. In addition, clients will suffer from high legal fees, delayed completion and occupation and general dissatisfaction. The contractor's profits will diminish and to these will be added additional legal fees. There are no winners under these circumstances.

The British construction industry offers its customers great flexibility. It is envied in many parts of the world. But, due to the pressures involved in getting a project completed as quickly as possible, many construction projects are inadequately or not fully designed before the construction work starts on-site. Changes to the design are therefore both inevitable and welcome in achieving satisfactory solutions to meet the needs of clients. But such changes have the result of changes to the contract programme, and this in turn affects the price originally quoted by the contractor to complete the project works. Disputes between the various parties that are involved are an ordinary and everyday occurrence. Thankfully, most of these differences of opinion or interpretation are resolved in an amicable and agreeable way. However, if the dispute is not managed efficiently, it leads onwards to conflict and the project begins to suffer in terms of time, cost and quality. This chapter considers the potential for escalation and the alternative ways in which differences of opinion or judgment can be resolved.

5.1 The reasons why disputes arise

The construction industry is a risky business. It generally does not build many prototypes, with each different project being individual in many respects. Even identical buildings that have been constructed on different sites create their own special circumstances, are subject to the vagaries of different site and weather conditions and use labour that may have different trade practices from one site to another. Even the identical building constructed on an adjacent site by a different contractor will have different costs and different problems associated with its construction. The introduction of new building materials and designs, changes to the procurement and organisation of the project and the poor margins of profitability provide a good platform for disputes. Disputes are therefore likely to arise under the best circumstances, even where every possibility has been potentially eliminated. Disputes between parties, it should be remembered, are really in no one's best interests. Here are some of the main areas where disputes might occur:

5.1.1 General

- Adversarial nature of construction contracts.
- Poor communication between the parties concerned.
- Proliferation of forms of contract and warranties.
- Fragmentation in the industry.
- Tendering policies and procedures.

5.1.2 Clients

- Poor briefing.
- Changes and variation requirements.
- Changes to standard conditions of contract.
- Interference in the contractual duties of the contract administrator.
- Late payments.

5.1.3 Consultants

- Design inadequacies.
- Lack of appropriate competence and experience.
- Late and incomplete information.
- Lack of coordination.
- Unclear delegation of responsibilities.

5.1.4 Contractors

- Inadequate site management.
- Poor planning and programming.
- Poor standards of work.
- Disputes with subcontractors.
- Delayed payments to subcontractors.
- Coordination of subcontractors.

5.1.5 Subcontractors

- Mismatch of subcontract conditions with main contract.
- Failure to follow and adopt agreed procedures.
- Poor standards of work.

5.1.6 Manufacturers and suppliers

- Failure to define performance or purpose.
- Failure of performance.

5.2 Issues for the resolution of disputes

The following matters need to be resolved in order to reduce the possibility of future disputes occurring:

- Clarification of responsibilities.
- Need of single-point responsibility contracts.
- Allocation of risk to the parties who are best able to control it.
- Further investigation of insurance-based alternatives.
- Need to develop and extend non-adversarial methods of dispute resolution.
- Partnership sourcing (contractors and consultants working in a consortium).
- Quality management and quality assurance.

5.3 Claims

It is evident from society in general that as individuals we are becoming more claims conscious. Firms of lawyers are now touting their services, often on a no-win no-fee basis. Everyone wants their pound of flesh and what they rightfully believe belongs to them. Claims are seen by many to be a last-resort issue. Even in the construction industry this is true, although some would want to argue that some contractors prepare their claims alongside their tender submissions.

Contractual claims arise where contractors assess that they are entitled to additional payments over and above that paid within the general terms and conditions of the contract. For example, the contractors may seek reimbursement for some alleged loss that has been suffered for reasons beyond their control. On many occasions, the costs incurred lie where they fall and contractors will have recourse to recover them. Thus, losses and delays arising from the intervention of third parties who are unconnected with the contract almost invariably fall with the contractor. The fact that a loss has been sustained, without fault on the part of the loser, may merit sympathy, but does not in itself demand compensation. Where a standard form of contract is used, many attempts may be made by contractors to invoke some of the compensatory provisions of the contract in order to secure further payment to cover the losses involved.

The details of such claims will be investigated by the quantity surveyors and a report made to the architect, engineer or other lead consultant. The report should summarise the arguments involved and set out the possible financial effect of each claim. Quantity surveyors frequently end up negotiating with contractors over such issues in an attempt to solve the financial problems and to arrive at an amicable solution, wherever this is possible. This is preferable to a lengthy legal dispute.

As with many issues in life, contractual claims are rarely the fault of one side only. If the claim cannot be resolved in this way, then some form of legal proceedings may be initiated. Particular care therefore needs to be properly exercised in the conduct of the negotiations since they may have an effect upon the outcome of any subsequent legal proceedings. Claims may be classified in several different ways. They usually reflect a loss and expense to a contractor.

5.3.1 Contractual claims

Contractual claims have a direct reference to conditions of contract. When the contract is signed by the two parties, the contractor and the employer, there is a formal agreement to carry out and complete the works in accordance with the information supplied through the drawings, specification and contract bills. Where the works constructed are of a different character or executed under different conditions, then it is obvious that different costs will be involved. Some of these additional costs may be recouped under the terms of the contract, through, for example, remeasurement and revaluation of the works, using the appropriate rules from the contract. Other additional costs that an experienced contractor had not allowed for within the tender may need to be recovered in a different way. This is usually under the heading of a contractual claim.

5.3.2 Ex gratia payments

Ex gratia payments are not based upon the terms or conditions of contract. However, the carrying out of the works has nevertheless resulted in some loss and expense to the contractor. The contractor has completed the project on time, to the required standards and conditions and at the price agreed. Perhaps, due to a variety of different reasons, and at no fault of the contractor, a loss has been sustained that cannot be related to the contractual conditions. On rare occasions, a sympathetic employer may be prepared to make a discretionary payment to the contractor. Such payments are made out of grace and kindness. They may be made because of a long-standing relationship and trust between employer and contractor, or because of outstanding service and satisfaction provided by the contractor. Nevertheless, they are rare occasions.

5.3.3 Forms of construction contracts

The standard forms of building and civil engineering contracts seek to clarify the contractual relationship between the employer and the contractor. As far as possible, ambiguities have been removed, but some nevertheless remain. If such forms or conditions of contract were not available, then the uncertainty between the two parties would be even greater. This could have the likely effect of increasing tender totals. Under the present conditions of contract, the contractual risks involved are shared between the employer and the contractor. Claims may arise, and these are known as loss and expense claims. They may also arise due to a breach of contract. The contractor must make a written application to the architect, in the first place, stating that a direct loss and expense has occurred or is likely to occur in the execution of the project. The contractor must further state that any reimbursement under the terms of the contract is unlikely to be sufficient. This information should be given to the architect or engineer promptly in order to allow as much time as possible to plan for other contingencies.

As soon as reasonably possible, the contractor should provide the architect or engineer with a written interim account providing full details of the particular claim and the basis upon which it is made. This should be amended and updated when necessary or when required. If the contractor fails to comply with this procedure, this might prejudice the investigation of the claim and any subsequent payments by the employer to the contractor.

The contractor is entitled to have such amounts included in the payment of interim certificates. However, in practice a large majority of claims are not agreed until the completion of the contract. In these circumstances, the contractor is entitled to receive part of the claim included in an interim certificate where this can be substantiated.

5.3.4 Contractors

Many contractors have well-organised systems for dealing with claims on construction projects and the recovery of monies that are rightly due under the terms of the contract. They are likely to maintain good records of most events, but particularly those where difficulties have occurred in the execution of the work. However, some of the difficulties may be due to the manner in which the contractor has sought to carry out the work and thus remain the entire responsibility of the contractor.

Claims that are notified or submitted late will inevitably create problems in their approval. In these circumstances, the architect might not have the opportunity to check the details of the contractor's submission. Such occurrences will not be favourably looked upon by the architect or the employer.

The contractor must prepare a report on why a particular aspect of the work has cost more than expected, substantiate this with appropriate calculations and support it with reference to the architect's instructions, drawings, details, specifications, letters, etc. The contractor must be able to show that, as an experienced contractor, they could not have foreseen the difficulties that occurred. They will also need to show that the work was carried out in an efficient, effective and economic manner.

Claims are for additional payments that cannot be recouped in the normal way simply through measurement and valuation. They are based on the assumption that the works constructed differed considerably from the works for which the contractor originally submitted a tender. The differences may have changed the contractor's preferred method of working and this in turn may have altered or influenced the costs involved. The rates inserted by the contractor in the contract bills are not now a true reflection of the work that has been executed.

Example

The construction of a major new factory on a greenfield site requires a large earth-moving contract. The quantities of excavation and its subsequent disposal have been included in the contract bills and priced by the contractor. It is found that the quantities of materials to be taken to tips has increased by 25 per cent by volume. This is due to variations to the contract and the unforeseen nature of some of the ground conditions of the construction site.

The contractor's pre-tender report indicates that a variety of tips at different distances from the site will be used for the disposal of the excavated materials. The contractor's tendering notes indicated that the tips nearer the site would be filled first. In this case they result in lower haulage costs, but also in this case are shown to have lower tipping charges. The disposal of the excavated material therefore includes two separate elements:

- The haul charges to the tip.
- The costs of the tip.

In the contract bills, the rate used by the contractor for disposal of excavated materials represents an average rate. This is based upon calculated average haul distances and average tipping charges based upon the weighted quantities in each tip. A revised average rate for the disposal of the excavated materials can be calculated similarly for the actual quantities of excavated materials that are involved. The data is shown in Box 5.1.

To simply continue to apply the contract bill rates in similar situations to that shown in Box 5.1 is erroneous. The rates no longer reflect the work to be carried out and the contractor's method of working, which have been changed by the variation to the contract. Other factors that might also need to be considered are the excavation of different types of construction materials that have been encountered, whether they bulk at different rates or whether they are more difficult to handle.

The increase in the amount of excavated materials on this scale may have other repercussions, such as an extension of time, which might also need to be considered. The method of carrying out the works might now be different from that originally envisaged by the contractor. Different types of mechanical excavators may be required or the plant originally selected to do the work might no longer be the most appropriate. This is especially so where cut and fill excavations are considered. The contractor may also be involved in hiring additional plant at higher charges and employing workpeople at overtime rates in order to keep the project on schedule.

The contractor must, as a matter of good practice, always put in writing:

- Applications for instructions, drawings, etc.
- Application for the nomination of subcontractors.
- Progress of the works and any delays.
- Notification of any claims under the contract in respect of:
    - variations
    - extensions of time
    - loss and expense.
- Confirmation of any oral instructions from the architect.

The contractor should also ensure that any certificates that are required under the terms of the contract are issued at the appropriate time. These may have some effect upon the validity or otherwise of a contractor's claim at a later stage.

5.4 Adjudication

The dispute resolution techniques that are described in the forms of contract (see Chapter 6) include adjudication, arbitration and litigation. The courts may also need to be called upon to

Box 5.1 Earth-moving claim

Contract bills

Excavated materials for disposal in the contract bills = 1,000,000 m^3. Contract bill rate is based upon:

Tip A	500,000 m^3	distance =	1 km	tip charge =	£0.10/m^3
Tip B	300,000 m^3	distance =	2 km	tip charge =	£0.20/m^3
Tip C	200,000 m^3	distance =	4 km	tip charge =	£0.30/m^3

$$\text{Average distance} = \frac{500{,}000 + 600{,}000 + 800{,}000}{1{,}000{,}000} = 1.9\text{km per m}^3$$

$$\text{Average tip charge} = \frac{(500{,}000 \times 10\text{p}) + (300{,}000 \times 20\text{p}) + (200{,}000 \times 30\text{p})}{1{,}000{,}000}$$

$$= £0.17 \text{ per m}^3$$

Final account

Actual quantities in the final account:

Tip A	500,000 m^3	distance = 1 km	tip charge = £0.10/m^3
Tip B	300,000 m^3	distance = 2 km	tip charge = £0.20/m^3
Tip C	300,000 m^3	distance = 4 km	tip charge = £0.30/m^3
Tip D	150,000 m^3	distance = 6 km	tip charge = £0.50/m^3

$$\text{Average distance} = \frac{500{,}000 + 600{,}000 + 1{,}200{,}000 + 900{,}000}{1{,}250{,}000} = 2.56\text{ km per m}^3$$

Average tip charge

$$= \frac{(500{,}000 \times 10\text{p}) + (300{,}000 \times 20\text{p}) + (300{,}000 \times 30\text{p}) + (150{,}000 \times 50\text{p})}{1{,}250{,}000}$$

$$= £0.22 \text{ per m}^3$$

enforce settlements that are reached by other methods. The parties may, however, decide to agree amongst themselves to use other alternative methods to settle their differences, such as alternative dispute resolution. This is considered towards the end of this chapter. It is claimed to be a non-adversarial technique, although its rise in popularity in the construction industry appears now to have waned in favour of more established techniques.

Adjudication was first introduced into the UK in the mid 1970s. Its application was restricted to disputes that occurred between the main contractor and the directly employed or domestic subcontractors. The process involved using an independent third party, an adjudicator, to help resolve a dispute that had arisen. The adjudicator could be appointed as part of the subcontract conditions, but invariably was only appointed after the dispute had occurred. The main advantage of using an adjudicator was the rapid response of the decision. The decision was binding, although, as in all disagreements, the parties had the right to take the dispute to a higher authority. Adjudication was subsequently introduced into the Joint Contracts Tribunal (JCT) form with contractor's design and more recently into JCT 2005. This followed one of the principles of better practice recommended by the Latham Report.

Adjudication is described in clause 41A of JCT 2005. The referral of a dispute to an adjudicator must be made within seven days and a decision must be given to the parties concerned, in writing, within twenty-eight days. The period for the decision can be extended for a further fourteen days if the parties to the dispute agree. The adjudicator, like anyone in hearing disputes, must act impartially to determine the facts and law that are applicable to the dispute.

The appendix to the form of contract seeks to identify who should nominate the adjudicator. Unless the parties agree to the contrary, this shall be the president or vice-president of the Royal Institute of British Architects (RIBA). As an alternative, the adjudicator may be a president or vice-president of the Royal Institution of Chartered Surveyors (RICS), the Construction Confederation (CC) or the National Specialist Contractors Council (NSCC). The adjudicator can be named in the contract in order to save time should a dispute occur. Whilst this is useful, the appointed adjudicator might not be suitable to review all disputes. The *JCT adjudication agreement* is a standard form that has been produced by JCT.

Within seven days of the notice to refer a matter to adjudication, a referral document should be provided that includes the particulars of the dispute, a summary of the issues involved and the remedy and relief that the adjudicator should consider. The powers of the adjudicator are described as follows:

- Using the adjudicator's own knowledge and expertise.
- Opening up, reviewing and revising certificates, opinions, decisions or notices.
- Requiring the parties to provide additional information.
- Requiring the parties to carry out tests or open up work.
- Visiting the site and workshops.
- Obtaining information from the employees of the parties concerned.
- Obtaining information from other third parties.
- Determining the payment of any interest within the terms of the contract.

The parties are normally responsible for their own costs, but the adjudicator may direct that, in fairness, the unsuccessful party can recover their costs from the successful party. There are several identified advantages of using adjudication in preference to other methods of settling disputes:

- It seeks to eliminate conflicts as quickly as possible by resolving disputes as they arise.
- It is recognised that the adjudicator's decision will be provided in the fastest possible way.
- It is intended to be the least expensive process for settling disputes, by reducing lawyers' charges.
- It can act as a referral system, preceding arbitration or litigation where the dispute is not resolved at this stage.
- It is anticipated that many disputes will be resolved and terminated at this stage, rather than proceeding towards more litigious action.

However, in practice the complexity that sometimes occurs with construction disputes cannot be dealt with effectively, and more rigorous methods of settlement may need to be employed. Sometimes adjudication may simply be considered as a temporary solution to a problem.

5.5 Arbitration

Arbitration is the main alternative to legal action in the courts, in order to settle an unresolved dispute. No one is compelled to submit a dispute to arbitration unless they have agreed to do so within the terms of the contract. Care should be taken, therefore, when considering the completion of the appendix. Once a person has agreed to this method of settling a disagreement, they cannot then take legal proceedings prior to arbitration. If they attempt to do so, the courts will stay such proceedings. All of the standard forms of contract used in the construction industry include an arbitration provision. It is therefore the procedure that is most commonly used for dealing with disputes that arise between the various parties concerned.

Arbitration is described as a private procedure for settling a wide range of disputes in a diverse range of industries. The dispute is settled by an impartial individual or panel of arbitrators appointed solely for that purpose. The decision of the arbitrator is described as an award. Arbitrators do not need to have any particular skills or qualifications. In the majority of cases, an individual is appointed to serve as an arbitrator on the basis of experience or expertise in the subject of the dispute. However, most practising arbitrators are fellows of the Chartered Institute of Arbitrators (FCIArb). There is a preference in the construction industry for arbitrators who have qualifications in construction and law.

The arbitration agreement in the JCT 2005 is covered in article 7A of the articles of agreement. The parties, under article 5.2, agree that proceedings will not take place until after practical completion has been achieved, or unless the contractor's termination of the contract has been made or if the project is abandoned. There are exceptions, however, to these rules where matters can be taken to arbitration during the progress of the works. Of course, where the two parties otherwise agree, the guidelines can be amended as required. The following matters can therefore be dealt with prior to the issue of the certificate of practical completion:

- *Article 3*: contractor's objection to the appointment following the death of the architect.
- *Article 4*: contractor's objection to the appointment following the death of the quantity surveyor.
- *Clause 4*: dispute over the power to issue an instruction.
- *Clause 30*: a certificate being improperly withheld or not being in accordance with the conditions.
- *Clause 25*: dispute over the difference of an extension of time.
- *Clauses 32/33*: disputes concerning outbreak of hostilities or war damage.

An arbitrator's powers are very wide. They may review and revise certificates and valuations. They may also disregard opinions, decisions or notices that have already been given.

Here are the essential features of a valid arbitration agreement:

- The parties must be capable of entering into a legally binding contract.
- The agreement should whenever possible be in writing.
- It must be signed by the parties concerned.
- It must state clearly those matters which will be submitted to arbitration, and when the proceedings will be initiated.
- It must not contain anything that is illegal.

5.5.1 Arbitration Act 1996

The Arbitration Act 1996 extended the previous Act of 1979. It came into force in 1996 and it applies to England, Wales and Northern Ireland. Scotland is excluded because different laws apply in Scotland. The footnote in JCT 2005 reminds the reader of this. The Act consists of a number of provisions, some of which will apply to all arbitrations, and others to arbitrations provided by the parties in their agreement. The majority of arbitration agreements adopted by the construction industry accept the Act in its entirety.

5.5.2 Arbitration versus litigation advantages

- Arbitration is generally less expensive than court proceedings.
- Arbitration is a more speedy process than an action at law; a year awaiting a case to come before the courts is not uncommon.
- Arbitration hearings are usually held in private; this therefore avoids any bad publicity that might be associated with a case in the courts.
- The time and place of the hearing can be arranged to suit the parties concerned; court proceedings take their place in turn amongst the other cases and at law courts concerned.
- Arbitrators are selected for their expert technical knowledge in the matter of dispute; judges do not generally have such knowledge.
- In cases of dispute which involve a building site or property, it can be insisted that the arbitrator visit the site concerned; although a judge may decide upon a visit, this cannot be enforced by the parties concerned.

5.5.3 Disadvantages

- The courts will generally always be able to offer a sound opinion on a point of law; the arbitrator may seek the opinion of the courts, but this could easily be overlooked and then a mistake could occur.
- An arbitrator does not have the power to bring into an arbitration a third party against their wishes; the courts are always able to do this.

5.5.4 Terminology

arbitrator	The person to whom the dispute is referred for settlement. Arbitrators are often appointed by, for example, the president of the RIBA, RICS, CC, NSCC or CIArb (JCT 2005 appendix). In practice, they may be selected because of their expert technical knowledge regarding the subject matter in dispute.
umpire	It may occasionally be preferable to appoint two arbitrators. In this event, a third arbitrator, known as an umpire, is appointed to settle any dispute over which the two arbitrators cannot agree.
reference	The actual hearing of the dispute by the arbitrator.
award	The decision on the matter concerned made by the arbitrator.
respondent	This is the equivalent of the defendant in a law court.
claimant	The equivalent of the plaintiff.
expert witness	This is a special type of witness who plays an important part in arbitrations. Ordinary witnesses must confine their evidence strictly to the statement of facts. Expert witnesses may, however, forward their opinion based upon technical knowledge and practical experience. Prior to presenting evidence, they must show by experience and academic and professional qualifications that they can be recognised as an expert in the subject matter.

5.5.5 *Appointment of the arbitrator*

An arbitrator or umpire should be a disinterested person who is quite independent from the parties that will be involved in the proceedings. The person appointed should be someone who is sufficiently qualified and experienced in the matter of the dispute. However, it is for the parties concerned to choose the arbitrator, and the courts will not generally interfere even where the person appointed is not really the most appropriate person to settle the dispute. Arbitrators may, however, be disqualified if it can be shown that:

- They have a direct interest in the subject matter of the dispute (e.g. where the decision may have direct repercussions on their own professional work).
- They may fail to do justice to the arbitration by showing a bias towards one of the parties concerned (e.g. it could be argued that architects might show favour to an employer since they are normally employed on this 'side' of the industry).

Each of the parties to the arbitration agreement must be satisfied that the arbitrator who is appointed will give an impartial judgment on the matter of the dispute. The remarks expressed by the arbitrator during the conduct of the case will probably reveal any favouring of one of the parties in preference to the other. This may lead to removal by the courts of the arbitrator on the grounds of misconduct.

Once an arbitrator has been appointed, the general matters relating to the dispute will need to be established. It will also be necessary to determine:

- the facilities available for inspecting the works
- whether the parties will be represented by counsel
- the matters the parties already agree upon
- the time and place of the proposed hearing.

5.5.6 *Outline of the procedure: the pleadings*

Pleadings are the formal documents which may be prepared by counsel or a solicitor. The arbitrator will first require the claimant to set out the basis of the case. This will be included in a document, termed the *points of claim*, which will then be served on the respondent. The respondent then submits a reply in answer to the points of claim, termed the *points of defence*. The respondent may also submit points of counterclaim, which will be served on the claimant at the same time. This may raise relevant matters that were not referred to in the points of claim. The claimant, in reply to the matters raised in the counterclaim, will submit points of reply and defence to counterclaim.

The purpose of the above documents is very important, since they will make clear to the arbitrator the matters that are in dispute. Furthermore, the parties involved cannot stray beyond the scope of the pleadings without leave from the arbitrator.

5.5.7 *Discovery*

Once the pleadings have been completed, the precise issues which the arbitrator is to decide should be clear. Every fact to be relied upon must be pleaded, but the manner in which it is to be proved need not be disclosed until the reference. The term *discovery* means the disclosure of all documents which are in the control of each party and which are in any way relevant to the issues of arbitration. Each party must allow the other to inspect and to take copies of all or any of the documents in their list, unless they can argue on the grounds that it is privileged. The most important of these types of document are the communications between a party and their solicitors for the purpose of obtaining legal advice. A party who refuses to allow inspection may be ordered to do so by the arbitrator.

In fixing the date and place of the hearing, arbitrators have the sole discretion, subject to anything laid down in the arbitration agreement. They must, however, be seen to act in a reasonable manner. A refusal to attend the hearing by either party, after reasonable notice has been given, may empower the arbitrator to proceed without that party, i.e. *ex parte*.

5.5.8 The hearing

The procedure of the hearing follows the rules of evidence used in a normal court of law. The parties may or may not be represented by counsel or other representatives.

The claimant sets out the case, and calls each of the witnesses in turn. Once they have given their evidence, they are cross-examined by the respondent (or counsel/representative where appropriate). The claimant is then allowed to ask these witnesses further questions on matters which have been raised by the cross-examination.

A similar procedure is then adopted by the respondent, who sets out the details of the counterclaim if this is necessary. Witnesses are examined and these in turn are cross-examined by the claimant. The claimant then replies to the respondent's defence and counterclaim, and presents a defence to the counterclaim. When this has been completed, the respondent sums up the case in an address to the arbitrator known as the respondent's closing speech. The claimant then has the right of reply, or the last word in the case.

The trial is now ended and both parties await the publication of the award. The arbitrator may now decide to inspect the works, if this has not already been done, or re-examine certain parts of the project in more detail. The arbitrator will usually make the decision in private, and will set the decision out in the award. This is served on the parties after the appropriate charges have been met. It is enforceable in much the same way as a judgment debt, where the successful party can reclaim such costs as the arbitrator has awarded.

5.5.9 Evidence

To enable the arbitrator to carry out justice between the parties, the evidence must be carefully considered. This is submitted in turn by the claimant and the respondent. Evidence is the means by which the facts are proved. There are rules of evidence which the arbitrator must ensure are observed. These have been designed to determine four main problems:

- Who is to assume the burden of proving facts?
  Generally speaking, the person who sets forth a statement has the burden of determining its proof. The maxim 'innocent until proved guilty' is appropriate in this context.
- What facts must be proved?
  A party must give proof of all material facts which were relied upon to establish the case, although there can be exceptions to this rule. For example, the parties may agree on formal 'admissions' in order to dispense with the necessity of proving facts which are not in dispute.
- What facts will be excluded from the cognisance of the court?
  In order to prevent a waste of time or to prevent certain facts from being put before juries,

which might tend to lead them to unwarranted conclusions, English law permits proof of facts which are in issue and of facts which are relevant to the issue.

- How proof is to be effected?
  The law recognises three kinds of proof. Oral proofs are statements made verbally by a witness in the witness box. Documentary proof is contained in the documents that are available. Real proof could include models or a visit to the site in order to view the subject matter.

It is usual in arbitration for the evidence to be given on oath. The giving of false evidence is perjury and is punishable accordingly by fine or imprisonment. The arbitrator has no right to call a witness, except with the consent of both parties. Witnesses will, however, usually answer favourably to the party by whom they are called, in order to further that party's case. They must not, in general, be asked leading questions which attempt to put the answer in the mouth of the witness. For example, this is a leading question: Did you notice that the scaffold was inadequately fixed? It must be rephrased: Did you notice anything about the scaffolding? Leading questions can, however, be used to the opposing party's witnesses. The arbitrator must also never receive evidence from one party without the knowledge of the other. Where communications are received from one party, the other party must immediately be informed. An arbitrator must always refuse to admit evidence on the grounds that:

- The witness is incompetent, e.g. refuses to take the oath, too young, mentally incapacitated.
- The evidence is irrelevant, e.g. evidence which the arbitrator considers has no real bearing upon the facts.
- The evidence is inadmissible, e.g. hearsay evidence not made under oath and not capable of cross-examination. (There are, however, exceptions to this rule such as statements made on behalf of one party against their own interests.)

Where documents are used as evidence, it is the arbitrator's responsibility to ensure their authenticity. Where one party produces a document, this must be proved unless the other side accepts it as valid. Documents under seal must be stamped, and generally the original document must be produced wherever possible.

5.5.10 Stating a case

A question of law may arise during the proceedings, and the arbitrator may deal with this in one of three ways:

- Decide the matter personally.
- Consult counsel or a solicitor.
- State a case to the courts.

Having decided on the third course of action, a statement is prepared outlining clearly all the facts in order that the courts may decide a point of law. Once the court has given its decision, the arbitration proceedings can continue. The arbitrator must then proceed in accordance with the court's decision. Failure to do so results in misconduct on the part of the arbitrator. The arbitrator may voluntarily take this course of action or be required to do so by one of the parties.

The arbitrator may also state a case to the courts upon the completion of the arbitration. In this case, the award will be based upon the alternative findings to be resolved by the courts. The decision that the courts reach will depend upon which alternative is to be followed. For example, the arbitrator may state that a particular sum should be paid by one of the parties to the other if the courts approve the arbitrator's view of the law. Where the courts do not confirm this opinion, the arbitrator will have also indicated the course of action to be taken.

5.5.11 The award

The arbitrator's award is the equivalent of the judgment of the courts. The award must be made within the terms of reference, otherwise it will be invalid and therefore unenforceable. The essentials of a valid award can be summarised as follows:

- It must be made *within the prescribed time limit* that has been set by the parties.
- It should *comply with any special agreements* regarding its form or method of publication that has been laid down in the arbitration agreement.
- It must be *legal and capable of enforcement* at law.
- It must *cover all the matters* which were referred to the proceedings.
- It must be *final* in that it settles all the disputes which were referred under the arbitration.
- It must be *consistent and not contradictory or ambiguous*; its meaning must be clear.
- It must be *confined solely to the matters in question*, and not matters which are outside the scope of dispute.
- The award should *generally be in writing*, in order to overcome any problems of enforcing it in practice.

5.5.12 Publication of the award

The usual practice is for the arbitrator to notify both the parties that the award is ready for collection upon the payment of the appropriate fees to the arbitrator. If the successful party pays the fees, then they are able to sue the other party for that amount, assuming that the costs follow the event. If the award is defective or bad, or it can be shown that there have been irregularities in the proceedings, application may be made to the courts to have it referred back for reconsideration or set aside altogether.

5.5.13 Referring back the award

Here are the grounds on which the court is likely to refer back an award:

- Where the arbitrator makes a mistake so that it does not express the true intentions.
- Where it can be shown that the arbitrator has misconducted the proceedings, for example, in hearing the evidence of one of the parties in the absence of the other.
- Where new evidence, which was not known at the time of the hearing, comes to light and as such will affect the arbitrator's award.

The application to refer back an award should be made within six weeks after the award has been published. The courts have the full discretion regarding the costs of an abortive arbitration.

There is, of course, the right of appeal against the court's decision. The arbitrator's duty in dealing with the referral will depend largely upon the order of the court. As a rule, fresh evidence will not be heard unless new evidence has come to light. The amended award should normally be made within three months of the date of the court's order.

5.5.14 Setting aside the award

When the courts set aside an award it becomes null and void. The situations where the courts will do this are similar to those for referring back an award, but much more serious:

- Where the award is void, for example, if the arbitrator directs an illegal action.
- The discovery of evidence that was not available at the time the arbitration proceedings were held.
- Where the arbitrator has made an error on some point of law.
- Misconduct on the part of the arbitrator by permitting irregularities in the proceedings.
- Where the award has been obtained improperly, for example, through fraud or bribery.
- Where the essentials of a valid award are lacking, for example, the award is inconsistent or impossible of performance.

5.5.15 Misconduct by the arbitrator

Arbitrators must carry out their duties in a professional manner. Where they are guilty of misconduct, the award can be referred back by the courts, and in a serious case the effects of setting aside an award are to make it null and void. Misconduct may be classified as *actual* or *technical*.

Actual misconduct would occur where arbitrators have been inspired in their decisions by some corrupt or improper motive or have shown bias to one of the parties involved. Technical misconduct is when some irregularities in the proceedings occur, and may include:

- the hearing of evidence of one party in the absence of the other
- the examination of witnesses in the absence of both parties
- the refusal to state a case when requested to do so by one of the parties
- exceeding jurisdiction beyond the terms of the reference
- failing to give adequate notice of the time of the proceedings
- delegating authority
- an error of law.

Although an arbitrator may be removed because of misconduct, this will not terminate the arbitration proceedings in favour of, say, litigation. The courts, on the application of any party to the arbitration agreement, may appoint another person to act as arbitrator. The courts may also remove an arbitrator who has failed to commence the proceedings within a reasonable time or has delayed the publication of the award.

If the arbitrator dies during the proceedings, the arbitration will not be revoked. The parties must agree upon a successor.

The parties may at any time, by mutual agreement, decide to terminate the arbitrator's appointment. Where duties have been commenced, there will be entitlement to some remuneration.

5.5.16 Costs

An important part of the arbitrator's award will be the directions regarding the payment of costs. These costs, which include the arbitrator's fees, can sometimes exceed the sum which is involved in the dispute. The arbitrator must exercise discretion regarding costs, but should follow the principles adopted by the courts.

The agreement generally provides for the 'costs to follow the event', which means that the loser will pay. Where an arbitration involves several issues, and the claimant succeeds on some but fails on others, the costs of the arbitration will be apportioned accordingly. If the award fails to deal with the matter of costs, then any party to the reference may apply to the arbitrator for an order directing by whom or to whom the costs shall be paid. This must be done within fourteen days of the publication of the award. A provision in an arbitration agreement that a party shall bear their own costs or any part of them is void.

The entire costs of the reference, which will include not only the costs of the hearing but also the costs incurred in respect of preliminary meetings and matters of preparation, are subject to taxation by the courts. This involves the investigation of bills of costs with the objective of reducing excessive amounts and removing improper items. For example, if a party instructs an expensive counsel which the issues involved did not justify, then the fees paid will be substantially reduced. If witnesses have been placed in expensive hotels, then this may be struck from the claim as an unnecessary expense. Such items are known as 'solicitor and client charges' and have usually to be borne by the successful party.

5.5.17 JCT arbitration rules

The Joint Contracts Tribunal has published a set of arbitration rules for use with all of its various forms of building contracts. The rules contain stricter time limits than those prescribed by some arbitration rules or those frequently used in practice, largely in an attempt to avoid unnecessary costs.

The rules follow much of what has previously been described, but also seek to capitalise on what might be termed good practice. There are twelve rules:

- *Rule 1*: arbitration agreements in each of the JCT forms of contract.
- *Rule 2*: interpretation and provisions as to time.
- *Rule 3*: service of statements, documents and notices – content of statements.
- *Rule 4*: conduct of the arbitration – application of rule 5, rule 6 or rule 7.
- *Rule 5*: procedure without hearing.
- *Rule 6*: full procedure with hearing.
- *Rule 7*: short procedure with hearing.
- *Rule 8*: inspection by arbitrator.
- *Rule 9*: arbitrator's fees and expenses – costs.
- *Rule 10*: payment to trustee-stakeholder.
- *Rule 11*: the award.
- *Rule 12*: powers of arbitrator.

5.6 Litigation

Litigation is a dispute procedure which takes place in the courts. It involves third parties who are trained in the law, usually solicitors and barristers, and a judge who is appointed by the courts.

This method of solving disputes is often expensive and can be a very lengthy process before the matter is resolved, sometimes taking years to arrive at a decision. The process is frequently extended to higher courts, involving additional expense and time (Chapter 1). Also, since a case needs to be properly prepared prior to the trial, a considerable amount of time can elapse between the commencement of the proceedings and the trial, as noted above.

A typical action is started by the issuing of a writ. This places the matter on the official record. A copy of the writ must be served on the defendant, either by delivering it personally or by other means, such as through the offices of a solicitor. The general rule is that the defendants must be made aware of the proceedings against them. The speed of a hearing in most cases depends upon:

- availability of competent legal advisers to handle the case, i.e. its preparation and presentation
- expeditious preparation of the case by the parties concerned
- availability of courts and judges to hear the case.

The amount of money involved in the case will determine whether it is heard in the county court or High Court. Where the matter is largely of a technical nature, the case may be referred in the first instance to the Official Referees Court (now referred to as the Technology and Construction Court). An official referee is a circuit judge whose court is used to hearing commercial cases, and hence handles most of the commercial and construction disputes. Under these circumstances, a full hearing does not normally take place, but points of principle are established. The outcome of this hearing will determine whether the case then proceeds towards a full trial.

Under some circumstances, the plaintiff may apply to the court for a judgment on the claim (or the defendant for a judgment on the counterclaim) on the ground that there is no sufficient defence. Provided that the court is satisfied that the defendant (or plaintiff) has no defence that warrants a full trial of the issues involved, judgment will be given, together with the costs involved.

Every fact in a dispute that is necessary to establish a claim must be proved to the judge by admissible evidence whether oral, documentary or of other kind. Oral evidence must normally be given from memory by a person who heard or saw what took place. Hearsay evidence is not normally permissible.

In a civil action, the facts in the dispute must be proved on a balance of probabilities. This is unlike a criminal case, where proof beyond reasonable doubt is required. The burden of proof usually lies upon the party asserting the fact.

5.7 Alternative dispute resolution

Alternative dispute resolution (ADR) is a non-adversarial technique which is aimed at resolving disputes without resorting to the traditional forms of either litigation or arbitration. The process was developed in the USA but has also been widely used elsewhere in the world. It is claimed to be less expensive, fast and effective. It is also less threatening and stressful. ADR offers the parties who are in dispute the opportunity to participate in a process that encourages them to solve their differences in the most amicable way possible. Table 5.1 illustrates a comparison between litigation, arbitration, adjudication and ADR.

Before the commencement of an ADR negotiation, the parties who are in dispute should have a genuine desire to settle their differences without recourse to either litigation or arbitration. They must therefore be prepared to compromise some of their rights in order to achieve a

Table 5.1 Characteristics of dispute resolution

Characteristics	*Litigation*	*Arbitration*	*Alternative dispute resolution*
Place/conduct of hearing	Public court: unilateral initiation; compulsory	Private (with few exceptions); bilateral initiation; voluntary (subject to statutory provisions)	Private: bilateral initiation; voluntary
Hearing	Formal: before a judge	Formal: conforming to rules of arbitration; before an arbitrator	Informal: before a third party (a neutral)
Representation	Legal: lawyers influence settlement	Legal: lawyers influence settlement	Legal only if necessary; disputants negotiate settlement
Resolution/ disposal	Imposed by a judge after adjudication; limited right of appeal	Award imposed by an arbitrator; limited right of appeal	Mutually accepted agreements; option of arbitration if dissatisfied
Outcome	Unsatisfactory: legal win or lose	Unsatisfactory: legal win or lose	Satisfactory: business relationship maintained
Time	Time-consuming	Can be time-consuming	Fast
Cost	Expensive; uneconomic	Uneconomic	Economic

Source: Based upon *Alternative Dispute Resolution (ADR) in Construction* (Kwayke 1993).

settlement. Proceedings are non-binding until a mutually agreed settlement is achieved. Either party can therefore resort to arbitration or litigation if the ADR procedure fails.

5.7.1 Common forms of ADR

Conciliation

Conciliation is a process where a neutral adviser listens to the disputed points of each party and then explains the views of one party to the other. An agreed solution may be found by encouraging each party to see the other's point of view. With this approach, the neutral adviser plays the passive role of a facilitator. Recommendations are not made by the adviser, and any agreement is reached by the parties agreeing to settle their differences. Where an agreement is achieved, the neutral adviser will put this in writing for each of the parties to sign.

Mediation

Mediation is often defined as facilitated negotiation. It enables the party to a dispute to arrive at a mutually acceptable settlement. It is a system of 'give and take'. No judgment is passed and the outcome of the dispute is in the hands of those who are party to it. This is fundamentally different from the standard form of dispute resolution, with most regarding adjudication as the first step in a construction dispute. Mediation can have significant savings in both time and cost. The hours that can be spent on progressing a claim, submitting a referral notice, replying to evidence submitted by the other party, attending meetings and the possibility of a lengthy court process all have a financial impact on a business. Where 'experts' become involved, this all adds to the cost. Mediation, by comparison, can be extremely flexible, can be arranged at short notice and the time involved can be limited. The process is also completely confidential, so any commercial agreements remain private. This can be significant where agreements made will not be disclosed and thus set a precedent for other similar claims. Creative solutions can be found that are usually not possible through adjudication or the courts.

A disadvantage may be where the strength of one party tests out the other in attempting to assess at what level of settlement they can be persuaded to agree. Mediation is not suitable for all occasions. Parties, it should be remembered, are also able to settle their differences without involving any third party such as a mediator.

The mediation process can take many formats, but the preferred method is for the parties to meet and to have the opportunity to briefly state their case. Only strict rules of courtesy apply. Each party is given the opportunity to reply to the points made by the other party. The mediator will then spend time with both of the parties separately. As the process evolves, both parties will move towards forming a binding agreement, highlighting the points that are important to themselves and conceding the lesser issues. Mediation is generally accepted as a good process for the construction industry by allowing the parties involved to fully air their grievances and work towards a mutually agreed settlement.

Executive tribunal

An executive tribunal is a more formal arrangement undertaken by a group comprising a neutral adviser together with representatives of the parties involved in the dispute. This group has the authority to settle the matters of dispute that created the conditions of ADR in the first place. At the hearing, each party makes representations to the chairperson and each party can raise questions and seek points of clarification. Witnesses may be called upon to give evidence, although this is unusual. Each of the parties in dispute then attempts to settle the matter in private in order to achieve a negotiated settlement. Again, the agreement is put in writing for endorsement by the parties concerned.

Combination

Whilst the above arrangements may appear to be separate approaches to ADR, in practice a combination of the different aspects may be used in order to solve the dispute.

5.7.2 ADR advisers

The neutral advisers are typically construction professionals or lawyers. Lawyers are likely to have some knowledge of the construction industry and will probably have been involved in settling such disputes in the traditional way. Persons appointed as ADR advisers will need to be trusted by the different parties and acceptable to both with known impartiality and fairness. They will seek to develop a process that suits the particular circumstances involved, seek out the truth of the dispute through questioning and debate and be the sorts of individuals who are able to solve problems. They will have good communication skills and be able to put each of the parties and the witnesses at ease in order to build their trust and confidence. Additionally, they will be able to appreciate all of the relevant issues involved and direct the parties towards the matters that are of crucial importance to the case.

5.7.3 Role of advisers in ADR

- Educating the parties involved of the procedures to be employed and the outcomes expected.
- Arranging the meetings, setting agendas and outlining the protocol to be used.

- Managing the process.
- Clarifying the issues involved and seeking conciliation rather than encouraging confrontation.
- Allowing each party to state and explain their case.
- Encouraging each party to see the other party's point of view.
- Preparing a report and assisting the parties to accept the agreement as binding.

Part 2
Procurement

6 Forms of contract

The UK is fortunate to have the benefit of a wide variety of published institutional standard forms of contract available for use in the building and civil engineering industries. However, a number of government-sponsored reports have also highlighted that this has major disadvantages, identifying duplication of effort and a wasteful use of resources at almost every level of activity. It has been suggested that many of these forms also probably help to fuel the adversarial nature of the construction industry in which they are applied. It can also be argued that to write and interpret the clauses of the various forms alone represents an industry in itself.

The widespread use of standard forms within the construction industry is also partly accounted for by the practical impossibility of writing a set of new contract conditions for every project, even if this were in any way desirable.

The construction of building and civil engineering projects represents a major investment for any client. In some cases, this will represent the single highest purchase ever made in an organisation. For clients who undertake construction projects as a regular part of their activities, the correct choice of a form of contract is more important, since the application of the principles involved may be seen as precedents in the administration of their contracts. Since large sums of money are likely to be involved in these activities, it is important that the contractual arrangements should always be formal and legal from the outset of the project. Where a client allows consultants or contractors to begin their work on an informal basis, then the client's bargaining position is thereafter weakened or even eliminated. Under these circumstances, a worst-case scenario for clients is that they may expect to spend years, often at substantial legal expense, in arguing over the precise nature of the contractual arrangements which should have been clarified from the start of the project.

6.1 The standard form of contract for building works

A standard form of contract for building projects in the UK was introduced in the latter part of the nineteenth century. It consisted of only nineteen clauses, compared with the forty-two clauses plus supplemental provisions in JCT 98. The form was commonly described as the RIBA form until 1977, when the term JCT (Joint Contracts Tribunal) contract was adopted.

The RIBA form was originally written by the Royal Institute of British Architects (RIBA), the National Federation of Building Trades Employers (NFBTE) and the then Institute of Building (IOB). In 1931, the IOB withdrew and much later the NFBTE became known as the Construction Confederation (CC) after a number of different name changes. The Royal Institution of Chartered Surveyors (RICS) became involved in the preparation of the form in 1952, and by 1963 the JCT consisted of ten bodies from the construction industry. Subcontracting bodies also eventually became members of the JCT.

The JCT has existed since 1931 and has produced forms of contract, guidance notes and other standard documentation for use in the construction industry. The standard form was substantially rewritten in 1939, 1963 and 1980. The last of these was described as JCT 80. Since then there have been major revisions in 1998, 2005, 2011 and 2016.

The present body responsible for drafting the current form includes public and private sector employers, architects, quantity surveyors, contractors and subcontractors.

6.2 Types of contract envisaged

The building or civil engineering contract typically refers to the contractual arrangement between the client (employer in building forms of contract and promoter in civil engineering) and contractor. A large part of this book has been written with this in mind. However, it should be remembered that, with the construction of most building and civil engineering projects, contracts also need to be formed between other individuals or firms. Besides the main contract, these include:

- engagement of the different consultants
- nominated subcontractors and suppliers
- trade or domestic subcontractors.

In addition, the different contractual arrangements also require collateral warranties and performance bonds, and in some cases personal and parent company guarantees.

6.3 Employer and contractor

As long ago as 1964, the Banwell Report recommended the use of a single form of contract for the whole of the construction industry, this being both desirable and practicable. In more recent years, the Latham Report (1994) has reiterated these comments. *Rethinking Construction*, the Egan Report (1998), makes further comments. Chapter 12 considers some of the issues raised from these reports in more detail. Unfortunately, since the Banwell Report of 1964 this apparently good suggestion has been thwarted, with just the opposite taking place. Since that time there has been a plethora of different forms of contract designed to suit the individual interests of particular employers and changes in the way that construction work is now often procured. The different forms of contract also have the vested interest of different parties and institutions, who, for a variety of reasons, whilst incorporating good practices, will at the same time wish to retain their separate identity. The better or fairer forms of contract have incorporated the views of the different interested groups within the construction industry, such as employers and contractors. Such forms have then been prepared as a joint effort between the various different parties involved. Many of the current forms of contract include wide representation from different organisations, sectors and interests in the construction industry.

The widespread use of different forms of contract is, of course, exacerbated where international projects are concerned. Not only are further additional forms required, but the procurement methods used can also be considerably different from those used in the UK and different laws may exist. Note that aspects of the laws of England and Scotland themselves continue to remain at variance. There are common threads that run throughout all of the forms of contract regarding payments, variations, quality, time, etc. However, whilst the general layout and content of the various forms may appear somewhat similar, the details may vary considerably. The interpretation of the individual clauses will also differ. In some cases, these have been clarified through the application of case law resulting from differences of opinion being settled in a court of law. The principles of the case law may also only apply to the form of contract in question and thus may not be applied universally across all of the different forms.

The selection of a particular form of contract depends upon several different circumstances, such as the following:

- *Type of work to be performed*: building, civil engineering, process plant engineering.
- *Size of project*: forms are available for major and minor works and those of an in-between nature.
- *Status of designer*: architects are more likely to prefer JCT or Association of Consultant Architects (ACA), whereas civil engineers will opt for an Institution of Civil Engineers (ICE) or New Engineering Contract (NEC) form.
- *Public or private sector*: different forms are available for use by private clients and local and central government. In addition, large industrial corporations may have their own forms of contract.
- *Procurement method to be used*: procurement methods such as design and build necessitates the use of forms more amenable to peculiarities of the procurement method.

A major advantage of using a standard document is that those who use it regularly become familiar with its contents and can apply them more easily and more consistently in practice. Individuals become aware of the strengths and weaknesses of a form and are able to identify the potential areas where disputes may arise, and take corrective action where possible. They are also able to identify the form's suitability for projects with which they are concerned. The range of forms adopted by consultants is often more restricted than those faced by contractors.

6.4 Main contract forms

According to the *Construction Contracts and Law Survey* (2015) (Chapter 7), the JCT form of contracts remains the most popular form of contract for building contracts in the UK. This is commonly referred to as the Standard Building Contract (SBC). The Institution of Civil Engineers (ICE) form remains the most popular form for civil engineering contracts, although the NEC form is gaining in popularity. There is a trend towards moving to more collaborative forms of contract such as NEC. Whilst JCT is in common use, its original introduction into the industry in 1980 faced considerable opposition due to the new procedures that were being introduced. It could also be argued that the introduction of JCT 80 also encouraged the ACA to prepare their own form of contract. This had great similarities to what was referred to as the 1963 RIBA form. Due to the overt complexity of administering aspects of JCT 80, particularly in respect of nominated subcontractors, the Joint Contracts Tribunal introduced a new intermediate form of contract in 1984. This has since been revised to bring it in line with the other current JCT forms. The reason for introducing this form was to provide contract conditions that were more appropriate for use on 'medium-sized' building projects. In practice, this form has received a more widespread use, often on the sort and size of projects that should have adopted the SBC form as the preferred form of contract.

6.5 Joint Contracts Tribunal (JCT) forms 2016

The JCT is widely represented from within the construction industry. Its constituent bodies in 2016 were:

- British Property Federation.
- Contractors Legal Grp Limited.
- Local Government Association.
- National Specialist Contractors Council Limited.
- Royal Institute of British Architects.
- Royal Institution of Chartered Surveyors.
- Scottish Building Contract Committee Limited.

In 1967, when standard forms were issued, there were eleven constituent bodies, but these subsequently reduced to the current configuration due to industry changes and mergers of several constituent bodies.

The different forms of contract available from JCT are discussed below (see Figure 6.1).

6.5.1 JCT Standard Building Contract (SBC) family

The JCT standard form of contracts are intended for traditionally procured large and complex construction projects. There are three main options to select from, as described in the following sections. The SBC family also contains a series of standard subcontract agreements, with subcontract conditions and subcontractor design portion of works.

Standard Building Contract with Quantities

This form of contract is appropriate for larger projects that have been designed by the employer where drawings and bills of quantities have been provided. A contract administrator and quantity surveyor are normally employed to administer the conditions. It can be used where the contractor might design discrete parts of the works. It is suitable for those projects that are carried out in sections and can be used for both private and local authority employers.

Standard Building Contract without Quantities

With this form of contract, the employer provides the contractor with drawings and either a specification or work schedules (schedule of works) to define and adequately scope the quality of the works. It is used where the degree of complexity is not such to require bills of quantities.

In addition, there are also a number of additional supplements to cover. It can be used where the contractor might design discrete parts of the works. It is suitable for those projects that are carried out in sections and can be used for both private and local authority employers.

Standard Building Contract with Approximate Quantities

This is similar to the form described above. It is used in those circumstances where the full extent of the works cannot be properly described at the tender stage. This may be because there

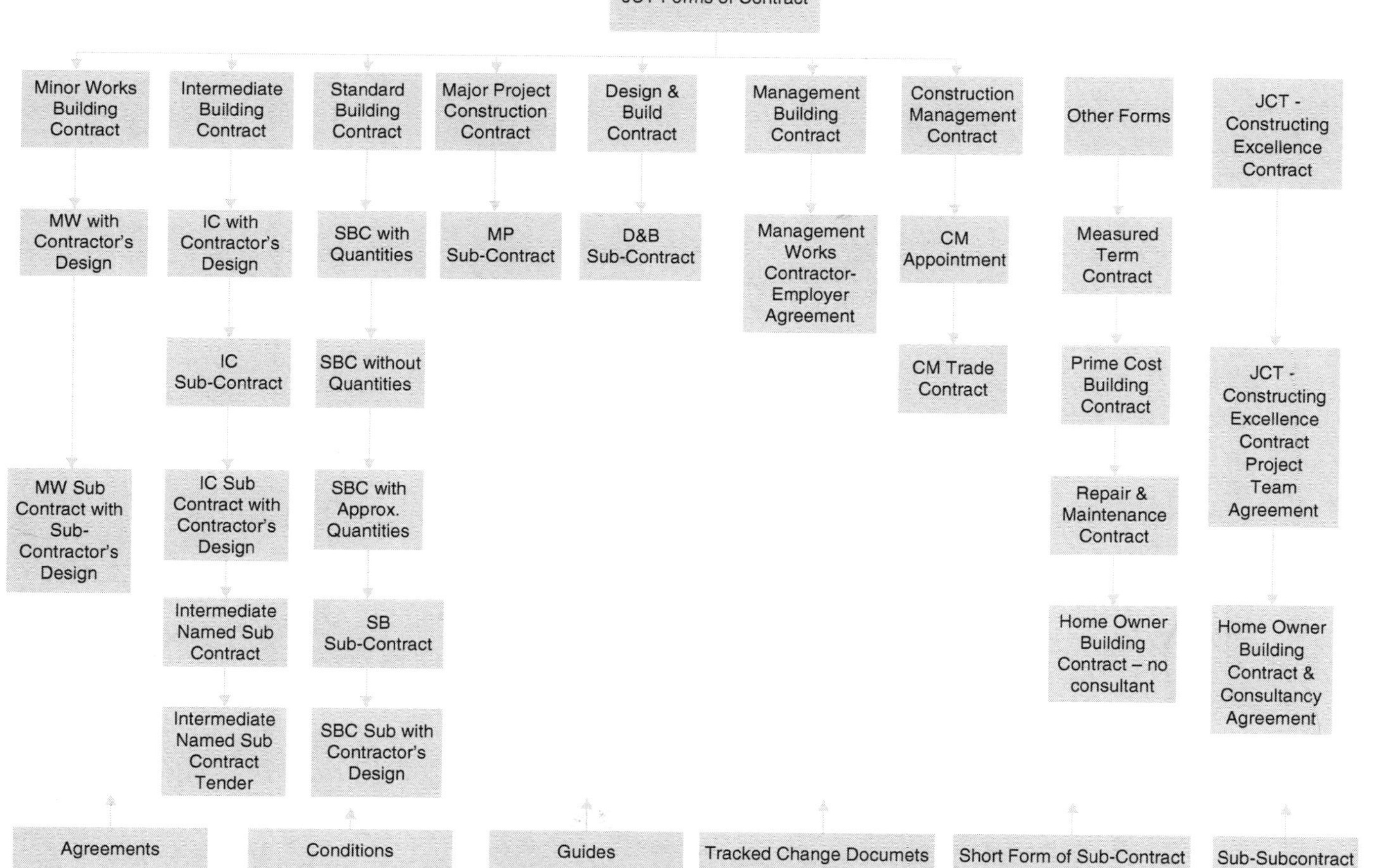

Figure 6.1 JCT forms of contract 2016.

Source: Adapted from Joint Contracts Tribunal (2016) *The Standard Form of Building Contract.*

is insufficient time available in which to prepare the detailed drawings that are necessary for accurate bills of quantities to be produced. The works are subject to remeasurement on completion. It can be used where the contractor might design discrete parts of the works. It is suitable for those projects that are carried out in sections and can be used for both private and local authority employers.

Standard Building Contract subcontracts

There are subcontract forms available to be used with the main contract forms described above. These forms can only be used with the SBC forms of contract. Different forms are available depending upon whether the subcontractor is required to specially design any discrete part of the subcontract works. This is then referred to as the subcontractor's design portion. The options available are:

- Standard Building Sub-Contract Agreement (SBCSub/A).
- Standard Building Sub-Contract Conditions (SBCSub/C).
- Standard Building Sub-Contract with sub-contractor's design Agreement (SBCSub/D/A).
- Standard Building Sub-Contract with sub-contractor's design Conditions (SBCSub/D/C).

The following two forms of subcontract can be used with all other JCT forms:

- Short Form of Sub-Contract (ShortSub).
- Sub-subcontract (SubSub).

6.5.2 JCT Intermediate Building Contract family

This form of contract family is available where the proposed building works are of simple content involving the normal recognised basic trades and skills of the industry, without building service installations of a complex nature or other complex specialist work. The main form is the Intermediate Building Contract (IC). The works will be designed on behalf of the employer, where fairly detailed contract provisions are necessary. The employer will provide the contractor with drawings and a specification, work schedules or bills of quantities. These will define adequately the quantity and quality of the works. A contract administrator and quantity surveyor will administer these conditions.

This form of contract is not suitable where the contractor is to design discrete parts of the works, even though all other criteria are met. There is a separate version of this form called the Intermediate Building Contract with contractor's design (ICD), where the contractor is required to design discrete parts of the works. This form is not suitable for a design and build contract.

There are a range of intermediate subcontract conditions and agreements that should be used with the IC. These are:

- Intermediate Sub-Contract Agreement (ICSub/A).
- Intermediate Sub-Contract with sub-contractor's design Agreement (ICSub/D/A).
- Intermediate Sub-Contract Conditions (ICSub/C).
- Intermediate Sub-Contract with sub-contractor's design Conditions (ICSub/D/C).
- Intermediate Named Sub-Contract Tender and Agreement (ICSub/NAM).
- Intermediate Named Sub-Contract Conditions (ICSub/NAM/C).

6.5.3 JCT Minor Works Building Contract family

The Minor Works Building (MW) form of contract should only be used where the work is simple in character. The work will be designed by or on behalf of the employer and drawings, specifications or work schedules will be provided to adequately define the quantity and quality of the works. A contract administrator should be used to administer the conditions. It can be used by both private and local authority employers and is suitable where bills of quantities are not required. It is not suitable where provisions are required to govern the work carried out by named specialists or where detailed control procedures are needed. Neither should it be used where the contractor is required to design a discrete portion of the works. In this case, the Minor Works Building Contract with contractor's design (MWD) should be used. Where subcontractor design is involved, the Minor Works Sub-Contract with sub-contractor's design (MWSub/D) should be used. Other standard documents available for minor works include:

- Minor Works Building Contract (MW) Tracked Change Document.
- Minor Works Building Contract with contractor's design (MWD) Tracked Change Document.

6.5.4 JCT Major Project Construction Contract family

This family of forms of contract is appropriate for works where the employer regularly procures large-scale construction work and where the contractor to be appointed is experienced and is able to take greater risks than would arise under other JCT contracts. The main form is Major Project Construction Contract (MP). The parties are likely to have their own detailed procedures and therefore only a limited set of procedures need to be set out in the contract conditions. The contractor is not only required to carry out and to complete the works, but also to complete the design. The employer employs a representative to exercise the powers and functions of the employer under the contract. It is also a suitable form of contract where the work is carried out in sections. There is a dedicated Major Project Sub-contract (MPSub).

6.5.5 JCT Design and Build Contract family

This form of contract family is to be used where detailed contract provisions are necessary and where the employer's requirements have been prepared and provided to the contractor. The main form of contract is Design and Build Contract (DB). It is used in those circumstances where the contractor is not only to carry out and complete the works, but also to be fully responsible for their design. The employer may appoint an agent who may be an external consultant or employee to administer the conditions. It can be used where the works are to be carried out in sections and is suitable for both public and private employers. It should not be used where the contractor is only responsible for designing small, discrete parts of the works. In these circumstances, the main contract form described above should be used. Bespoke subcontract agreements and conditions are provided to be used with this form of contract. Where subcontractors are employed, the following subcontract forms can be used:

- Design and Build Sub-Contract Agreement (DBSub/A).
- Design and Build Sub-Contract Conditions (DBSub/C).

JCT also provides the following guide documents for administering design and build projects:

- Design and Build Contract Guide (DB/G).
- Design and Build Sub-Contract Guide (DBSub/G).
- Design and Build Contract (DB) Tracked Change Document.

6.5.6 JCT Management Building Contract family

This form of contract family, and its associated subcontract conditions, are appropriate for large-scale projects requiring an early start on-site. The works will be designed by or on behalf of the employer. It is used where it is not possible to prepare full design information before the works commence and where much of the detail design may be of a sophisticated or innovative nature requiring proprietary systems or components designed by specialists. The employer will provide the drawings and specification, and a management contractor will administer the conditions. The management contractor does not carry out any construction work, but manages the contract for a fee. The management contractor employs works contractors to carry out the construction works. This form can be used where the works are to be carried out in sections and is able to be used by both private and public organisations. The main forms of contract and supporting documents include:

- Management Building Contract (MC).
- Management Works Contract Agreement (MCWC/A).
- Management Works Contract Conditions (MCWC/C).
- Management Works Contractor/Employer Agreement (MCWC/E).

6.5.7 JCT Construction Management family

This form of contract family is to be entered into between the client and the trade contractor to carry out the works. It is an appropriate form of contract where a client is to enter into direct separate trade contracts and where a construction manager is to administer the conditions on behalf of the client. It can also be used where the works are to be carried out in separate sections. There are two main forms in this family of contracts:

- Construction Management Appointment (CM/A).
- Construction Management Trade Contract (CM/TC).

The forms are used where separate contractual responsibility for design, management and construction is desired whilst apportioning the works in to different sections.

6.5.8 JCT Constructing Excellence Contract family

This forms of contract family is specifically designed for use in partnering where parties intend to follow collaborative and integrated working practices. The forms were developed through collaboration with UK Constructing Excellence to procure a range of construction services that uses partnering approaches. It also can be used in conjunction with framework agreements. The main forms available are:

- JCT – Constructing Excellence Contract (CE).
- JCT – Constructing Excellence Contract Project Team Agreement (CE/P).

6.5.9 JCT Measured Term Contract (MTC)

This form of contract is used by employers who have a regular flow of maintenance and minor works, including improvements. A single contractor is normally appointed over a fixed period of time under a single contract. The work will be measured and valued usually on the basis of an agreed schedule of works. A contract administrator or quantity surveyor will administer the conditions.

6.5.10 JCT Prime Cost Building Contract (PCC)

This form is suitable for projects requiring an early start on-site, where the works are designed by or on behalf of the employer. It is used when it is not possible to prepare full design information before the works commence on-site. Detailed contract conditions are provided along with a specification. Drawings may or may not be available. A contract administrator or quantity surveyor will administer the conditions. The work will proceed on a basis of a brief specification and an estimate of cost. The contractor is paid reasonable costs plus either a percentage of the cost or a lump sum as a fee. It can be used where the works are carried out in sections and by private and public organisations.

6.5.11 JCT's Repair and Maintenance Contract (RM)

This form of contract is primarily aimed at organisations that have a defined programme of repair and maintenance works to specified buildings or sites. As such, it is useful for local authorities and larger organisations that have a portfolio of buildings or structures that require periodic maintenance works. Often, such organisations will have their own maintenance department managing such maintenance works and therefore do not require the appointment of a contract administrator.

6.5.12 JCT's home owner contracts

This is primarily aimed at self-build or owner-occupier type home owners who are seeking contractual protection through a formal contract. There are two forms available:

- Building Contract for a Home Owner/Occupier who has not appointed a consultant to oversee the work (HO/B).
- Building Contract and Consultancy Agreement for a Home Owner/Occupier (HO/C and HO/CA).

The first document is for building contracts, and the latter serves as both building contract and consultancy agreement for appointment of consultants. These documents are fairly comprehensive and cover all aspects from commencement to completion of works.

6.6 JCT agreements

The JCT agreements provide an important element in the contractual administration process of a construction process. They deal with the appointment of consultants, specialists and adjudicators and such other parties in relation to a construction project. They help in maintaining relatively long-term agreements with such appointments including the pre-construction phase.

6.6.1 Framework Agreement (FA)

This agreement is appropriate for the procurement of construction or engineering related works on a regular basis over a period of time. It is used by clients with contractors or suppliers. It is used by contractors, subcontractors or suppliers in subletting to others in the supply chain. It can be used appropriately with most standard forms of construction and engineering contracts and subcontracts. The form enables good compliance with UK public procurement rules. The JCT current version was published in 2016.

6.6.2 Pre-construction Services Agreement

This agreement enables the contractor to be brought in to collaborate in the early stages of design in a construction project. It is designed for appointing a contractor to carry out pre-construction services under a two-stage tender process, facilitating contractors' input in collaborating with the employer and/or team of consultants responsible for design development. The agreement can be used where the main contract is Standard Building Contract, Design and Build Contract, Major Project Construction Contract or Intermediate Building Contract. The advantages of early contractor involvement include specialist knowledge about construction and aspects of buildability, programme and cost related issues, amongst others.

There are two forms of standard agreements available (both published in 2016):

- Pre-Construction Services Agreement (General Contractor) (PCSA).
- Pre-Construction Services Agreement (Specialist) (PCSA/SP).

The former is for the appointment of a general contractor and the latter is for appointment of a specialist contractor, at the pre-construction stage of a project.

6.6.3 Consultancy Agreement (Public Sector) (CA)

The appointment of consultants happens at a very early stage of a construction project, when the notion of a construction project (some form of building or structure or facility) is envisaged. The *National Construction Contracts and Law Survey* (2015) indicates that more than 50 per cent of respondents still use bespoke consultancy agreements. Nevertheless, standard forms are gaining acceptance in the industry, and the JCT forms designed specifically for use by the public sector employers are amongst the leading standard forms.

Other commonly used consultancy service agreements include the NEC Professional Service Contract, RIBA agreements, the Association of Consulting Engineers (ACE) Agreement, the RICS form, FIDIC client/consultant model services, the ACA form, the CIC Consultants' Contract, amongst others.

6.6.4 Adjudication Agreement

This form is used where the adjudicator is to be appointed in respect of a construction dispute that has arisen under a JCT contract, subcontract or agreement. There are two forms available (published 2016):

- Adjudication Agreement (Adj).
- Adjudication Agreement Named Adjudicator (Adj/N).

The latter is where an adjudicator named in the contract is to be appointed under the contract conditions.

6.6.5 Project Bank Account Documentation (PBA)

This documentation has been created as part of the government's fair payment guidelines, a part of the construction strategy. The documentation is useful for parties who wish to adopt the use of project bank accounts as part of their fair payment policy implementation. The documentation is in line with the JCT 2011 suite of contracts and as such a revision can be expected to match with the 2016 suite of contracts.

6.7 JCT contract guides

JCT has produced a number of guides, for main contracts and the subcontracts, to help those completing the contracts by advising on unfamiliar terms and specifying who should complete it. JCT has also published collateral warranty forms for contractors and subcontractors, and there are forms available for home owner contracts.

Figure 6.2 provides a guide on which JCT form of contract to use on an individual project, based upon a series of yes/no possibilities.

6.8 Other main forms of contract

6.8.1 General Conditions of Government Contracts for Building and Civil Engineering Works (GC/Works/1 and GC/Works/2)

These are published by the Stationery Office. They are used almost exclusively by central government departments. GC/Works/1 is used on major projects and GC/Works/2 on minor building projects. In addition, there are other related forms for both mechanical and electrical services in buildings. The body responsible for content, style and updating is the Department of the Environment, Transport and the Regions (DETR). The forms were revised in 1998 and are now compliant with the Latham Report and the Housing Grants, Construction and Regeneration Act 1996. GC/Works/1 is available in three formats: lump sum with quantities; lump sum without quantities; and single-stage design and build.

The family of GC/Works contracts include the following:

GC/Works/1	Contract for building and civil engineering major works – this is in seven parts that include the typical variants of the form.
GC/Works/2	Contract for building and civil engineering minor works.
GC/Works/3	Contract for mechanical and electrical engineering major works.
GC/Works/4	Contract for mechanical and electrical engineering minor works.
GC/Works/5	General conditions for the appointment of consultants.
GC/Works/6	Standard form of daywork contract.
GC/Works/7	Standard form of measured contract that is based upon a schedule of works.
GC/Works/8	Specialist term contract for use where specified maintenance of equipment required can be costed per task.
GC/Works/9	Lump sum maintenance contract (one to five years) for operation, maintenance of fixed mechanical and electrical plant, equipment and installations.
GC/Works/10	Facilities management contract.
GC/Works/11	Minor works term contract.

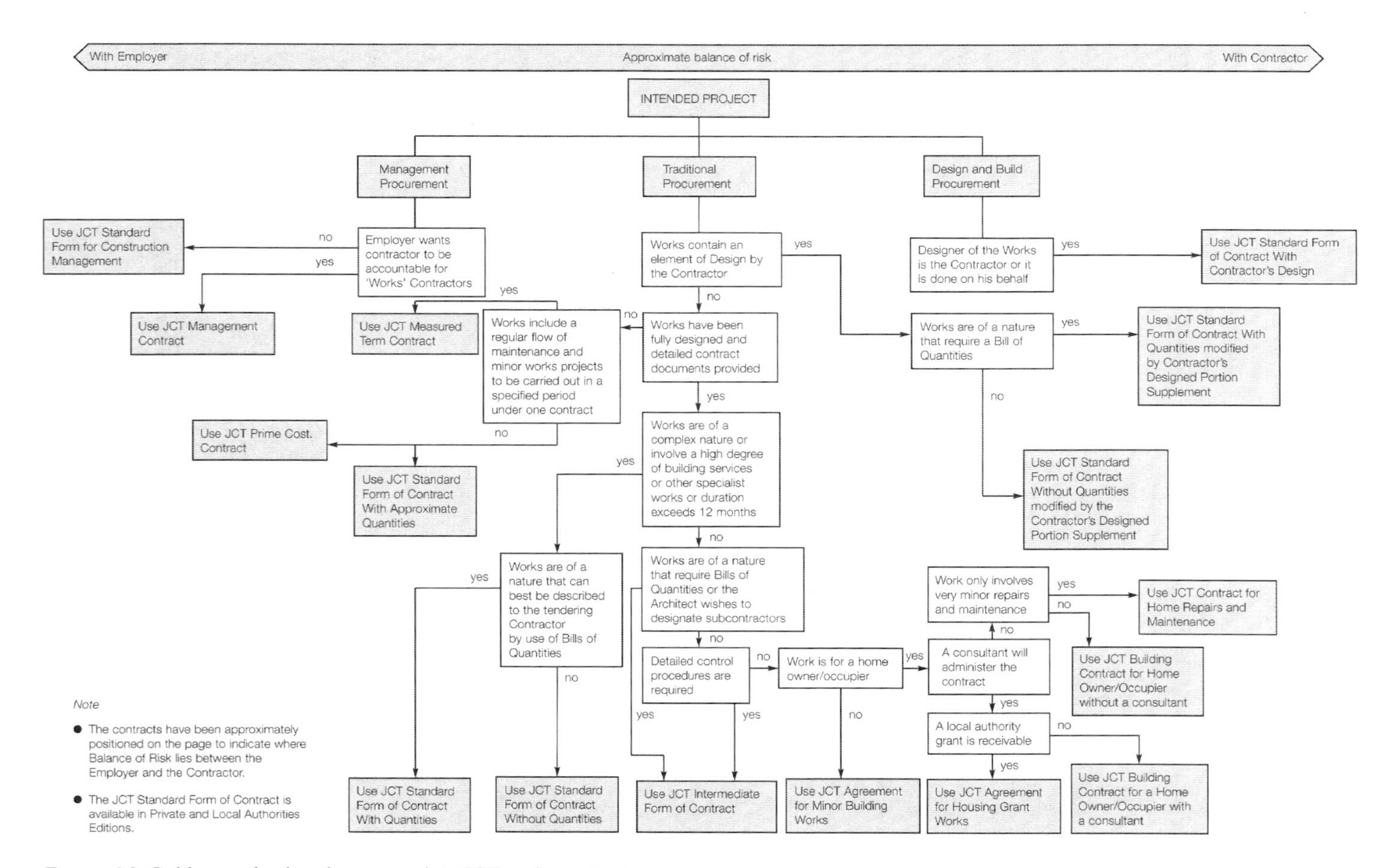

Figure 6.2 Guide to selecting the appropriate JCT main contract.

Source: The Joint Contracts Tribunal, *Practice Note: Deciding on the Appropriate JCT Contract 2011*.

The GC/Works suite of contracts are no longer updated by the government, although still available for use. The public sector is now adopting the NEC forms of contracts for their projects.

6.8.2 Association of Consultant Architects Form of Building Agreement

ACA publish a range of key contract forms for use by the building professions. These are alternatives to those published by JCT. ACA have been unhappy with the complexities of the JCT forms and claim that their own documentation is short, clear, legally precise and has been prepared with the advice of legal experts and in consultation with clients, contractors and the professions.

ACA also publishes a suite of partnering related contracts:

- PPC2000 (Amended 2013) – ACA Standard Form of Contract for Project Partnering.
- TPC2005 (Amended 2008) – ACA Standard Form of Contract for Term Partnering.
- SPC2000 (Amended 2008) – ACA Standard Form of Specialist Contract for Project Partnering.
- SPC2000 Short Form (Issued 2010) – ACA Standard Form of Specialist Contract for Project Partnering.
- STPC2005 (Issued 2010) – ACA Standard Form of Specialist Contract for Term Partnering.
- Guide to ACA Project Partnering Contracts PPC2000 and SPC2000.
- Guide to ACA Term Partnering Contracts TPC2005 and STPC2005.
- Introduction to Pricing Under PPC2000.
- Introduction to Pricing Under TPC2005.

In addition, they publish forms of architectural appointment, forms of building agreement and forms of subcontract, amongst other documentation.

6.8.3 ICE Conditions of Contract and Form of Tender, Agreement and Bond for Use in Connection with Works of Civil Engineering Construction

Produced by the Conditions of Contract Standing Joint Committee (CCSJC), the ICE Conditions of Contract are jointly sponsored by the Institution of Civil Engineers (ICE), the Civil Engineering Contractors Association (CECA) and the Association of Consulting Engineers (ACE). The ICE Conditions of Contract have been in use for more than sixty years (first edition 1945) and were designed to standardise the duties of contractors, employers and engineers and to distribute the risks inherent in civil engineering to those best able to manage them. The seventh edition of this form was published in 1999. The following represent the current editions of ICE Conditions of Contract, each with their corresponding guidance notes:

- Measurement Version, 7th edition.
- Design and Construct, 2nd edition.
- Term Version, 1st edition.
- Minor Works, 3rd edition.
- Partnering Addendum.
- Tendering for Civil Engineering Contracts.

- Agreement for Consultancy Work in Respect of Domestic or Small Works.
- Archaeological Investigation, 1st edition.
- Target Cost, 1st edition.
- Ground Investigation, 2nd edition.
- Amendments to ICE Conditions of Contract.

In addition to the main form, appendices cover an arbitration procedure, a conciliation procedure and contract price fluctuations. It is recommended that all clauses are incorporated unaltered because they are closely interrelated and any changes made in some may have unforeseen effects on others.

Guidance notes have been prepared specifically to assist the users of the ICE Conditions of Contract in the preparation of contract documents and the carrying out of the contract works. They do not purport to provide a legal interpretation, but they do represent the unanimous view on what constitutes good practice in the execution of civil engineering works. Guidance notes are available for:

- Measurement Version.
- Design and Construct.
- Minor Works.
- Term Version.
- Ground Investigation.

The ICE Conditions of Contract has not been published since 2009, when the ICE withdrew its support for its Conditions of Contract in favour of the NEC.

6.8.4 New Engineering Contract (NEC)

The ICE commissioned the development of a new form of contract in 1986 as it was felt that there was a need for a form that had clearer language, clearer allocation of responsibilities and reduced opportunities for contractual gamesmanship. This resulted, in 1991, in a consultative form of the NEC form of contract, with a first edition published in 1993. Wider use of the NEC form was recommended by the Latham Report in 1994. It promotes a gravitation towards more use of plain language in the formulation of conditions of contract as a means of promoting greater collaboration in construction.

The NEC is a family of standard contracts for use on building and civil engineering projects, each of which has these characteristics:

- Its use stimulates good management of the relationship between the two parties to the contract and, hence, of the work included in the contract.
- It can be used in a wide variety of commercial situations, for a wide variety of types of work and in any location.
- It is a clear and simple document – using language and a structure which are straightforward and easily understood.

In 2005 NEC3 was developed as a suite of contracts that included Term Service Contract and Framework Contract. By 2013 the NEC3 suite included thirty-nine forms and contract guides, including Professional Services Short Contract, so the NEC3 April 2013 edition consists of eighteen forms of contract and twenty-one guidance notes and flow charts. The supporting documents are still being added to the suite. The main forms are listed below:

- NEC3 Engineering and Construction Contract (ECC).

  a NEC3 ECC Option A: Priced contract with activity schedule.
  b NEC3 ECC Option B: Priced contract with bill of quantities.
  c NEC3 ECC Option C: Target contract with activity schedule.
  d NEC3 ECC Option D: Target contract with bill of quantities.
  e NEC3 ECC Option E: Cost reimbursable contract.
  f NEC3 ECC Option F: Management contract.

- NEC3 Engineering and Construction Subcontract (ECS).
- NEC3 Engineering and Construction Short Subcontract (ECSS).
- NEC3 Engineering and Construction Short Contract (ECSC).
- NEC3 Professional Services Contract (PSC).
- NEC3 Professional Services Short Contract (PSSC).
- NEC3 Term Service Contract (TSC).
- NEC3 Term Service Short Contract (TSSC).
- NEC3 Supply Contract (SC).
- NEC3 Supply Short Contract (SSC).
- NEC3 Framework Contract (FC).
- NEC3 Adjudicator's Contract (AC).

These options offer a framework for tender and contract clauses that differ primarily in regard to the mechanisms by which the contractor is reimbursed and motivated to control costs. The core clauses of the main options listed above are used in conjunction with the secondary options and the additional conditions of contract. The clauses of these options can be adapted by tenders for low risk projects and with:

- The Engineering and Construction Short Contract.
- The Engineering and Construction Short Subcontract Contract.

NEC3 is the fastest growing form of contract in terms of usage according to the *National Construction Contracts and Law Survey* (2015) with an 8 per cent increase in usage from 2012 to 2015. The suite has been endorsed by the UK public sector as the preferred form, and there is significant acceptance from the private sector as well. NEC contracts are used internationally for construction projects.

NEC3 is a highly project management oriented form of contract that promotes collaboration between parties involved in a construction project. Recent land mark projects that adopted NEC3 include the London 2012 Olympic Games, Crossrail and Heathrow Terminal 5, amongst others.

6.8.5 FIDIC: Conditions of Contract (International) for Works of Civil Engineering Construction

These conditions of contract for construction cover both building and civil engineering works. They are prepared by the International Federation of Consulting Engineers

(Fédération Internationale des Ingénieurs-Conseils, FIDIC). The forms of contract are colour coded:

Red Book: Conditions of Contract for Construction
Harmonised Red Book: Conditions of Contract for Construction (Multilateral Development Bank)
Green Book: Short Form of Contract
Yellow Book: Conditions of Contract for Plant & Design-Build
Silver Book: Conditions of Contract for EPC Turnkey Projects

The Conditions of Contract for Construction (Multilateral Development Bank Harmonised Ed. Version 3: June 2010) for Building and Engineering Works is the latest edition in the series. The form is approved by several other organisations representing the construction interests of various other countries and is available in many different languages. It provides a framework that can be adapted to suit local laws and has attempted to harmonise definitions. FIDIC also includes a subcontractor agreement form, and publishes a range of other information, for example:

- Client–Consultant Agreement.
- Joint Venture Agreement.
- Plant and Design and Build Contract.
- EPC/Turnkey Projects.
- Consultant Selection.
- Contracts Guide.

The FIDIC forms of contract have significant worldwide usage for international construction projects, and have been adopted by the World Bank for its own form of contract for World Bank-funded projects.

6.8.6 Other forms of contract

Other forms of contract are also available, most notably those prepared by the large industrial corporations who undertake a substantial amount of construction work. Even in cases where such forms are not used, and one of the more conventional forms of contract is applied, it is not uncommon to find a few pages being added as supplementary conditions. These conditions generally place the standard forms in a more favourable position with the building owner or employer. A much greater risk is therefore placed with the contractor than is usually the case. Whether this offers any real advantages to the employer is open to speculation. Under 'normal' tendering circumstances, the contractors' prices will reflect the more onerous conditions.

It is much more unusual to tamper with the contract conditions themselves, since there is always the danger of making the entire contract null and void, or failing to alter the conditions consistently throughout the particular form being adapted.

6.9 The Housing Grants, Construction and Regeneration Act 1996

This Act is also known as the Construction Act. Part II of this Act is concerned with construction contracts. Included in the Act is a definition of such contracts as an agreement to do architectural, design or surveying work or to provide advice on building, engineering, interior or exterior decoration or on the laying out of landscape in relation to construction operations. Construction operations include:

- construction, alteration, repair, maintenance, extension, demolition or dismantling of a building or structure, both temporary and permanent
- works commonly considered to be of a civil engineering nature
- installation in any building or structure fittings forming part of the land
- external or internal cleaning of buildings or structures
- operations which form an integral part of, or are preparatory to, such work
- painting or decorating the internal or external surfaces of any building or structure.

Within the meaning of the Act, the following operations are not defined as construction operations:

- drilling for, or extraction of, oil or natural gas
- extraction of minerals
- assembly, installation or demolition of plant and machinery
- manufacture or delivery to site of:
    - building or engineering components or equipment
    - material plant or machinery
    - components for engineering services in buildings
- making or repair of artistic works such as murals or sculptures, unless they are installed in the works.

The provisions of the Act apply only to agreements in writing and there are detailed provisions as to what this includes. The Act requires that all construction contracts must include the following provisions:

- *Adjudication*: each party must have the right to refer disputes to adjudication with the object of achieving a decision within twenty-eight days.
- *Stage payments*: if the project runs for longer than forty-five days, there must be provision for stage payments.
- *Date for payment*: all contracts must have methods of working out the value of payments, the date on which these are due and the final date for payment.
- *Set-off*: payments cannot be withheld nor money set-off unless notice has been given detailing the amounts withheld and the reasons why it has been withheld.
- *Suspension of the works*: a party has the right to suspend the works if payments are not made within the terms of the contract.
- *Pay when paid*: this is intended to outlaw the principle of the pay when paid clause.

The standard forms of contract that are used in the construction industry comply with the Act.

The Act was amended in 2011, creating a greater focus on construction contracts, creating greater clarity as to payment terms in construction contracts, delivering a fairer payment regime and making adjudication more mainstream and accessible in construction contracts.

The Act stipulates that if adjudication clauses or payment related clauses are not clearly defined in the contract, the scheme provided in the Act applies. This makes clear payment provisions as well as adjudication procedures mandatory in construction contracts.

6.10 Other contractual documentation

6.10.1 National Joint Consultative Committee

The National Joint Consultative Committee (NJCC) is an advisory body concerned with practice and procedures in building contracts. It has a good practice panel and a wide range of publications concerned with contracts. It provides a range of codes of procedure, the most common of which is *Code of Procedure for Single Stage Selective Tendering* (1996) (Chapter 8). In addition, it provides codes on tendering procedure for industrialised building projects, two-stage selective tendering, selective tendering for design and build, management contracting and the letting and management of domestic subcontract works. The NJCC also offers good practice on preparing lists of approved contractors. It provides guidance notes on issues such as performance bonds, collateral warranties and alternative dispute resolution, and produces a set of procedure notes covering topics such as financial controls and cash flow, placing contracts with substantial building services engineering content, and the use and completion of nominated subcontracts.

6.10.2 Definition of prime cost for daywork carried out under a building contract

The Royal Institution of Chartered Surveyors, the Building Cost Information Service and the Construction Confederation jointly published this new definition in 2007. There are some significant differences in the new definition, which now offers two options for dealing with the prime cost of labour:

- *Option A*: the percentage addition is based upon the traditional method of pricing labour in daywork. This allows for a percentage addition for incidental costs, overheads and profit; to be made to the prime cost of labour applicable at the time the daywork is carried out.
- *Option B*: this provides all-inclusive rates. It includes not only the prime cost of labour but also an allowance for incidental costs, overheads and profit. The all-inclusive rates are deemed to be fixed for the period of the contract. However, where a fluctuating-price contract is used, or where the rates in the contract are to be index-linked, the all-inclusive rates shall be adjusted by a suitable index in accordance with the contract conditions.

Whoever prepares the contract documents will select which of the above methods is the most appropriate. Consideration should be given, for example, to length of contract, whether the contract is fixed, firm or fluctuating price, or whether the costs are to be index-linked. Using Option B gives the client price certainty in terms of the labour rate to be used in any daywork during the contract. However, there is the potential that the rate will be higher, as the contractor is likely to build in a contingency to cover any unknown increases in labour rates that may occur during the contract period. If Option B is used, and the rates in the contract are to be index-linked, BCIS publish a number of cost indices that may be suitable in these circumstances.

The new definition gives guidance on how to build up the prime cost of labour in daywork for non-productive overtime purposes. Some model documentation is also included, illustrating how the Definition of Prime Cost may be applied in practice.

6.10.3 Formula rules

These rules are derived from the Price Adjustment Formulae for Building Contracts (Series 2) Guide to Application and Procedure, prepared on behalf of the National Consultative Council Standing Committee on Indices for Building Contracts and published by The Stationary Office in 1977. The current JCT rules were published in 2006. A similar price adjustment system also exists for works of civil engineering construction, published by the ICE.

6.10.4 Construction Industry Model Arbitration Rules (CIMAR)

These have been introduced for the conduct of arbitration that will be contractually binding on the parties. The rules apply to the settlement of all disputes coming within the terms of the arbitration agreements in all forms of JCT contract and subcontract and other tribunal agreements.

6.11 Appointment of consultants

The building or civil engineering contract is between the client (employer or promoter) and the contractor. The designers or consultants for the project, whilst referred to in this contract, are not

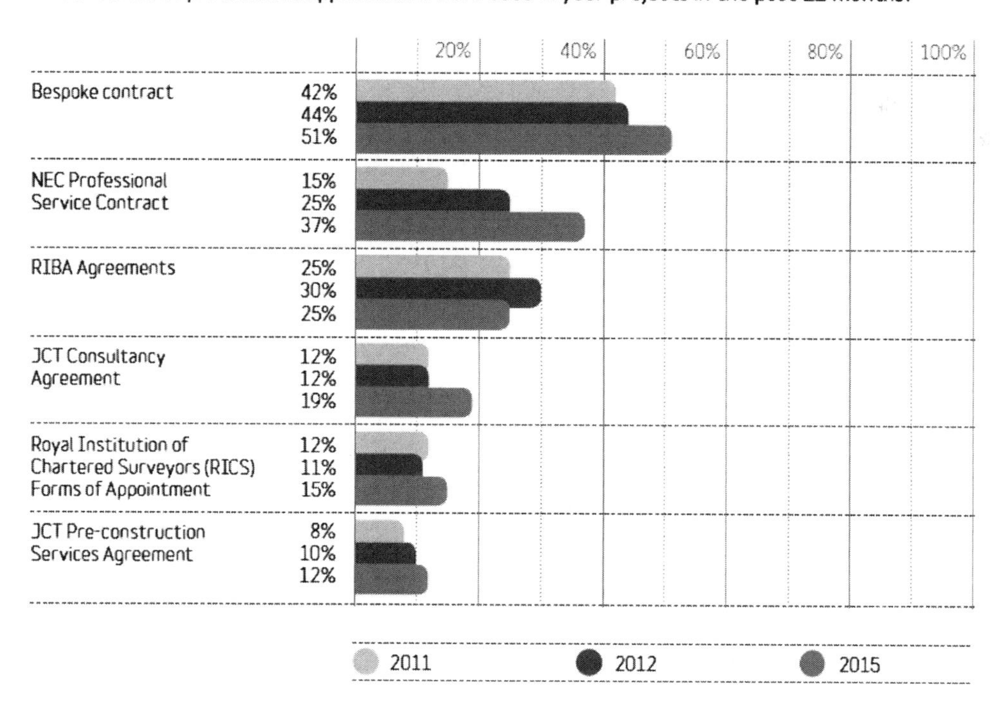

Figure 6.3 Trends in forms of professional appointments.

Source: The NBS National Construction Contracts and Legal Issues Survey (2015).

a part of what is commonly described as the construction contract. The various consultants engaged in a project will therefore have their own individual contracts with the person who employs them. Under the normal arrangements this will be the client (or employer, as referred to in the JCT forms), but where a consultant is involved as part of a design and build arrangement, then the contractual relationship may well be with the contractor and the consultant. Different forms of consultant contracts have been prepared to suit the particular needs of the different consultants, and are used for the appointment of:

- architects – Royal Institute of British Architects
- quantity surveyors – Royal Institution of Chartered Surveyors
- structural engineers – Association of Consulting Engineers
- building services engineers – Association of Consulting Engineers
- project managers – Royal Institution of Chartered Surveyors
- civil engineers – Association of Consulting Engineers.

The contracts in these circumstances are for the professional services that are required in the provision of a construction project. The different consultant forms will cover much of the same material, such as the duty of care and negligence, the amounts to cover professional fees and when these will be paid, the duration of the contract, intellectual rights and copyright, collateral warranties and agreements, professional indemnity insurance, and so on.

The various forms identify materials or substances that should not be used in the construction process at the present time. These include high alumina cement in structural elements, wood wool slabs in permanent formwork to concrete, calcium chloride in admixtures for use in reinforced concrete and asbestos products and naturally occurring aggregates for use in reinforced concrete which do not comply with British Standards 882 or 8110.

The fees charged by these different consultants may be guided by the fee scales (or now more correctly described as fee guides) that are published by the different professional bodies. In common with the principles of competitive tendering and laws relating to restrictive practices, it is usual to invite consultants to compete for their work against each other with one of the criteria being the fees charged for the consultancy work.

The *National Construction Contracts and Law Survey* (2015) indicates that bespoke consultancy service agreements are becoming more popular, with a greater trend in the adoption of NEC contracts. Figure 6.3 indicates the current trends in the use of forms in appointment of construction professionals.

7 Contract strategy

In the early 1960s the clients of the construction industry had only a limited choice of procurement methods which they might wish to use for the construction of a new project. The RIBA form of contract, although even then not used universally, had yet to experience extensive rivalry from elsewhere or even from within its own organisation for a serious alternative. Contract bills were becoming the preferred document in place of the specification, and the mistrusted cost-plus type contracts, which had been a necessity a few years earlier for the rapid repair of war-damaged property, were already in decline.

Society is now a world of endless change: in recent years there have been many developments in the technologies that are used, the types of groups in which we live and amongst professional attitudes. The causes of these changes are not difficult to find. Society, generally, is on the search for something new, and the construction industry in a way mirrors that change to avoid being thought of as outdated or old-fashioned. The search is also for improvement, knowing that we have not as yet found the perfect system – even if one exists. In this context, largely due to improved geographical communications of the twentieth century, the construction industry has looked with interest to both East and West, to the way that others around the world arrange the procurement of capital works projects. Not surprisingly, it is the countries which appear to have been the most successful that have received most of the attention. The methods adopted in the USA or Japan for their construction industries have been at the forefront of our ideas for change.

The apparent failure of the construction industry to satisfy the perceived needs of its customers, particularly in the way it organises and executes its projects, has been another catalyst for change, and various pressure groups, too, have evolved to champion the causes of organisations who have particular axes to grind. The 1970s oil price shocks, which had a massive influence on inflation and hence industries' borrowing requirements, also motivated the construction industry to improve its own efficiency through the way it managed and organised its work. The depression of the 1970s, like previous slumps, encouraged firms of all kinds to attempt to persuade employers to build by using what were then innovative approaches to contract procurement.

7.1 Industry analysis

The construction industry can be measured in several different ways; for example, turnover, profitability, number of firms and number of employees (see Harvey and Ashworth 1997). In 2014 its annual worth was about £103 billion, which represents 6.5 per cent of total output. The construction industry at the start of the twenty-first century has been buoyant. It is also diverse: about 60 per cent of the total output is repairs and maintenance, less than 25 per cent is public sector projects and less than 20 per cent is civil engineering projects.

During the past decades, changes in the methods of construction procurement have been one of the most fundamental driving forces within the construction industry. Greater comparisons have been made with procurement methods used by other industries and in other countries around the world. Research has been undertaken to better inform clients, consultants and contractors. The different methods each have their own peculiar advantages and disadvantages and there is still no panacea that applies across the whole industry. The choice of method depends upon the following characteristics:

- familiarity amongst parties
- type of client
- size of project
- type of project
- risk allocation
- form of contract to be used
- major objectives of the client
- status of the designer
- relationships with contractors and consultants
- type of contract documentation.

7.2 The use of construction contracts in the industry

7.2.1 The Contracts in Use Survey

The use of construction contracts in the UK construction industry often reflects the current thinking and trends in the overall construction strategy employed by the UK public and private sectors. There are a number of different firms and organisations in the UK construction industry who monitor trends in procurement systems. The *Contracts in Use Survey*, published by the RICS, has been the industry wide accepted indicator in the past. The survey has been carried out by the construction consultants Davis Langdon and published by the RICS Construction Faculty. As with any other survey, it is meant to provide a representation of practice. Construction consultancies will reflect their own client base, and those may present a different picture from this survey. The first *Contracts in Use Survey* was published in 1985. The last reported survey, the twelfth edition, was published in 2012 reporting on use of contracts in 2010. This survey reported only 3.9 per cent of the value of new orders in 2010 (with eighty-six responses). As such, its representative nature has diminished to some extent.

The survey indicated that more than 97 per cent used a standard form of contract, whilst 88 per cent used JCT forms, followed by NEC forms at 7.4 per cent. In terms of value of contracts, the NEC portion moved to 26 per cent, up by 12 per cent from 2007, indicating its gaining popularity, especially in the larger (high-value) projects. In terms of procurement methods, the survey indicated that specifications and drawings were used in 52 per cent of projects, firm bill of quantities (BQ) in 24 per cent followed by design and build (D&B) in 17 per cent of projects. However, in terms of value of projects awarded and procurement systems, D&B leads, with 39 per cent, followed by specifications and drawings, 23 per cent, and firm BQ, 19 per cent.

Whilst smaller projects continue to be dominated by plan and specification procurement routes and lump sum contracts, larger projects show a preference for construction management or a version of design and build. There is therefore a distinctive difference between the way in which smaller and large projects are procured. This difference is probably also reflected in the size of firms that undertake such work.

7.2.2 The National Construction Contracts and Law (NCCL) Survey

In recent years the *National Construction Contracts and Law Survey*, conducted by the National Building Specification (NBS), has been gaining more popularity and recognition. Its latest survey, published in 2015, reports use of contracts between summer 2014 and summer 2015. The survcy is based on 981 responses received from across the industry.

The survey reports Traditional Procurement Systems as the most used procurement system, with 47 per cent, followed by D&B, at 39 per cent. All other procurement systems (such as management contracting, construction management, measured term, cost-plus, private finance initiative (PFI) or public private partnership (PPP), partnering/alliancing) were seen to have less than 3 per cent usage. It reports an interesting shift in procurement methods from Traditional to D&B, with contractors now reporting 49 per cent usage.

Single-stage tendering remains the main means of tendering (over 75 per cent) followed by two-stage tendering and negotiation. Although e-tendering is gaining greater usage, it is surprisingly underused, with 36 per cent reporting that they have not used e-tendering for projects in the year under review.

The survey finds that 30 per cent sign contracts after the commencement of work, and 2 per cent do not sign contracts at all. This is somewhat disturbing for an industry that deals with the construction of complex structures and buildings. JCT forms remain the most widely used contract, with 39 per cent usage, closely followed by NEC forms, 30 per cent, and bespoke contracts, 11 per cent. However, in comparison to 2012 data from the same survey, NEC forms have increased in usage by 8 per cent, whereas JCT forms have reduced in usage by 21 per cent during the same period. This is a major decline in use for the JCT forms, although seen as the most tried and tested in the industry. The adoption of NEC forms for public sector contracts, and the increased emphasis on project collaboration, has propelled the use of NEC forms (see Figure 7.1).

The *NCCL Survey 2015* also provides a picture of the forms used for professional appointments (see Figure 7.2); that is, the appointment of architects, surveyors, engineers and other

Which suite of contracts have you/your organisation used most often?

	2011	2012	2015
JCT contracts	60%	48%	39%
NEC contracts	16%	22%	30%
Bespoke contracts	10%	9%	11%
FIDIC contract Agreement	3%	4%	7%

Figure 7.1 Use of popular standard forms of contract.

Source: NBS 2015.

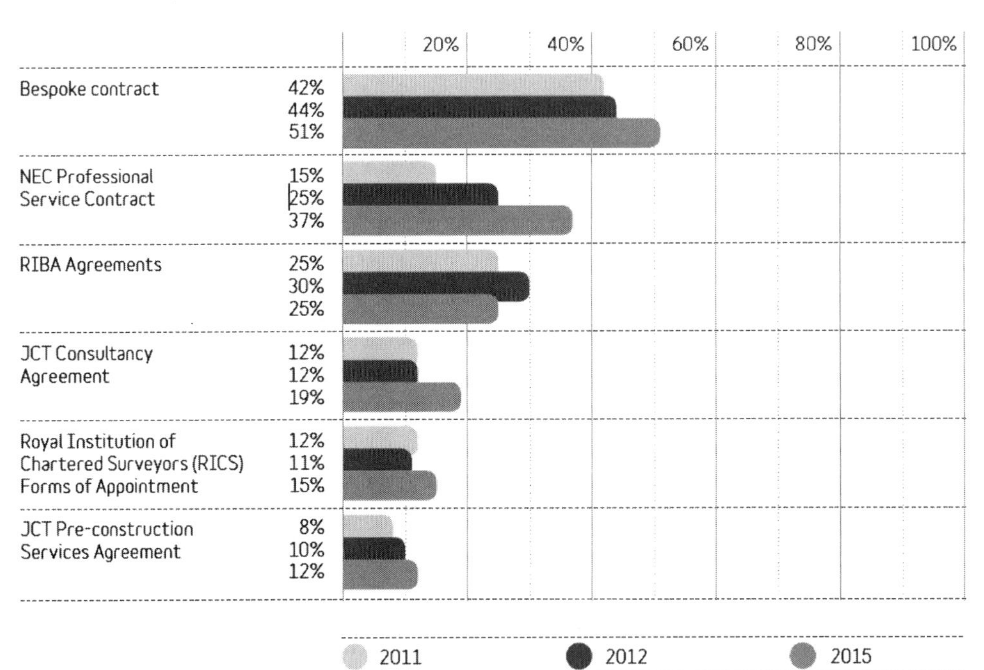

Figure 7.2 Forms of professional appointments.

Source: NBS 2015.

specialists and professional advisers. It is clear that bespoke forms are the overwhelmingly most popular forms to be used, and this indicates a trend in the departure from standard forms in more recent times.

7.2.3 Calculating the costs of construction

It is worth reminding ourselves briefly about the origins of the quantity surveyor and construction costs. Historically, architects were responsible for both the design and construction of the project. They employed a number of master craftsmen who carried out the work in each trade. Drawings were very sketchy and much of the design evolved as the job was being constructed. The craftsmen submitted accounts to the architect for the work that was carried out. Over time, the craftsmen employed measurers or surveyors to prepare these accounts. Some of the accounts appeared to be excessive and the architects began to employ their own surveyors to contest such claims. When general contractors became established, during the industrial revolution, they submitted estimates to cover all trades, and they engaged their own quantity surveyors to prepare bills of quantities for labour and materials. When competitive tendering became more common, instead of each firm employing a quantity surveyor to prepare a set of bills, contractors tendering for a project employed a single firm of quantity surveyors and the successful tenderer then paid the quantity surveyor's fees. The fees were included and recouped through the contractor's tender. Architects, on realising that the costs of the quantity surveyor were really being paid for by the client, persuaded clients to employ the quantity surveyor directly on their behalf and to use them for calculating the costs of interim payments and final accounts. Contractors themselves employed quantity surveyors to contest the amount of such payments in order to arrive at a fair sum for the work that they carried out. Today, calculating the costs of construction has almost gone full circle in the context of the history of quantity surveying. The use of firm BQs is declining, as is evident by both the *Contracts in Use Survey* and the *NCCL Survey 2015*.

7.2.4 Lump sum design and build v. traditional (bills of quantities)

In 1985, the use of bills of quantities was the preferred method for obtaining tenders in the UK construction industry. There was only a relatively small use of other methods (see Tables 7.1 and 7.2). Design and build was less than 10 per cent of the total value of all contracts and, even then, this method was used more on larger projects. In terms of the total number of contracts, design and build represented less than 5 per cent. Over the next thirty years a rapid increase in the use of design and build across all sizes of project have been recorded (39 per cent in 2015). This shift in practices was largely due to the transfer of risk, where clients wanted more certainty of risk transfer. The 2010 *Contracts in Use* survey (RICS and Davis Langdon 2012) reinforces the dominance of design and build as a preferred procurement strategy along with a continued decline in the use of bills of quantities, which is then further reinstated in the *NCCL Survey 2015* (NBS 2015). However, this is only part of the picture since the use of bills or measured quantities refuse to die. It is recognised that someone somewhere in the supply chain needs to measure quantities since it is largely on these that real price is determined. Whilst the use of the 'with quantities' forms have declined, there are still Standard Method of Measurement of Building Works (SMM7) (and now RICS *New Rules of Measurement* (NRM)) bills being measured, often by professional quantity surveying practices. However, instead of preparing these for the client, they are now being prepared on behalf of contractors in support of design and build tenders. Knowing this, the RICS has recently published its *New Rules of Measurement* (RICS 2009, 2012), and it will be interesting to observe in future years how the market responds to this suite of documents. These new rules were introduced as a replacement for SMM7 in 2012. The early indications are that NRM is being adopted, but at much slower pace than desired.

7.2.5 Negotiation and two-stage tendering

The use of negotiation and two-stage tendering methods both showed an increase in general use over the earlier surveys. This reflects a market that, at the time of the survey, was booming. This is evident in the *Contracts in Use Survey* where up to 2007 there is steady growth and in the 2010 survey it dropped (especially the use of negotiated contacts) as the UK entered a recessive period (RICS 2007; RICS and Davis Langdon 2012). It may, from other anecdotal evidence at the time, possibly underestimate what was actually happening in the construction industry. It has often been suggested that, in times of boom, two-stage tendering often increases when contractors are reluctant to price single-stage tenders. This is evident in the NCCL surveys where growth in two-staged tendering is reordered from 2012 to 2015 (NBS 2015). A net increase in use is reported by clients, consultants and contractors. Another category that is becoming noticeable is the use of design competition as means of obtaining work. The *NCCL Survey 2015* reports around 10 per cent per cent use of this method (NBS 2015).

7.2.6 Drawings and specification

The use of drawings and specifications as a procurement route has a long history. Initially, it was used largely for minor or small works projects where the level of complexity was minimal. In terms of the number of contracts issued in the construction industry this method remains the most popular, at approximately 52 per cent in 2010 (see Table 7.2). Table 7.1 indicates that its use on larger projects is much less in evidence, although its use on these types of project has almost doubled over the past twenty years. There used to be a rule that contractors should not tender for large projects under this method because of the duplication of effort in measuring quantities. The costs of tendering under any method is not free. Drawings and specifications are the second most popular procurement method and, when considering the value of projects, overtaking BQs. Table 7.2 indicates that it continues to dominate the smaller projects market.

7.2.7 Guaranteed maximum price (GMP)

This is often a client-driven contractual amendment to standard forms. The use of GMP remains a fundamental part of the UK Health Service Procure 21 system for awarding contracts. It is possible that the *Contracts in Use Survey* analysed very few hospital and other health service projects and, as such, have under reported its use (RICS and Davis Langdon 2012). The *NCCL Survey 2015* reports 3 per cent use (NBS 2015).

7.2.8 Management contracting/construction management

The use of management contracting, which was a procurement method that showed real promise in the 1980s, peaked in 1991 and has declined ever since. The use of this now represents only a very small amount of work that is carried out in the construction industry. It is viewed by some as a blip in procurement history. Some argued in the 1980s that this method was to be the panacea and model for procurement in the future. Construction management was viewed in a similar way, but now remains in use only on larger and more complex projects.

7.2.9 Electronic tendering

The use of electronic tendering through EDI (electronic data interchange) was developed in the early 1980s. Electronic tendering has not yet reached mainstream centre stage, but there has been a steady considerable growth. E-tendering evolved from the use of spreadsheets for tendering on disks, CDs and memory sticks to the internet as a platform. The, now defunct, RCIS e-tendering service led the way in the private sector, and the public sector promoted the Bravo Solutions™ platform. Extranet-based tendering was the logical next step in the market, which is now led by Software as a Service (SaaS) software providers.

Increased digitisation of the construction industry seems not to have much of an impact in driving the adoption of e-tendering in construction projects. The *NCCL Survey 2015* indicates that, although one-third have either used e-tendering regularly or sometimes, a shocking 36 per cent still do not use it at all. The case is worst with consultants, where 40 per cent do not use e-tendering, and it drops to 30 per cent for contractors (see Figure 7.3). Unfortunately, there is no marked improvement in these figures from the *NCCL Survey 2012*.

7.2.10 Building information modelling (BIM)

BIM has moved the construction industry towards digitalisation and is providing the platform for next level of digital disruption in the construction industry. It also provides a platform for

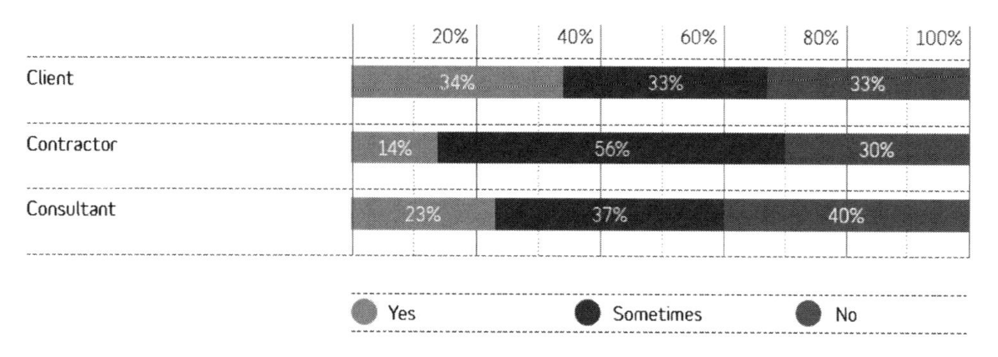

Figure 7.3 Use of e-tendering in the last twelve months reported by *NCCL Survey 2015*.

Source: NBS 2015.

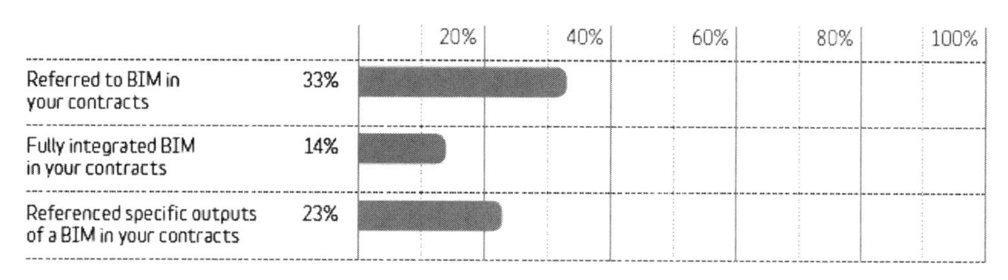

Figure 7.4 State of BIM in construction contracts.

Source: NBS 2015.

digital collaboration in construction procurement and administration. Many developed and rapidly developing countries are actively promoting the adoption of BIM in the construction industry. The UK government requires that all projects that are funded by central government adopt collaborative 3D BIM (Level 2) by 2016. It is interesting to see to what extent BIM has penetrated the industry. The *NCCL Survey 2015* indicates that over 58 per cent of respondents recognise BIM as contractually binding in the same way as drawings and specifications (NBS 2015). This is very positive and logical, but it would be interesting to see the full legal stance through case law. The survey further explores the linkages between BIM and contracts, as indicated in Figure 7.4.

7.2.11 JCT v. NEC

The *Contracts in Use Survey* continues to record the dominance of the use of the JCT suite of contracts on construction projects. They had a market share of 88 per cent by use and 67 per cent by value in 2010. There has been some decline in relation to previous years as the NEC (New Engineering Contract) documents are seeing an increase in use. Some evidence suggests that

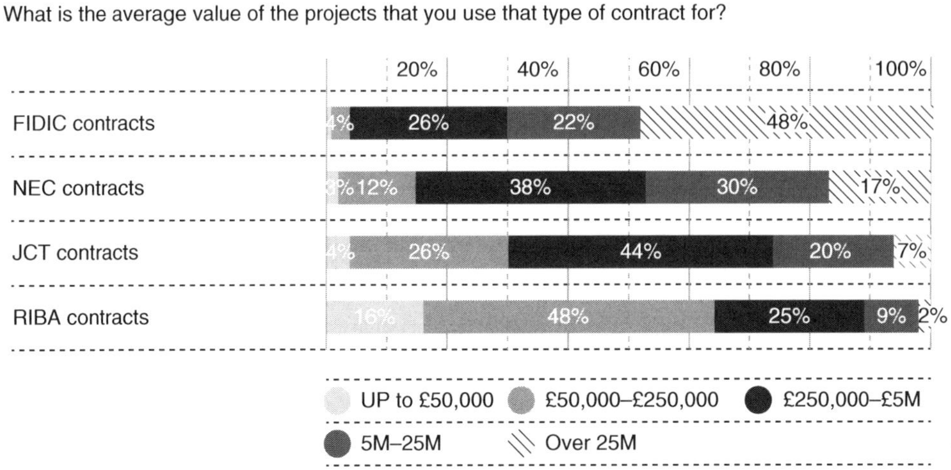

What is the average value of the projects that you use that type of contract for?

20% 40% 60% 80% 100%

FIDIC contracts 4% 26% 22% 48%

NEC contracts 3% 12% 38% 30% 17%

JCT contracts 4% 26% 44% 20% 7%

RIBA contracts 16% 48% 25% 9% 2%

UP to £50,000 £50,000–£250,000 £250,000–£5M 5M–25M Over 25M

Figure 7.5 Value of contract v. type of contract.

Source: NBS 2015.

NEC is gaining greater use and recognition on the larger construction projects. The NEC users' forum suggests that the data from the survey does not fully reflect the wider use of NEC in practice. The *NCCL Survey 2015* indicates that there is only 9 per cent difference in use between JCT and NEC now (NBS 2015). The gap is closing up rapidly with the adoption of NEC for public sector projects (see Figure 7.1). Of the JCT forms of contract, the Design and Build Form remains the most widely used, followed by the Standard Form with Quantities. The Intermediate Form is widely represented on lower-value projects, as is the Minor Works Form.

The NEC was strongly supported and recommended in the Latham (1994) and Egan (1998) reports because of its non-adversarial features. Some lawyers dislike the NEC form for this reason! However, this form of contract has taken a long time to make any impact amongst many of the clients who were surveyed. Whilst it has a strong recommendation from the two reports above, it needs to be recognised that these reports were written some considerable time ago.

The *NCCL Survey 2015* also provides a good insight in to the type of contract used and the value of contract (NBS 2015). FIDIC standard form of contracts are the preferred format for high-value contracts, followed by NEC and JCT (Figure 7.5).

7.2.12 Partnering

Partnering contracts were recorded for the first time in the 2001 *Contracts in Use Survey*. The 2004 survey recorded only a small increase in the use of partnering agreements and then only for larger-value projects. More importantly, the 2007 *Contracts in Use Survey* shows a 0.4 per cent (to 2.3 per cent) decline in use, further reducing to 1 per cent by 2010 in the number of contracts on which partnering was used. Although there is a significant increase in terms of value, to 15.6 per cent in 2007, there is a dramatic reduction in 2010 to 0.9 per cent. There is a somewhat similar feeling to that of management contracting, which preceded it. Partnering as a contractual arrangement, however good its intentions and ideas, is not making the impact that was either predicted or expected.

7.2.13 PFI/PPP

There was no information in the *Contracts in Use Survey* about the use of PFI/PPP. Apparently, questions were asked about these, but no responses were forthcoming. This may suggest that the organisations returning the survey forms were not working in this field. Perhaps, and more likely, lawyers who were dealing with these contractual arrangements do not complete this survey. Whilst the value of these arrangements is not inconsiderable (see Chapter 14), their overall proportion in respect of total industry turnover remains small but significant. Now that PPP/PFI contracts are in use in many countries, it will be interesting to observe how their use develops in the future.

These processes are being evaluated by many different bodies, who sometimes come to conflicting views about the benefits of PPP/PFI. It is widely recognised that many of the projects would never have been built without these practices, but the question remains the same – at what cost?

7.2.14 Trends in methods of procurement

Tables 7.1 and 7.2 show the proportion by the value and number of contracts, respectively, of trends in procurement methods since 1985.

It is important to reflect on the data included in these tables and to compare the results with our own experiences. Had the survey been done fifty years ago, the picture would have been

Table 7.1 Trends in methods of procurement (by value of contracts)

Procurement method	*1985* %	*1987* %	*1989* %	*1991* %	*1995* %	*1998* %	*2001* %	*2004* %	*2007* %	*2010* %
Contract bills	59.3	52.1	52.3	48.3	43.7	35.5	20.4	23.6	13.2	18.8
Drawings/spec	10.2	17.7	10.2	7.0	12.2	10.0	20.2	10.7	18.2	22.6
Design and build	8.0	12.2	10.9	14.8	30.1	41.4	42.7	43.2	32.6	39.2
Target contracts	0.0	0.0	0.0	0.0	0.0	0.0	0.0	11.6	7.6	17.1
Approximate quantities	5.4	3.4	3.6	2.5	2.4	2.8	2.8	2.5	2.0	0.7
Prime cost	2.7	5.2	1.1	0.1	0.5	0.3	0.3	0.1	0.2	0.6
Management contract	14.4	9.4	15.0	7.9	6.9	2.3	2.3	0.8	1.0	0.0
Construction management	0.0	0.0	6.9	19.4	4.2	7.7	9.6	0.9	9.6	0.1
Partnering	0.0	0.0	0.0	0.0	0.0	0.0	1.7	6.6	15.6	0.9
Totals	100.0	100.0	100.0	100.0	100.0	100.0	100.0	100.0	100.0	100.0

Source: RICS and Davis Langdon 2012.

Table 7.2 Trends in methods of procurement (by number of contracts)

Procurement method	*1985* %	*1987* %	*1989* %	*1991* %	*1995* %	*1998* %	*2001* %	*2004* %	*2007* %	*2010* %
Contract bills	42.8	34.0	35.8	29.0	39.2	30.9	19.7	31.1	20.0	24.5
Drawings/spec	47.1	55.4	49.7	59.2	43.7	43.9	62.9	42.7	47.2	52.1
Design and build	3.6	5.2	9.1	9.1	11.8	20.7	13.9	13.3	21.9	17.5
Target cost	0.0	0.0	0.0	0.0	0.0	0.0	0.0	6.0	4.5	3.7
Approximate quantities	2.7	1.9	2.9	1.5	2.1	1.9	1.7	2.9	1.7	0.3
Prime cost	2.1	2.3	0.9	0.2	0.7	0.3	0.2	0.2	0.5	0.6
Management contract	1.7	1.2	1.4	0.8	1.2	1.5	0.6	0.2	0.7	0.0
Construction management	0.0	0.0	0.2	0.2	1.3	0.8	0.4	0.9	1.1	0.3
Partnering	0.0	0.0	0.0	0.0	0.0	0.0	0.6	2.7	2.4	1.0
Totals	100.0	100.0	100.0	100.0	100.0	100.0	100.0	100.0	100.0	100.0

Source: RICS and Davis Langdon 2012.

vastly different, with a huge emphasis on bills of quantities as a contract document, a limited use of design and build and virtually nothing on the more alternative and more recently introduced methods of procurement. In the 1970s, some predicted that design and build would eventually become the norm in the construction industry. This has been realised somewhat, but not to the extent of being the norm. The question always would be on how to publicly account for the funds used in public sector projects. Is full value for money achieved from the procurement system used?

Conversely, despite the apparent dislike of some of the more traditional methods of procurement, these appear to have survived better than other modern methods that have been introduced and then discarded. Neither the various types of management methods or partnering have made the impact that many expected. Management contracts reached their peak twenty-six years ago (in 1991), when, in terms of value, they accounted for over a quarter of the construction industry workload. Today, they have very much disappeared, even amongst the larger contracts.

It is worth remembering that, for the construction industry as a whole, the figures are also slightly distorted. A majority of housing projects are not, by definition, contract work and thus are excluded from this analysis.

7.2.15 Use of JCT standard forms

Tables 7.3 and 7.4 show the proportion of use by the value and number of contracts, respectively.

Variants of the JCT forms of contract continue to be used on the majority of building projects. The 91 per cent quoted in 1998 was the largest proportion, by number, since the survey was first introduced. Since then, its use, according to the survey, has declined, although still representing

Table 7.3 Trend in use of JCT standard forms

	1984 %	*1991* %	*1993* %	*1995* %	*1998* %	*2001* %	*2004* %	*2007* %	*2010* %	*2011* %	*2012* %	*2015* %
Number	84	78	82	85	91	91	78	79	88	60	48	39
Value	75	61	80	76	68	79	70	62	67			

Source: RICS and Davis Langdon 2012; NBS 2015.

Table 7.4 Use of JCT standard forms

Form	*By number of contracts (%)*				*By value of contracts (%)*			
	1995	*2004*	*2007*	*2010*	*1995*	*2004*	*2007*	*2010*
JCT with quantities	22.4	14.9	11.7	12.5	32.3	18.4	10.2	11.2
JCT without quantities	5.2	5.3	6.2	4.8	3.6	3.6	7.8	5.0
JCT with approx. quantities	1.1	2.6	0.9	0.2	1.5	2.5	1.5	0.2
Design and build	9.3	11.2	19.4	15.7	20.4	35.6	35.3	33.4
IF with quantities	12.0	11.8	6.7	10.7	7.0	3.5	1.1	6.2
IF without quantities	9.9	8.1	8.5	9.3	3.5	2.4	1.4	3.8
Minor works	22.7	23.5	23.5	33.0	2.3	2.4	0.9	4.9
Prime cost	0.5	0.1	0.4	0.1	0.3	0.1	0.1	0.1
Management	0.8	0.2	0.6	0.0	4.5	0.7	0.6	0.0
Total JCT forms	83.9	77.7	77.9	86.3	75.4	69.2	58.9	64.8

Source: RICS and Davis Langdon 2012.

Table 7.5 Use of standard forms of contract on building contracts

	By number		*By value*	
	2007	*2010*	*2007*	*2010*
Prime contracting agreement	0.1%	0.0	9.4%	0.0
ACA	2.2%	0.9	5.5%	0.8
GC/Works/1	6.1%	0.8	2.9%	2.2
NEC	7.7%	7.4	14.0%	26.4
ICE	1.2%	0.4	2.1%	1.6
Other standard forms	0.9%	0.0	2.8%	0.0
Other contracts	2.5%	2.5	1.8%	1.9
JCT	79.3%	88.0	61.5%	67.2

Source: RICS and Davis Langdon 2012.

a large proportion. In respect of the value of projects, it is now used on almost two-thirds of the contracts from this survey. In the 2001 survey, it was reported that 95 per cent of all building projects used a standard form of contract of one kind or another.

The following are the main findings of the *Contracts in Use Survey* (RICS and Davis Langdon 2012) regarding forms of contract (see Table 7.5):

- JCT in all of its variants remains the most commonly used form of contract for the procurement of building contracts, although it shows a steady decline in use over other forms (especially NEC and FIDIC).
- The use of JCT with quantities has declined considerably, in line with the reduction in the use of bills of quantities, as shown in Table 7.1.
- There is a high use of the design and build form.
- The use of IF contracts appears to have peaked in terms of the number of contracts on which it is used. In terms of the value of work using this form, the decline is more pronounced where contract bills are used as a part of the contract.
- The vast majority of construction contracts continue to use one of the standard forms of contract, albeit sometimes with client or consultant amendments.
- Only 2.5 per cent by number, and 1.9 per cent by value, have used a non-standard form of contract, down from the figures recorded in the previous two surveys.

Within the survey sample, 88 per cent of all contracts by number employed a JCT standard form, indicating an increase from the 2007 survey, though the *NCCL Survey 2015* indicates a steady decline in use, dropping to 39 per cent (NBS 2015). By value, the proportion of contracts employing one of the JCT family of contracts dropped to 67 per cent (in 2010), indicating a marginal increase. However, this is bound to have decreased judging by the decline in use (actual figures by value not available in *NCCL Survey 2015*).

7.3 Major issues to be resolved

There are four main aspects that needs to be considered in deciding and developing a contract procurement strategy:

- consultant or contractor
- price competition or negotiation

- measurement or reimbursement
- traditional or alternative procurement.

These are discussed in the following subsections.

7.3.1 Consultants versus contractors

The *Contracts in Use Survey* (RICS and Davis Langdon 2012) underestimates the total amount of D&B projects because it is almost entirely a survey of consultant organisations. The arguments for engaging a consultant rather than a contractor as the main employer's adviser are inconclusive. The respective advantages and disadvantages may be summarised as follows. Advantages of a contractor-centred approach are said to be:

- better time management
- single-point responsibility
- inherent buildability
- certainty of price
- teamwork
- inclusive design fees.

Disadvantages may be:

- problems of contractor proposals matching with employer requirements
- payment clauses
- emphasis may be directed away from design towards other factors
- employers may still need to retain consultants for payments, inspections, etc.

7.3.2 Price competition versus negotiation

There are a variety of ways in which a contractor may seek to secure business. These include speculation, invitation, reputation, rotation arrangement, recommendation and selection. Irrespective of the final contractual arrangements which are made by the employer, the method of choosing the contractor must first be established. The alternatives which are available for this purpose are either competition or negotiation.

Some form of competition on price, time or quality is desirable. All of the available evidence suggests that, under the normal circumstances of contract procurement, the employer is likely to strike a better bargain if an element of competition exists. There are, however, a number of circumstances that can arise in which a negotiated approach may be more beneficial to the employer. Some of these include:

- business relationship
- early start on-site
- continuation contract
- state of the market
- contractor specialisation
- financial arrangements
- geographical area.

This list is not exhaustive, nor should it be assumed that negotiation would be preferable in all of these examples. Each individual project should be examined on its own merits, and a decision made bearing in mind the particular circumstances concerned and specific advantages to the employer.

Certain essential features are necessary if the negotiations are to proceed satisfactorily. These include equality of the negotiators in either party, parity of information, agreement as to the basis of negotiation and a decision on how the main items of work will be priced.

7.3.3 Measurement versus reimbursement

There are only two ways of calculating the costs of construction work. Either the contractor adopts some form of measurement and is paid for the work on the basis of quantity multiplied by a rate, or the contractor is reimbursed the actual costs. A drawing and specification contract, for example, relies upon the contractor measuring and pricing the work, even though only a single sum is disclosed to the employer. The measurement contract allows payment for risk to the contractor, the cost reimbursement approach does not. Many measurement contracts may include for a small proportion of the work to be paid for under dayworks (a form of cost reimbursement), but it is more unusual to find cost reimbursement contracts with any measurement aspects. Here are some points to bear in mind when choosing between measurement or cost reimbursement contracts:

- *Contract sum*: this is not available with any form of cost reimbursement contract.
- *Final price forecast*: this is not possible with any of the cost reimbursement methods or with measurement contracts which rely extensively on approximate quantities.
- *Incentive for contractor efficiency*: cost reimbursement contracts can encourage wastage that must then be passed on to the employer.
- *Price risk*: measurement contracts allow for this, employers may therefore pay for such non-events.
- *Cost control*: the employer has little control over costs where any form of cost reimbursement contract is used.
- *Administration*: cost reimbursement contracts require a large amount of clerical work.

7.3.4 Traditional versus alternatives

Until recently, the majority of the major building projects were constructed using single-stage selective tendering. This method of procurement has many flaws, so alternative procedures have been devised in an attempt to address them. The newer methods, or alternative procurement paths, overcame the failures of the traditional approach, but they created their own particular problems. In fact, if a single method was able to be devised that addressed all of the problems, then the remaining methods would quickly fall into disuse. In choosing a method of procurement, therefore, the following issues are important (they are more fully described in Chapter 9):

- project size
- costs inclusive of the design
- time from brief to handover
- accountability
- design, function and aesthetics
- quality assurance
- organisation and responsibility
- project complexity

- risk placing
- market considerations
- financial provisions.

7.4 The framework of society

The correct application of contractual procurement systems is influenced by a number of factors that are present within any society. Such factors, although external to the construction industry, do have implications for the successful completion of each project. Also, the appropriate recommendations today may have different implications in the future, because the framework of society is constantly evolving. In selecting the right method, the following factors should be considered and evaluated:

- *Economic*: interest rates, inflation, land costs, investment policies, market levels, taxation opportunities, opportunity costs.
- *Legal*: contract law, case law, arbitration, discharge of contracts, remedies for breach of contract.
- *Technological*: new techniques, off-site manufacture, production processes, use of computers.
- *Political*: government systems, public expenditure, export regulations and guarantees, planning laws, trade union practice, policies.
- *Social*: demography, ageing population, availability of potential employees, retraining, the environment.

Many of these factors are unstable, even in a well-run economy, or for the duration of the contract period, owing to influences from other world markets. An employer's satisfaction with the project relies upon the procurement adviser's skill and, to some extent, the intuition which can be provided in being able to reflect on how these aspects are likely to influence the method recommended for contract procurement.

7.5 Employers' essential requirements

When buying a particular service, employers seek to ensure that it fully meets their needs. The consultant or contractor employed needs to identify with the employer's objectives, within the context in which the employer has to operate, and particularly with any constraints that may be present. *A Study of Quantity Surveying Practice and Client Demand* (RICS 1984) identified some of the following criteria as important requirements for the majority of employers:

- impartial and independent advice
- trust and fairness in all dealings
- timely information ahead of possible events
- implications on the inter-reactions of time, cost and quality
- options from which the employer can select the best possible route
- recommendations for action
- good value for any fees charged
- advice based upon a skilled consideration of the project as a whole
- sound ability and general competence
- reliability of advice
- enterprise and innovation.

7.6 Procurement management

It is of considerable importance to employers who wish to have buildings erected that the appropriate advice is provided on the method of procurement to be used. The advice offered must be relevant and reliable and based upon skill and expertise. There is, however, a dearth of objective and unbiased advice available. It is often difficult to elicit the relevant facts appropriate to a proposed building project. Construction employers will tend to rely on the advice from their chosen consultant or contractor. The advice provided is usually sound, and frequently successful, according to the criteria set by the employer. It may, however, tend to be biased and even in some cases tainted with self-interest. It may also sometimes be given on the basis of 'who gets to the client first'. Methods and procedures have now become so complex, with a wide variety of options available, that an improvement in the management approach to the procurement process is now necessary to meet the employer's needs. For instance, the need to match the employer's requirements with the industry's response is very important if customer care and satisfaction are to be achieved. The employer's procurement manager must consider the characteristics of the various methods that are available and recommend a solution that best suits the employer's needs and aspirations. The manager will need to discuss the level of risk involved for the procurement path recommended for the project under review.

The process of procurement management may be broadly defined to include the following:

- Determining the employer's requirements in terms of time, cost and quality.
- Assessing the viability of the project and providing advice in terms of funding and taxation.
- Advising on an organisational structure for the project as a whole.
- Advising on the appointment of consultants and contractors bearing in mind the criteria set by the employer.
- Managing the information and coordinating the activities of the consultants and contractor through the design and construction phases.

The simplistic view is that architects design and contractors build; those are their strengths. It is important, however, that someone is especially responsible for the contractual matters surrounding the contract.

The National Economic Development Corporation publication, *Thinking About Building* (1985), makes the following suggestions on procurement:

- Do make one of your in-house executives responsible for the project.
- Do bring in an outside adviser if the in-house resources and skills are inappropriate.
- Do take special care to define the needs for the project.
- Do choose a procurement path to fit the defined priorities.
- Do go to some trouble to select the organisations and individuals concerned.
- Do ensure that a professional appraisal is done before the scheme becomes too advanced.

The effectiveness of a procurement path is a combination of three things:

- The correct advice and decision on which procurement path to use.
- The correct implementation of the chosen path.
- The evaluation during and after its execution.

7.7 Coordinated project information (CPI)

Research at the Building Research Establishment on fifty representative building sites has shown that the biggest single cause of events which stop site managers, architects or tradespeople from working together is *unclear or missing project information*. Another significant cause is uncoordinated design, and at times the entire effort of site management can be directed to searching for missing information or reconciling inconsistencies in the information that is available. To overcome these weaknesses, the Coordinating Committee for Project Information (CCPI) was formed with the task of developing a common arrangement for work sections that could be used throughout the various forms of documentation.

The CCPI has consulted widely to ensure that the proposals are practicable and helpful, and its work has been performed by practising professionals who are themselves fully aware of the pressures involved. Is it really more economical to produce incomplete, inconsistent, if not contradictory, project information during the design process, and then spend time wrestling with and trying to rectify the problems on-site, with the attendant waste of time in dealing with claims? Is it not better to produce complete and coordinated information once and thus avoid such problems? If properly coordinated project documentation is produced, much abortive work will be avoided and costs will be so much lower.

Regardless of the chosen procurement method, it is important that the principles of the CCPI are adopted and put into practice. Procurement methods that are unable to incorporate such principles may be severely flawed in practice.

The CCPI, originally established as the Building Project Information Committee (BPIC), was set up on 17 February 1987 under the joint sponsorship of the RIBA, RICS, Building Employers Confederation (BEC) and ACE, which represented both Chartered Institute of Building Services Engineers (CIBSE) and ICE. It is now the Construction Project Information Committee (CPIC), responsible for major advancements in standardisation of project information. These recent developments include the UNICLASS 2015 (which is embedded in the NBS BIM toolkit) and PAS1192 standards.

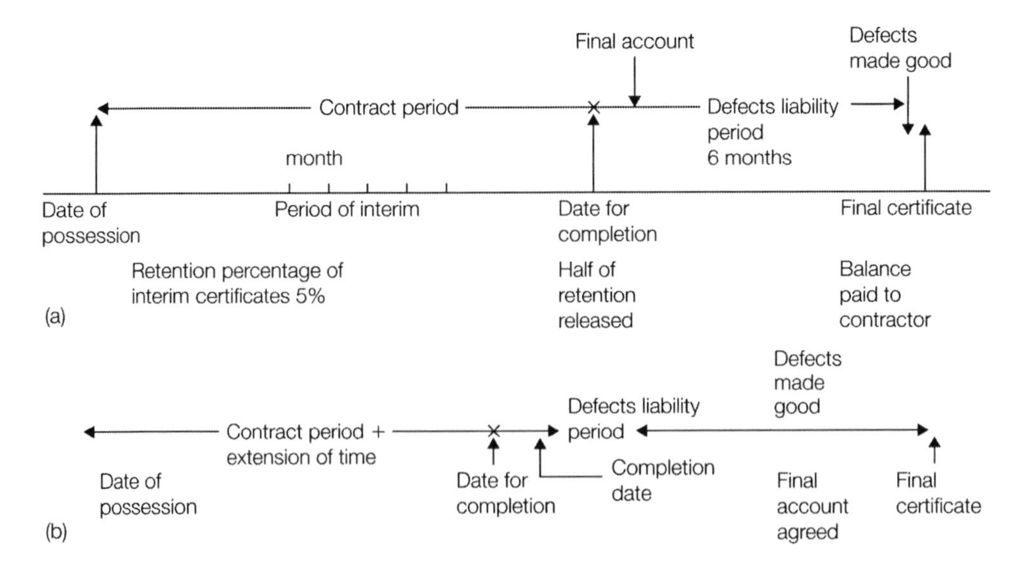

Figure 7.6 Contractual terminology: (a) expected contractual position, (b) likely contractual position.

7.8 Added value

Employers are rightly concerned with obtaining value for money. Cheapness is in itself no virtue. It is well worthwhile to pay a little more if the gain in value exceeds the extra costs. Value for money, in any context, is a combination of subjective and objective viewpoints. There are some items that can be measured, but there are other items that can only be left to opinion or, at best, expert judgement. Measurable items can largely be proven, or at least they could if our knowledge were fully comprehensive. But the professional's skill is largely in the area of judgment. The procurement method recommended to the employer needs to be the method which offers the best value for money. The need for careful assessment is required to obtain the desired results. Figure 7.6 illustrates the contractual terminology in both the expected and the likely positions.

7.9 Conclusions

Procurement procedures today are a dynamic activity. They are evolutionary, to suit the changing needs of society and the considerations in which the industry finds itself operating. There are no standard solutions, so each project, which invariably represents a one-off, needs to be considered independently and analysed accordingly. A wide variety of factors have to be taken into account before any practical decisions can be made. The various influences concerned need to be weighed carefully, and always with the best interests of the employer in mind.

8 Contract procurement

The execution of a construction project requires both design work and the carrying out of construction operations on-site. If these are to be carried out successfully, resulting in a satisfactorily completed project, then some form of recognised procedure must be employed at the outset to deal with their organisation, coordination and procedures. Procuring buildings is expensive. New buildings are one of the highest single-cost items for any firm or organisation, be they developers or owner-occupiers. It is therefore important to get this aspect as right as possible.

Traditionally, a client who wished to have a building constructed would invariably commission an architect to prepare drawings of the proposed scheme, and if the scheme was sufficiently large, employ a quantity surveyor to prepare appropriate contract documentation on which the building contractor could prepare a price. These would all be based upon the client's brief, and the information used as a basis for competitive tendering. This was the common system in use at the turn of the century and still continues to be widely used today.

However, particularly since the mid 1960s, a small revolution has occurred in the way designers and builders are employed for the construction of buildings. To some extent, this is the result of initiatives taken by the then Ministry of Works in the early 1960s. The Banwell Committee report in 1967 recommended several changes in the way that projects and contracts were organised, one of which was an attempt to try to bring the designers and the constructors closer together. The construction industry continues to examine and evaluate the methods available and devise new procedures which address the shortfalls and weaknesses of the current procedures. Some of these procedures remain fashionable for a short space of time and then fall into disuse.

There is, however, no panacea that will suit all projects. In fact, it may be argued that change is occurring so fast that today's solution may be quite inappropriate for tomorrow. Each of the methods described has its own characteristics, advantages and disadvantages. All have been used in practice, some more than others, largely due to familiarity and ease of application. New methods will continue to evolve to meet the rapidly changing circumstances of the construction industry. These will occur in response to current deficiencies and to changes in the culture of the construction industry.

8.1 Methods of price determination

Building and civil engineering contractors are paid for the work they carry out on the basis of one of two methods: measurement, i.e. payment against given criteria; or cost reimbursement, i.e. actual costs involved.

8.1.1 Measurement

The work is measured in place on the basis of its finished quantities. The contractor is paid for the work on the basis of quantity multiplied by a rate. Measurement may be undertaken by the employer's quantity surveyor, or by the contractor's surveyor or estimator. In the first example, an accurate and detailed contract document can be prepared. In the second example, the document prepared will be sufficient only to satisfy the particular builder concerned. The work may be measured as accurately as the drawings allow prior to the contract being awarded, in which case it is known as a lump sum contract. Alternatively, the work may be measured or remeasured after it has been carried out. In the latter case, it is referred to as a remeasurement contract. All contracts envisage some form of remeasurement to take into account variations to the original plans. In such cases, both the employer's and the contractor's surveyors measure the work together. Measurement contracts, unless they are entirely of a remeasurement type, allow for some sort of final cost to be recalculated. This offers advantages to the client for budgeting and cost control. Building contracts are more often lump sum contracts, whereas civil engineering projects are typically of the remeasurement type. Measurement is usually done against agreed criteria, such as a method of measurement. In the UK, the Standard Method of Measurement of Building Works (SMM7) was designed and compiled for this purpose. This has now been replaced by the New Rules of Measurement (NRM) from 2012.

The Civil Engineering Procedure, published by the Institution of Civil Engineers (ICE 2015), introduces another similar term 'Admeasurement'. This is likely to originate from the ICE Conditions of Contract measurement clauses where it is defined as the difference between the final quantity of work (actual work) carried out and the originally anticipated quantity of work (as per bills). It simply is the difference between the actual and estimated quantity of work.

8.1.2 Cost reimbursement

Cost reimbursement arrangements allow the contractor to recoup the actual costs of the materials which have been purchased, the actual time spent on the work by the operatives and the actual time used by mechanical plant. An agreed additional amount, often expressed as a percentage, is added to cover the contractor's overheads and profit. Daywork accounts, used on many contracts in conjunction with measurement, are costed and valued on much the same basis.

8.2 Measurement contracts

8.2.1 Drawings and specification

This is the simplest type of measurement contract and is really only suitable for small works or simple projects. In recent years there has been a trend towards using this approach on schemes much larger than it was originally intended for. Each contractor must then measure the quantities from the drawings and interpret the specification during pricing in order to calculate the tender sum. The method is wasteful of the contractor's estimating resources, since all contractors need to prepare their own measurements. It does not easily allow for a fair comparison of the tender sums received by the employer. Interpreting the specification can be a difficult job, even for the more experienced estimator. The contractor has also to accept a greater proportion of the risk, being responsible for the prices as well as for the measurements that have been based upon interpretation of the contract information. In order to compensate for possible errors or omissions, evidence suggests that sensible contractors will tend to overprice this type of work.

8.2.2 Performance specification

This method is a much more vague and imprecise approach to tendering and the evaluation of the contractors' bids. The use of a performance specification requires the contractor to prepare a price for the work based only upon the employer's brief and user requirements alone. The contractor is then left to select the materials to be used and to determine the method of construction that suit these broad requirements. In practice, the contractor will select materials and methods of construction which satisfy the prescribed performance standards in the cheapest possible way. Great precision is required in formulating a performance specification if the desired results of the employer are to be achieved. Performance specifications are sometimes used with mechanical equipment that is used in buildings.

8.2.3 Schedule of rates

With some projects it is not possible to predetermine the nature and full extent of the proposed building works. In these circumstances, where it is desirable to form some direct link between quantity and price, a schedule of rates may be used. This schedule is similar to a bill of quantities, but without any actual quantities being included. It should be prepared using the rules of a recognised method of measurement. Contractors are invited to insert their rates against these items, and these are then used in comparison against other contractors' schedules in selecting the best tender. Upon completion of the work, this is remeasured and the rates that have been provided by the contractor are used to calculate the final cost. This method does not allow either the prediction of a contract sum, or an indication of the probable final cost of the project from the outset. Contractors also find it difficult to price the schedule realistically in the absence of any quantities, since the amount of work to be performed has a direct influence upon the costs incurred.

8.2.4 Schedule of prices

An alternative to the schedule of rates is to provide the contractors with a ready-priced schedule, similar to the former Property Services Agency's schedule of rates. This is now referred to as the PSA schedule of rates, published by Carillion™. An alternative to this is the rates schedule published by the Building Costs Information Service Online Rates Database (BCIS-ORDB).

In this case, the contractors adjust each rate by the addition or deduction of a percentage. In practice, a single percentage adjustment is normally made to all of the rates. This standard adjustment is unsatisfactory, since the contractor will view some of the prices in the schedule as being high, and others as being too low in terms of covering the contractor's own costs. However, a schedule of prices does have the advantage of producing fewer pricing errors in the tender documents when compared with the contractor's own price analysis of the work.

8.2.5 Bill of quantities

Even with all the new forms of contract arrangement, the bill of quantities continues to remain the most common form of measurement contract. It also remains the most common contractual arrangement for major construction projects in the UK (see Chapter 7). The use of a schedule of quantities has many advantages over the alternative systems that are presently available (see Chapter 10). The tender sums that are received from contractors are able to be judged almost against the criteria of cost, since all contractors are using the same qualitative and quantitative information. This type of documentation is still recommended as the most appropriate for all but the smallest and simplest types of building projects. Bills of quantities are referred to in JCT standard forms as the contract bills.

8.2.6 Bill of approximate quantities

In some circumstances it is not possible to premeasure the work accurately because parts of the design remain incomplete. In these situations a bill of approximate quantities can be prepared, with the entire project being remeasured upon completion of the works. The JCT standard form of building contract (SBC/AQ 2016) with approximate quantities would then be used. Whilst an approximate cost of the project can be obtained, the uncertainty in the design information makes any reliable forecast of cost impossible.

8.3 Cost reimbursement contracts

Cost reimbursement contracts are not favoured by many of the industry's employers, since there is an absence of a tender sum and a forecasted final account cost. Some of these types of contract also provide little incentive for contractors to control their costs, although different varieties of cost reimbursement have attempted to build in incentives for the contractor to keep costs as low as possible. Evidence from industry also shows them to be an unpopular method (see Chapter 7). Cost reimbursement contracts are therefore used only in special or unusual circumstances. These might be:

- Emergency work projects, where time is not available to allow the traditional process to be used.
- When the character and scope of the works cannot be readily or easily determined.
- Where new technology is being introduced.
- Where a special relationship exists between the employer and the building contractor.

For example, part of a major highway between two important towns collapses and makes the route impassable. Diversions are put in place that increase journey times for commuters and freight traffic. It is important to reopen the highway as soon as possible. The design is started and a contractor is employed to commence preliminary works on-site. Early completion of the project is the important criteria. There is insufficient time available to fully complete the design prior to construction work commencing on-site. There is limited tender information and no time to obtain tenders in the usual way; this could further delay the work by anything up to three months, even for a modest scheme (documentation preparation and pricing). The project is thus awarded to a firm who has successfully worked with the highways agency on previous occasions. On this project the cost reimbursement is a reasonable solution.

Cost reimbursement contracts can take many different forms, and the following sections discuss three of the more popular types in use. Each of the methods repay the contractor's costs with an addition to cover profits. Prior to embarking on this type of contract it is especially important that all the parties involved are clearly aware of the definition of contractor's costs as used in this context.

8.3.1 Cost plus percentage

The contractor receives the costs of labour, materials, plant, subcontractors and overheads and to this sum is added a percentage to cover profits. This percentage is agreed at the outset of the project. A major disadvantage of this type of cost reimbursement is that the contractor's profits are directly related and geared to the contractor's own expenditure. Therefore, the more the contractor spends on the building works, the greater will be the contractor's profitability. Because it is an easy method to operate, this tends to be the selected method when using cost reimbursement. In the highway example described above, cost plus percentage would be the chosen method of cost reimbursement.

8.3.2 Cost plus fixed fee

With this method, the contractor's profit is predetermined by the agreement of a fee for the work before the commencement of the project. There is therefore some incentive for the contractor to attempt to control the costs, since the fee (contractor's profit) remains the same regardless of the actual costs involved. However, it is sometimes difficult to predict the costs of building with sufficient accuracy. Disagreement between the contractor and the employer's own professional advisers will occur where the predicted cost is widely at variance with the actual costs. The result is that the so-called *fixed fee* may thus need to be revised on completion of the project.

Examples

Cost plus percentage

Estimated cost	£500,000	Final cost	£511,964
Agreed 10% profit	£50,000	10% profit	£51,196
Tender	£550,000	Final account	£563,160

Cost plus fixed fee

Estimated cost	£500,000	Final cost	£511,964
Fixed fee	£60,000	Fixed fee	£60,000
Tender sum	£555,000	Final account	£571,964

Cost plus variable fee

Estimated cost	£500,000	Final cost	£511,964
Fixed fee	£50,000	Fixed fee	£50,000
Variable fee		Variable fee	
10% ± £500,000		10% × £11,964	–£1,196
Tender sum	£560,000	Final account	£560,768

8.3.3 Cost plus variable fee

The use of this method requires a target fee to be set for the project prior to the signing of the contract. The contractor's fee is then composed of two parts, a fixed amount and a variable amount. The total fee charged then depends upon the relationship between the target cost and the actual cost. This method provides a supposedly even greater incentive to the contractor to control the construction costs. It has the disadvantage of requiring a reasonably accurate target cost to be fixed on the basis of a very *rough* estimate of the proposed project.

The examples are for illustration only. Do not assume that the final accounts would necessarily follow these patterns. These are the simple explanations of the different calculations expected. In practice, the financial adjustments are often very complex and time-consuming. And, as explained earlier, the preferred choice with cost reimbursement contracts is frequently the *cost plus percentage* version for ease of administration.

8.4 Contractor selection

There are essentially two ways of choosing a contractor, either by competition or negotiation. Competition may be restricted to a few selected firms or open to almost any firm who wishes to submit a tender. The options described later are used in conjunction with one of these methods of contractor selection.

8.4.1 Selective competition

Selective competition is the traditional and most popular method of awarding construction contracts. It is also believed to be the fairest to all the parties concerned. In essence, a number of firms of known reputation are selected by the employer on the advice of the design team to submit a price. The firm who submits the lowest tender is then awarded the contract. The *Code of Procedure for Single Stage Selective Tendering* is a useful document of good practice on guidance about awarding of construction contracts (NJCC 1996). The code takes into account the changes in buildings procurement and the general principles of current recommendations. Here are some of the more important points from this code:

- The code assumes the use of a standard form of building contract with which the parties in the construction industry are familiar. If other forms of contract are used, some modification of detail may be necessary. There are clear advantages to all parties in the knowledge that a standard procedure will be followed in inviting and accepting tenders. Figure 8.1 illustrates the traditional contractual relationships.
- The code recommends that the number of tenderers should be limited to six. The number of tenderers is restricted because the cost of preparing abortive tenders will be reflected in prices generally throughout the building industry. There are no such things as *free estimates*.
- In preparing a shortlist of tenderers, the following must be borne in mind:
    - the firm's financial standing and record
    - recent experience of building over similar contract periods
    - the general experience and reputation of the firm for similar building types
    - adequacy of management
    - adequacy of capacity.
- Each firm on the shortlist should be sent a preliminary enquiry to determine its willingness to tender. The enquiry should contain:
    - job title
    - names of employer and consultants
    - location of site and general description of the works
    - approximate cost range
    - principal nominated subcontractors
    - form of contract and any amendments
    - procedure for correction of priced bill
    - contract under seal or under hand
    - anticipated date for possession

    - contract period
    - anticipated date for dispatch of tender documents
    - length of tender period
    - length of time tender must remain open for acceptance
    - amount of liquidated damages
    - bond
    - special conditions.

- Once a contractor has confirmed an intention to tender, that tender should be made. If circumstances arise which make it necessary to withdraw, the architect should be notified before the tender documents are issued or, at the latest, within two days thereafter.
- A contractor who has expressed a willingness to tender should be informed if not chosen for the final shortlist.
- All tenderers must submit their tenders on the same basis:

    - Tender documents should be dispatched on the stated day.
    - Alternative offers based on alternative contract periods may be admitted if requested on the date of dispatch of the documents.
    - Standard forms of contract should not be amended.
    - A time of day should be stated for receipt of tenders and tenders received late should be returned unopened.
    - The tender period will depend on the size and complexity of the job, but be not less than four working weeks, i.e. twenty days.

- If a tenderer requires any clarification, the architect must be notified and the architect should then inform all tenderers of this decision.
- If a tenderer submits a qualified tender, opportunity should be given to withdraw the qualification without amending the tender figure, otherwise the tender should normally be rejected.
- Under English law, a tender may be withdrawn at any time before acceptance. Under Scottish law, it cannot be withdrawn unless the words 'unless previously withdrawn' are inserted in the tender after the stated period of time the tender is to remain open for acceptance.
- After tenders are opened, all but the lowest three tenderers should be informed immediately. The lowest tenderer should be asked to submit a priced bill within four days. The other two contractors are informed that they might be approached again.
- After the contract has been signed, each tenderer should be supplied with a list of tender prices.
- The quantity surveyor must keep the priced bills strictly confidential.
- If there are any errors in pricing, the code sets out alternative ways of dealing with the situation:

    - The tenderer should be notified and given the opportunity to confirm or withdraw the offer. If it is withdrawn, the next lowest tenderer is considered. Where the offer is confirmed, an endorsement should be added to the priced bills that all rates, except preliminary items, contingencies, prime cost and provisional sums are to be deemed reduced or increased, as appropriate, by the same proportion as the corrected total exceeds or falls short of the original price.
    - The tenderer should be given the opportunity of confirming the offer or correcting the errors. Where it is corrected and is no longer the lowest tender, then the next tender should be examined. If it is not corrected then an endorsement is added to the tender.

- Corrections must be initialled or confirmed in writing and the letter of acceptance must include a reference to this. The lowest tender should be accepted, after correction or confirmation, in accordance with the alternative chosen. Problems sometimes occur because the employer will see that a tender will still be the lowest even after correction. If the first alternative has been agreed upon and notified to all tenderers at the time of invitation to tender, the choice facing the tenderer should clearly be to confirm or withdraw. The employer may require a great deal of persuading to stand by the initial agreement in such circumstances. The answer to the problem is to discuss the use of the alternatives thoroughly with the employer before the tendering process begins. The employer must be made aware that the agreement to use the code and one of the alternatives is binding on all parties. It is possible that an employer who stipulated the first alternative and subsequently allowed price correction could be sued by, at least, the next lowest tenderer for the abortive costs of tendering.
- The employer is not bound to accept the lowest or any tender and is not responsible for the costs of their preparation. There may be reasons why a decision is taken not to accept the lowest tender. Although the employer is entitled to do so, it will not please the other tenderers. The code is devised to remove such practices.
- If the tender under consideration exceeds the estimated cost, negotiations should take place with the tenderer to reduce the price. The quantity surveyor then normally produces what is called 'reduction or addendum bills'. They are priced and signed by both parties as part of the contract bills.
- The provisions of the code should be qualified by the supplementary procedures specified in EU directives which provide for a 'restrictive tendering procedure' in respect of public sector construction contracts above a specified value.

This method of contractor selection is appropriate for almost any type of construction project where a suitable supply of contractors is available.

8.4.2 Open competition

With open competition, the details of the proposed project are often advertised in the local and trade publications, or through the local branch of the Construction Confederation. Any contractor who then feels willing and able to carry out such a project can request the contract documentation. This method has the advantage of allowing new contractors, or contractors who are unknown to the design team, the possibility of submitting a tender for consideration. In theory, any number of

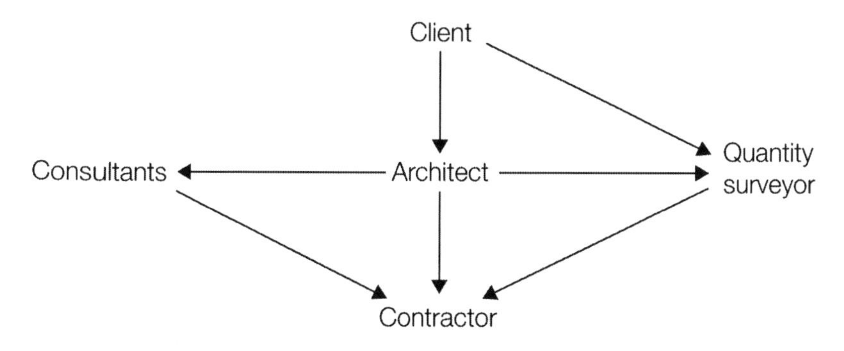

Figure 8.1 Traditional relationships.

firms are able to submit a price. In practice, there is usually a limit on the number of firms who will be supplied with the tender documents. Unsuitable firms are removed from the list where the number of firms becomes too large. The preparation of tenders is both expensive and time-consuming. The use of open tendering may relieve the employer of a moral obligation of accepting the lowest price, because firms are not generally vetted before tenders are submitted. Factors other than price must also be considered when assessing these tender bids, such as the capability of the firm who has submitted the lowest tender. There is no obligation on the part of the employer to accept any tender, should the employer consider that none of the contractors' offers is suitable. Prior to the Second World War, many major building contracts, particularly those in the public sector, were awarded using this method. Although it is still commonly used on minor works, its use on larger projects has been curtailed. There are no known records of the numbers of firms on an open tender list, although more than fifty firms tendering for a single project have been known.

8.4.3 Negotiated contract

This method of contractor selection involves the agreement of a tender sum with a single contracting organisation. Once the documents have been prepared, the contractor prices them in the usual way. The priced documents are then passed to the quantity surveyor, who will check the reasonableness of the contractor's rates and prices. The two parties then arrange a meeting to discuss the queries raised by the quantity surveyor and the negotiation process begins. It is important for both parties to arrive at reasonable rates and prices. Eventually, a tender price is agreed that is acceptable to both parties. There is an absence of any competition or other restriction, other than the social acceptability of the price. This frequently results in tender sums that are higher than might have been obtained by using one of the previous competitive procurement methods. It is believed that, under normal circumstances, negotiated tenders are approximately 5 per cent higher in price than competitive tenders. Negotiation does, however, have particular applications, as follows:

- Where a business relationship exists between the client and the building contractor.
- Where only one firm is capable of undertaking the work satisfactorily.
- On a continuation contract where the building contractor is already established on-site.
- Where an early start on-site is required by the client.
- Where it is beneficial to bring the contractor in during the design stage, to advise on constructional difficulties and how they might best be avoided.
- To make use of the contractor's buildability expertise.

Competitive tendering can create financial problems for both the industry and its clients, due to too keen pricing, which in the end benefits no one. Negotiated contracts do result in fewer errors in pricing since all of the rates are carefully examined and agreed by each party. This will result in fewer exaggerated claims in an attempt to recoup losses on cut-throat pricing. This type of procurement can also involve the contractor in some participation during the design stage, which can result in on-site time and cost savings, thus improving value for money. It should also be possible to achieve greater cooperation during the construction process between the design team and the contractor. However, public sector clients tend not to favour negotiated contracts because of:

- higher tender sums incurred
- public accountability
- suggestion of possible favouritism.

8.5 Contractual options

The following contractual options are an attempt to address the employer's objectives associated with the cost, time and quality of construction. They are not mutually exclusive. For example, a serial contract can be awarded using, in the first instance, a design and build (D&B) arrangement. Fast tracking may be used in conjunction with management contracting. All of these options will also need to include either a selective or negotiated arrangement in the selection of a contractor.

The approximate balance of risk is towards the employer where management procurement methods are selected. Where design and build procurement is used a greater proportion of the risk involved is transferred to the contractor.

8.5.1 Early selection

Early selection is also known as two-stage tendering. Its main aim is to involve the chosen contractor for the project as soon as possible so that the contractor can have an input at the design stage. It therefore seeks to succeed in getting the firm who knows what to build (the designer) in touch with the firm who knows how to build it (the constructor) before the design is finalised. The contractor's expertise in construction methods can thus be harnessed with that of the designer to improve, for example, the buildability criteria of the project. A further advantage is that the contractor may be able to start work on-site sooner than when the more traditional methods of procurement are used. In the first instance, an appropriate contractor must be selected for the project. This is often done through some form of competition and can be achieved by selecting suitable firms to price the major items of work connected with the project. A simplified bill of quantities can be prepared, which might include the following items:

- on-site costs on a time-related basis
- major items of measured works
- specialist items.

Specialist items allow the main contractor the opportunity of pricing the profit and attendance sums. The contractors should also be required to state their overhead and profit percentages. The prices of these items will then form the basis for the subsequent and more detailed price agreement as the project gets under way.

The NJCC have produced a *Code of Procedure for Two Stage Selective Tendering* (1979). This code is not concerned with aspects of the design, which may in this process involve the contractor. The code assumes the use of a standard form of building contract after the second stage has been completed. During the first stage it is important to:

- provide a competitive basis for selection
- establish the layout and design
- provide clear pricing documents
- state the respective obligations and rights of the parties
- determine a programme for the second stage.

Many of the conditions already outlined for single-stage selective tendering apply equally to two-stage tendering. Acceptance of the first-stage tender is a particularly delicate operation.

The employer does not wish to be in the position of having accepted a contract sum at this stage. The terms of the letter of acceptance must be carefully worded to avoid such an eventuality. Depending upon the circumstances, it may be that a contract has been entered into. The question could be: What are the terms of the contract? There are two pitfalls:

- No contract exists; this is likely in many cases.
- A contract binds the employer to pay and the contractor to build.

The existence of a binding contract could be the far worse situation if insufficient care is given to the drafting of the invitation to tender, the tender and the acceptance. After a contractor has been appointed, all unsuccessful tenderers should be notified and, if feasible, a list of first-stage tender offers should be provided. If cost was not the sole reason for acceptance, the reasons for this should also be stated.

It should be noted that public sector or publicly funded project procurement in the UK is governed by the regulations and EU directives as stated in the Official Journal of the European Union (EU nd) and the Public Contracts Regulations (Legislation.gov.uk 2015).

8.5.2 Design and build

Design and build projects aim to overcome the problem of having separate design and construction processes by providing for them within a single organisation. The method has gained in popularity (see Chapter 7). This single firm is generally the building contractor, who may employ the designers in-house or be responsible for employing consultants directly under their control. The major difference is that instead of approaching the designer for a building, the employer briefs the contractor direct (Figure 8.2). The employer may choose to retain the services of an architect or quantity surveyor to assess the contractor's design or to monitor the work on-site. The prudent employer will always want some form of independent advice.

The design evolved by the contractor is more likely to be suited to the needs of the contractor's organisation and construction methodology, and this should save construction time and construction costs. Some argue that the design will be more attuned to the contractor's construction capabilities, rather than to the design requirements of the employer. The final building should result in lower production costs on-site and an overall shorter design and construction period, both of which should provide price savings to the employer. There should also be some supposed savings on the design fees, even after taking into account the necessary

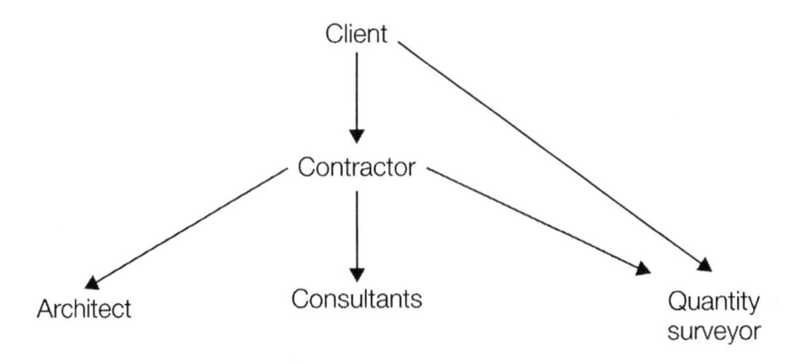

Figure 8.2 Design and build relationships.

costs of any independent architectural advice. A further advantage to the employer is in the implied warranty of suitability, because the contractor has provided the design as a part of the all-in service.

The contractor has a duty to use proper skills and care. An apparent disadvantage to the employer is the financial disincentive relating to possible changes to the design by the employer during construction. Where an employer considers these to be important to keep abreast of changing technologies or needs, excessive costs may be required to discourage them in the first place, or to allow for their incorporation within the partially completed building.

Design and build projects usually result in the employer obtaining a single tender from a selected contractor. When some form of competition in price is desirable, then both the type and quality of the design will need to be taken into account. This can present difficulties in evaluation and comparison of the different schemes. Many of these comments relate generally to design and build projects, not specifically to those that may use the JCT form of contract with contractor's design.

Here are some of the advantages claimed for the design and build approach:

- The contractor is involved with the project from inception and is thus aware of all of the employer's requirements.
- The contractor is able to use their specialised knowledge and methods of construction in evolving the design.
- The time from inception to completion should be able to be reduced due to the telescoping of the various parts of the design and construction processes.
- There can be no claims for delays due to a lack of design information, since the contractor is in total control of it.
- There is direct contact between the employer and the contractor.

The major criticisms against design and build have tended to be of an architectural or aesthetic nature. After all, most builders are not designers but constructors. In an attempt to combat this criticism, the building companies who are extensively involved in design and build work will have an architectural in-house office or employ an outside firm of architects. The rapid increase in this form of procurement indicates that architectural considerations are now given due recognition by the best design and build contractors.

Clients used to employ architects who would then employ building constructors. The trend at present is for clients to employ building firms who in turn employ designers. Design and build has changed the emphasis in order to reduce the possible waste in resources and to increase or add value on behalf of the client.

8.5.3 Package deal

In practice, the terms 'package deal' and 'design and build' are interchangeable. Design and build normally refers to a bespoke arrangement for a one-off project. But the package deal is strictly a special type of design and build project where the employer chooses a suitable building, often from a catalogue. The employer may also be able to view completed buildings of a similar design and type that have been constructed elsewhere. This type of contract procurement has been used extensively in the past for the closed systems of industrialised system buildings of timber or concrete. Multi-storey office blocks and flats, low-rise housing, workshop premises, farm buildings, etc., have been constructed on this basis. The building owner typically provides the package deal contractor with a site and supplies the user requirements or brief. An architect may be independently employed to advise on the building type selected, to inspect the works during construction or to deal with the contract administration. This architect's role is useful to the employer for those items which are outside the scope of the system superstructure. The type of building selected is an *off-the-peg* structure and often it can be erected very quickly. There is even less scope, however, for variations than with the more usual design and build approach, should the employer want to change aspects of the constructional detailing. It cannot be automatically assumed that this type of procurement will be a more economical solution to the employer's needs, either initially or in the long term. Some system buildings constructed in the 1960s are now very costly to maintain. Initial costs can also be high, but these buildings have the advantage that they can be constructed quickly on-site. This is due to the relative completeness of the design, the availability of standard components and the speed of construction on-site.

8.5.4 Design and manage

Design and manage is really the consultant's counterpart to the contractor's design and build. In this case, the design manager, who may be an architect, engineer or surveyor, has full control not just of the design phase but also of the construction phase. The design and manage firm effectively replaces the main contractor in this role, which is now largely management and organisation, administration and coordination of subcontractors. The design manager is responsible for all aspects of construction, including the programming and progressing and the rectification of any defects which may arise. The building contract is between the design and manage firms and the employer, giving the employer a single point of contractual responsibility. The work is generally let through competition in work packages to individual subcontracting firms. This method of procurement therefore offers many of the advantages of traditional tendering coupled with design and build. The design and manage firm will of course need to engage its own construction managers or develop existing staff who have potential in this direction. It will also need to consider continuity with this type of work.

It is suitable for all types and sizes of project, but employers undertaking large projects may, due to past experience, prefer to use a more traditional form of procurement using one of the larger contractors. A major disadvantage is with regard to the site facilities, which will need to be provided by the design and manage consultant, who may need to be hired in a similar way to the subcontractors.

This type of procurement method should be able to offer comparative completion times when compared with the other methods that are available. Since there is the traditional independent control of the subcontractor firms, this should ensure a standard of quality at least as good as that provided by the other contracting methods. In terms of cost, since the work packages will be sought through competition, this will be no more expensive.

8.5.5 Turnkey method

Turnkey contracting is still somewhat unusual in the UK and has thus not been used to any large extent. It has, however, had certain notable successes in the Middle and Far East. The true turnkey contract includes everything that is required and necessary. This normally means everything from inception up to occupation of the finished building. The method receives its title from the *turning-the-key* concept whereby, when the project is completed, the employer can immediately start using it since it will have been fully equipped by the turnkey contractor, including furnishings. Some turnkey contracts also require the contractor to find a suitable site for development. An all-embracing agreement is therefore formed with the one single administrative company for the entire project procurement process. It is an extension of the traditional design and build arrangements and, in some cases, it may even include a long-term repair and maintenance agreement.

On industrial projects, the appointed contractor is also likely to be responsible for the design and installation of the equipment required for the employer's manufacturing process. This type of procurement method can therefore be appropriate for use on highly specialised types of industrial and commercial construction projects.

The entire project procurement and maintenance needs can be handled by a single firm accepting sole responsibility for all events. However, it has been suggested that the employer's ability to control costs, quality, performance, aesthetics and constructional details will be highly variable and severely restricted by using this procurement method. A contractor who undertakes such an all-embracing project will require a variety of strengths and may well have definite ideas about the importance of the different aspects and outcomes of the scheme.

8.5.6 Management contract

Management contracting evolved at the beginning of the 1970s in the UK with an aim of building more complex projects in a shorter period of time and for a lower cost. It may therefore be argued that the more complex the project, the more suitable management contracting may be. This method is also appropriate to a wide range of medium-sized projects. In many ways, management contracting was ahead of its time.

The term *management contract* is used to describe a method of organising the building team and operating the building process. The main contractor provides the management expertise required on a construction project in return for a fee to cover the overheads and profit. The intention is to place the main contractor in a professional capacity to be able to provide the management skills and practical building ability for a fee. In theory, the contractor does not, therefore, participate in the profitability of the construction work. The construction work itself is not undertaken by the contractor, nor does the contractor employ any of the labour or plant directly, except with the possibility of setting up the site and those items normally associated with the preliminary works. Because the management contractor is employed on a fee basis, the appointment can be made early on in the design process. The contractor is then able to provide a substantial input into the design, particularly those aspects associated with the practical construction of the building. Each trade section required for the project is normally tendered for separately by subcontractors, either on the basis of measurement or a lump sum. This should result in the least expensive cost for each of the trades and thus for the construction works as a whole. The work on-site needs a considerable amount of planning and coordination, more so than a traditional procurement arrangement. This is the responsibility of the management contractor and an inherent part of the acquired skills.

In common with all procurement methods, there are advantages and disadvantages. It is somewhat open-ended, since the price can only be firmed up after the final works package quotation has been received. The later in the contract the work is let, the less time there will be for negotiating price reductions overall without seriously impairing a section of the works.

8.5.7 Management fee

Management fee contracting is a system whereby a contractor agrees to carry out building works at cost. In addition, a fee is paid by the employer to cover the overheads and profit.

Some contractors are prepared to enter into an agreement to offer an incentive on the basis of a target cost. This type of procurement is a similar approach to management contracting and cost-plus contracts and therefore has similar advantages and disadvantages.

An alternative approach can also be used where a bill of quantities can be prepared and priced net of the contractor's overheads and profit, or just the profit. These items are then recovered by means of a fee. The system can be as flexible and adaptable as the parties wish. Invariably, as with cost-plus contracting, the fees are percentage related unless some reasonably accurate forecast of cost can be made. Different contracting firms who use management fee contracting have different ways of determining the fee addition. In any case the total cost is largely unknown until completion is achieved and the records agreed. Overspending is therefore much more difficult to control, and any needed savings in cost tend to be required, and made, on the later sections of the project.

8.5.8 Construction management contracting

This offers a further alternative procedure to the management contract. The main difference is that the employer chooses to appoint a construction management contractor who is then responsible for appointing a design team with the approval of the employer (Figure 8.3). The employer chooses to instruct the constructor rather than the designer. The construction management contractor is thus in overall project control of both the design and the construction phases.

There are similarities with this method and the design and build method. A major difference is that the contractor would invariably appoint outside design consultants rather than choosing to employ an in-house service. The employer may thus form some contractual relationship with these consultants. In reality, it is a reversal of the traditional arrangements.

8.5.9 Project management

Descriptions of the different procurement arrangements may mean different things in different parts of the world, different industries or even in different sectors of the construction industry. They may even have different interpretations within a single country, such as the UK. Project management in this context is a function that is normally undertaken by the employer's consultants rather than by a contractor. Contractors do undertake project management, but in a different context from that associated with procurement. The title of project coordinator is similar terminology and perhaps better describes the role of a project manager. The employer

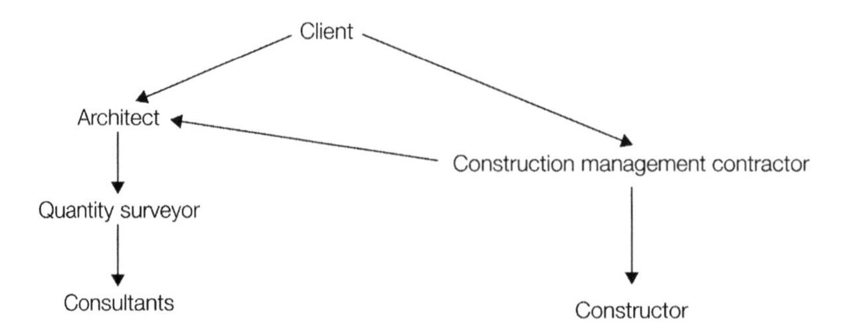

Figure 8.3 Construction management relationships.

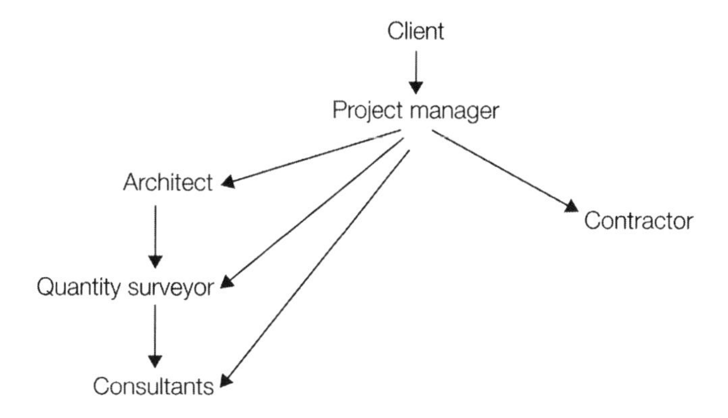

Figure 8.4 Project management relationships.

appoints the project manager who in turn appoints the various design consultants and selects the contractor (Figure 8.4). It is a more appropriate method for the medium- to large-sized project that requires an extensive amount of coordination owing to its complexity. The function of the project manager is therefore one of organising and coordinating the design and construction programmes. Any person who is professionally involved in the construction industry can become a project manager; it is the individual rather than the profession which is important.

In general terms, the need for a project manager is to provide a balance between function, aesthetics, quality control, economics and the time available for constructing. The project manager's aims are to achieve an efficient, effective and economic deployment of the available resources to meet the employer's requirements. The tasks to be performed include identifying those requirements, interpreting them as necessary and communicating them clearly to the various members of the design team and, through them, to the constructor. Programming and coordinating all of the activities and monitoring the work up to satisfactory completion are also a part of this role. A significant difference between this system and the majority of the others described is that the employer's principal contact with the project is through the project manager rather than through the designer.

8.5.10 The British Property Federation (BPF) system

The BPF represents substantial commercial property interests in the UK, and is thus able to exert some considerable influence on the building industry and its associated professions. The introduction of this system of building procurement to the industry created considerable interest, due largely to the fact that it proposed very radical changes to the status quo and practices and procedures that were being used. Some members of the design team felt threatened as their traditional roles and importance were questioned. The BPF system unashamedly makes the interests of the employer paramount. It attempts to devise a more efficient and cooperative method of organising the whole building process from inception up to completion by making genuine use of all who are involved with the design and construction process. Its development was due to employer dissatisfaction with the existing arrangements, where it was claimed that buildings on the whole cost too much, took too long to construct (when compared with other countries around the world) and did not always produce appropriate or credible results. The BPF manual is the only document which sets out in detail the operation of a system.

Whilst the system may appear to be a rigid set of rules and procedures, its originators claim that it can be used in a flexible way and in conjunction with other methods of procurement. Since the system has been devised almost entirely by only one party to the building contract, it lacks the considerations and compromises that are inherent in agreements devised by joint employer and contractor bodies such as the JCT. The system was designed as an attempt to help change outdated attitudes and to alter the way in which the various members of the professions and contractors dealt with one another. It seeks to create better cooperation between the various members of the building team, encourage motivation and remove any possible overlap of effort amongst the design team. Innovations, which were post-Latham, include single-point responsibility for the employer, utilisation of the contractor's buildability skills, reductions in the pre-tender period, redefinition of risks and the preference for specifications rather than bills of quantities. This is despite the desire of contractors for measured quantities of work.

8.5.11 Fast tracking

Fast tracking results in the letting and administration of multiple construction contracts for the same project at the same time, as outlined in Figure 8.5. It is appropriate to large construction projects where the employer's needs are to complete the project in the shortest possible time. The process results in the overlapping of the various design and construction operations of a single project. These various stages can therefore result in the creation of separate contracts or a series of phased starts and completions. When the design for a complete section of the works, such as foundations, are completed, the work is let to a contractor who will commence with this part of the construction work on-site, whilst the remainder of the project is still being designed. The contractor for this stage, or section, of the project will then see this work through to completion. At the same time as this is being done, another work section may be let, continuing

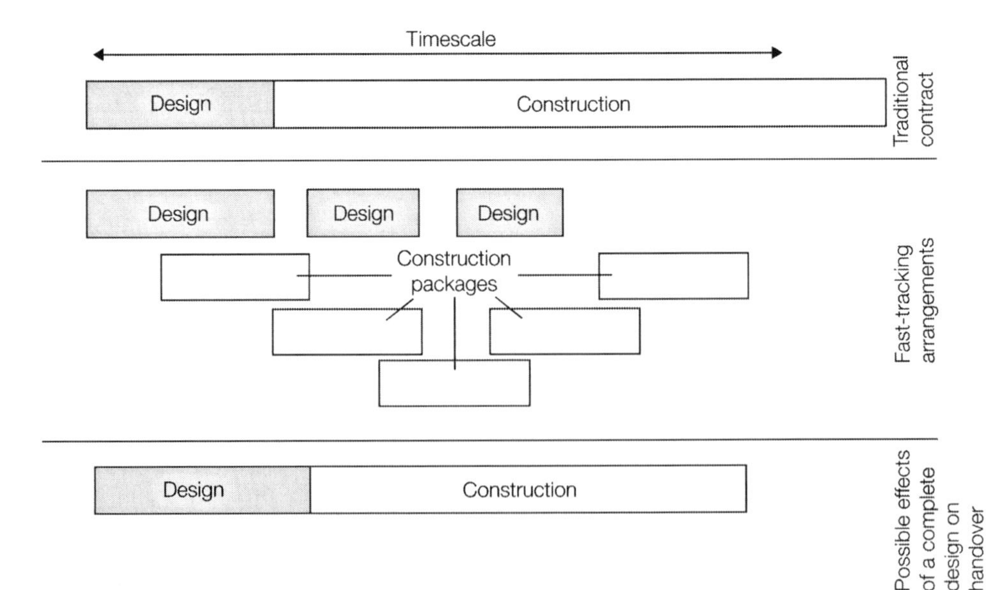

Figure 8.5 Different contractual arrangements and their effects upon time.

and building upon what the first contractor has already completed. The second contractor may or may not be the same as the first contractor, depending upon the amount of specialisation and competition that is involved. This staggered letting of the work has the objective of shortening the overall design and construction period from inception through to handover, as shown in Figure 8.5. This type of procurement arrangement requires a large amount of foresight, since the later stages of the design must take into account what has by now already been completed on-site. This type of arrangement also requires considerably more organisation and planning, particularly from the members of the design team. In practice, where an efficient application of this procurement procedure is envisaged, then a project coordinator will need to be engaged. Although the handover date of the project to the employer should be much earlier than any of the other methods that might be used, this might be at the expense of the other facets of cost and performance. These aspects may be much inferior to those achieved by the use of the more traditional methods of procurement and arrangement.

8.5.12 Measured term

A measured term contract is suitable for major maintenance work projects. It is often awarded not just for a single building project, but to cover a number of different buildings belonging to the same client. A contractor will usually be appointed for a specified period of time, although this may be extended depending upon the necessity of maintenance standards and requirements and the acceptability of the contractor's performance. The contractor is initially offered the maintenance work for a number of trades, such as painting and decorating, plumbing, etc. The work, when completed, will then be paid for using rates from an agreed priced schedule. This schedule may have been prepared specifically for the project concerned, or it may be a standard document similar to the Property Services Agency (PSA) schedule of rates. Where the employer has provided the rates for the work, the contractor is normally given the opportunity of quoting a percentage addition or reduction to these rates. The contractor offering the most advantageous percentage will usually be awarded the contract. An indication of the amount of work involved over a defined period of time would therefore need to be provided for the contractor's better assessment of the quoted prices.

8.5.13 Serial tender

Serial tendering is a development of the system of negotiating further contracts for work of a similar type where a firm has already successfully completed a project. Initially, the contractors tender against each other, possibly on a selective basis, for the first project. However, there is a contractual mechanism that several additional projects will automatically be awarded using the same contract rates. Some allowances are normally made for inflation, or perhaps more commonly increased costs are added to the final accounts. Contractors are therefore aware at the tender stage for the first project that they can expect to receive a further number of contracts in due course. This helps to provide some continuity in their workloads and incentive for investment. Conditions would be written into the documents to allow further contracts to be withheld where the contractor's performance was less than satisfactory.

Serial contracting should result in lower costs to contractors since they are able to gear themselves up for such work, perhaps purchasing mechanical plant and equipment that might otherwise be too speculative. They should therefore be able to operate to greater levels of efficiency. The employer will also achieve some financial gain since some of these lower production costs will find their way into the contractor's tenders.

A further advantage claimed for the use of serial tendering is that the same design and construction teams can remain involved for each of the projects that form part of the serial arrangement. This is a form of partnering where the skills of a team are utilised over a period of time. This regular and close working association aids and develops an expertise, which accelerates the production of the work and eases any anticipated problems. In turn, the operatives on-site improve their efficiency as they progress from project to project.

Serial projects are particularly appropriate for buildings of a similar nature, such as housing and school buildings in the public sector, where a large number of comparable schemes are constructed. This method may also be usefully employed in the private sector in the construction of industrial factory and warehouse units. It has also been successfully used with industrialised system building, where local authorities have worked together in a form of building consortium to gain marketing advantages from the construction industry and its material and component manufacturers.

8.5.14 Continuation contract

A continuation contract is an ad hoc arrangement to take advantage of an existing situation. Projects in an expanding economy are sometimes of insufficient size to cope with projected demands, even before they become operational. During their construction, an employer might choose to provide additional similar facilities, which may be constructed upon the completion of the original scheme. Such additions, often because of their size and scope, are beyond the definition of extra work or variations. For example, a housing project may be awarded to a contractor to build 300 houses. Due to perhaps an underestimation in demand, and where additional space permits, a continuation contract may be awarded to the same contractor to build an extra 100 houses on an adjacent site. Another example can occur where, during the construction of a factory building, it becomes apparent that it will be insufficient to meet the increased demands for the factory's product. Extensions to the factory building may therefore be agreed on the basis of a continuation contract. This will use the same designers and constructors, and the same contract prices adjusted only to allow for increased costs caused by inflation.

A continuation contract can be awarded as an add-on to the majority of the contractual arrangements. Continuing with the same building team, where this has already proved to be satisfactory to the employer, is a sensible arrangement under these circumstances. Each of the parties will already be familiar with each other's working methods and will be able to offer some cost savings since they are already established on the site. Both the employer and the contractor may wish to review the contractual conditions and the new market factors, which may have changed since the original contract was signed. It is also not unreasonable to expect that the contractor will want to share in the monetary savings available from the already established site organisation from the initial project.

8.5.15 Speculative work (develop and construct)

Because of market demand, many building firms have organised a part of their function into a speculative department. Some contractors are entirely speculative builders. These firms, or divisions, purchase land, obtain planning approvals and then design and construct buildings for sale, rent or lease. This is also described as develop and construct. Housing is the common type of speculative unit, but factories, whole industrial estates, new office buildings and refurbishment projects can also be constructed on a speculative basis to meet further and expected demand. The contractors may be in partnership agreements with financial institutions and developers who have already identified that demand for a particular type of development exists within a certain geographical area. Such assumptions are based upon detailed market research of the area, undertaken prior to development taking place. The developer or speculative builder may employ their own in-house design staff, or may employ consultants to do the work for them. Whilst design and build is done for a particular employer, speculative building is carried out on the basis that there is a need for certain types of buildings and that these will be purchased, leased or rented at some later date.

8.5.16 Direct labour

Some employers, notably local authorities and large industrial firms, have departments within their organisations that undertake building work. In some cases, they may restrict themselves to repair and maintenance works alone, or to the construction of minor projects. Other direct labour departments are much larger and are capable of carrying out any size of building project. Direct labour organisations within the public sector were numerous at one time. Successive governments, of different political persuasions, have for the time being abandoned many of these direct departments.

In order to execute construction projects in this way, the employer either needs to have an extensive building programme or be responsible for a large amount of repair and maintenance work. In-house services of this type allow the employer, at least in theory, much greater control than using some of the other methods which have been described. In practice, the level of control achieved may be less than satisfactory, because of accounting procedures that may be employed. Employers who choose to undertake their building works by this method often do so for either political or financial reasons.

8.6 Review

At the inception of any construction project, both the designers and the contractors should enquire about the employer's real objectives. Almost without exception, the employer's needs will be an amalgam of functional and aesthetic requirements, built to a good standard of construction and finish, completed when required, at an appropriate cost, offering value for money and easily and economically maintained (see early in this chapter). The contractual procurement arrangements must aim to satisfy all of these requirements, some in part and others as a whole, trading between them to arrive at the best possible solution.

The traditional approach has been to appoint a team of consultants to produce a design and estimate, and to select a constructor. The constructor would calculate the actual project costs, develop a construction programme to fit within the contract period, organise the workforce and material deliveries and build to the standards described in the contract documentation.

In practice, the quality of the work was often poor, costs were higher than expected and, even where projects did not overrun the contract period, they took much longer to complete than similar projects in the other countries around the world. Traditionally, little attention or consideration was given towards future repairs or maintenance aspects.

The obvious deficiencies in the above, supposedly due to the separation of the design role from the construction effort, have been known for some time. However, many employers have generally wished to retain the services of an independent designer, believing they would serve their needs better than the contractors with their own vested interests. Employers also wished to retain the competition element in order to keep costs to a minimum and to help improve the efficiency of the contractor's own organisation. In more recent years, they have seen some benefits from buildability, benefits which have been achieved through the early and integrated involvement of the contractor during the design stage. This has also had a spin-off effect in reducing costs and the time the contractor is on-site. However, employers have still been loath to commit themselves to a single contractor at the design stage when costs might still be imprecise, and where their bargaining position might be seriously affected before the final deal could be struck.

Increasingly, contractors have attempted to market their design and build methods as possible solutions to the employer's increasing dissatisfaction and frustration with existing procedures. The benefits of a truly fixed price, a single point of responsibility and having the contractor assume full responsibility for design and construction of the building were all attractive propositions. Remedies open to the employer in the case of default or delay were now no longer a matter for discussion between the designer and the constructor. If necessary, the employer could always employ professional advisers to oversee the technical and financial aspects of the project.

The entrepreneurs of the construction industry are always seeking new ways of satisfying the increasing demands of clients. Some are busy examining the methods that are in use in the USA, and thus seek to import systems which they believe are superior and will help to improve the failing image of the construction industry. Others feel that the employers' interests would be better served by an overarching project manager or construction manager. Contractors are, themselves, continuing to make changes in the way in which they employ and organise their workforce. An increase in subcontracting has occurred, as fewer and fewer operatives are directly employed by main contractors. Management contracting recognises and encourages these developments as trends which are desirable.

The construction industry will continue to evolve and adapt its systems and procedures to meet the new demands. New procurement methods will be developed which utilise new technologies and new ways of working. Construction projects are different from the majority of manufactured goods because they are procured in advance of their manufacture. The majority of the projects are also different from the previous ones and often incorporate some new characteristics. Some projects are almost wholly original, and all depend upon their particular site characteristics. In order to achieve value for money and client satisfaction, the appropriate procurement procedures must be selected or adapted to suit the individual needs of the employer and the project.

9 Contract selection

The selection of appropriate contract arrangements for any but the simplest type of project is difficult owing to the diverse range of options and professional advice. Much of the advice is in conflict. For example, professional design and build contractors are unlikely to recommend the use of an independent designer. Such organisations are also likely to believe that the integration of design and construction will result in better buildings, at a more economical cost and improve overall client satisfaction. Many of the design professions hold the opposite viewpoint, and believe that their independent approach will produce the best solution, particularly in the long term.

In the past two decades there has been a significant shift in the way that construction projects are procured. This is partially in response to the structural changes which have occurred within the construction industry. Personal experiences, prejudices, vested interests, the general desire of many to improve the system and the familiarity with particular methods are factors that have influenced capital project procurement recommendations. Those who have had a bad experience with a particular procurement method will be reluctant, at worst, and cautious, at best, in recommending such an approach again. When suggested, they will resist such a course of action and, if the results are then as expected, this further reinforces the view that such a method of procurement has but limited use.

In reality, too little is known and understood. Too little research has been undertaken properly to evaluate the various procurement options. In practice, this is difficult and complex because there are both successful and unsuccessful projects that have used identical means and methods of contract procurement. It is now believed that personal factors of all those involved have a major influence upon the possible outcomes. Nevertheless, it is essential to have more reliable information, based on empirical evidence rather than on subjective intuition and bias. It is also desirable to at least attempt to eliminate hearsay and folklore opinion.

The advice, wherever it originates, should always recommend a course of action which best serves the client's interests, not the advice giver's. This is true professionalism, even though it may mean losing a commission for the party or organisation concerned.

The employer may need to be convinced that a particular method and recommended procedures should be adopted as the most appropriate approach. This may at times be against the employer's own better judgement or preferences, particularly where there is some familiarity with the other options that are being suggested. Some clients are resistant to change and may require persuasion to adopt what might appear to be radical proposals. It may also be necessary to accept that, because of the mechanisms employed, a better course of action cannot easily be explained until after the event, and that such a recommendation may be rejected at the outset as being speculative.

Overseas clients working in different cultural environments may find that the traditional contractual procedures are wholly alien to their needs and expectations. They may require even greater assurances that their project should be constructed using a particular procurement method. Individuals working within the same team will also have different expectations and views on the optimum procurement path.

Procurement advisers should offer advice in the absence of any vested interests or personal gains. Any associated interests which might influence the judgments should be declared in order to avoid possible repercussions at a later stage. Whilst the professional institutions retain codes of conduct and disciplinary powers governing their members' activities, there has been a blurring of commercial and professional attitudes, with the former vying for position with the latter. Outmoded approaches and procedures should be removed when they no longer serve any really useful purpose. Sound, reliable and impartial advice is necessary from those who have the proficient skills, knowledge and expertise.

9.1 Employers' requirements

A confirmed cynic once described the construction industry as 'the design and erection of buildings that satisfy the architect and constructor alone and not the employer'. In addition, buildings often failed to function correctly, were too expensive and took too long to construct. As in most cynicism, the case is vastly overstated, yet there is enough truth not to dismiss it immediately.

One of the procurement adviser's main problems is in separating the employer's wants and needs. The employer's main trio of requirements are shown in Figure 9.1. The numerical values result from research and attempt to weigh these broad objectives in order of their relative importance. They reflect the 'average', or typical, client and offer some guidance on the main issues associated with a construction project. The weightings depend upon the employer's own aims and objectives. The attempted optimisation of these factors may not necessarily achieve the desired solution. For example, an employer may desire exceptional quality standards and future performance, and they may be prepared to incur extra costs or a longer contract period. Alternatively, a client may require the early use of the project as the main priority and be prepared to sacrifice a bespoke design or even high-quality standards in order to achieve this. Rarely are clients prepared to pay for the highest known quality. Clients may also choose to set contrasting objectives which are difficult to achieve using conventional procurement procedures. This may require the adviser to devise ingenious methods of procurement in order to satisfy such objectives.

In broad terms, the employer's objectives and the procurement path will need to identify the following priorities:

- An acceptable layout for the project in terms of function and use.
- An aesthetically pleasing design not only to the employer, but also as viewed by others.
- The final cost of the building should closely resemble the estimate.
- The quality should be in line with current expectations.
- The future performance of the building and the associated costs in use should fit within specified criteria.
- The project is available for handover and occupation on the date specified for completion of the works.

Client (100)

Performance (45)	**Cost (35)**	**Time (20)**
Function (25)	Initial (20)	
Technology (15)	Life cycle (15)	
Aesthetics (5)		
Factors to consider		
Quality	Budgets	Design length
Standards	Estimates	Start dates
Layout	Tender sums	Handover
Appearance	Cash flow	Completion
Maintenance	Final account	
	Costs in use	

Figure 9.1 Client's main requirements (%).

9.2 Factors in the decision process

The following need to be considered when choosing the most appropriate procurement path for a proposed project. Their importance and emphasis will differ on individual projects, types of clients and the relevant market conditions at the time of procurement:

- Size.
- Cost.
- Time.
- Accountability.
- Design.
- Quality assurance.
- Organisation.
- Complexity.
- Risk.
- Market.
- Finance.

9.2.1 Project size

Projects of a small size are not really suited to the more elaborate forms of contractual arrangements, since such procedures are likely to be too cumbersome and not cost-effective. Smaller-sized schemes therefore rely upon the traditional and established forms of procurement, such as a form of competitive tendering or design and build. The medium- to large-scale schemes can use the whole range of options available. On the very large schemes, a combination of different arrangements may be required to suit the project as a whole. On some of the very largest projects, bespoke arrangements might be introduced to meet particular circumstances.

9.2.2 Cost

Open tendering will generally secure for the client the lowest possible price from a contractor. This is the evidence that competition helps to reduce costs through efficiency, and lowers the price to the customer. There are limits to how far this theory can be applied in practice. For example, if a large number of firms have to prepare detailed tenders, then this increases the industry's costs, which must be either absorbed or passed on to the industry's employers through successful tenders. There is no such thing as a free estimate! Negotiated tenders supposedly add around 5 per cent to the contract price. In the absence of competition, contractors will price the work *up to what the traffic will bear*. Some will argue that this is a major factor in determining project costs. Projects which require unusually short contract periods incur cost penalties, largely to reflect the demands placed upon the constructor for overtime working and rapid response management. The imposition of conditions of contract which favour the client or insist upon higher standards of work than are usual also increase building costs. Under these circumstances, the employer may end up paying more for the stricter conditions that might not be necessary or for a quality of work that adds little to the overall ambience of the project and is thus not required.

In the past, employers have been overly concerned with a lowest tender sum, often at the expense of other factors such as the principles behind the indeterminate whole life cost. Cash flow projections which might have the effect of reducing the timing of expenditures are often ignored, except perhaps by the most enlightened employers. A cash flow analysis, for example, may be able to show that the lowest tender is not always the least expensive solution. The timing of cash flows on a large project may have a significant effect upon the real costs to the employer. With projects that include options on construction method, the cash flow analysis becomes even more important. Procurement and contractual arrangements which take into account these factors should be considered more frequently.

The cost of the project is a combination of land, construction, fees and finance, and the employer will need to balance these against the various procurement systems that are available. In terms of cost savings, a less expensive site elsewhere, rather than reducing the costs of quality, may be more appropriate.

Design and build projects show some form of cost savings in terms of professional fees, although the precise amount is difficult to calculate since fees are absorbed within other charges submitted by the contractor in the bid price. Where the building is of a relatively straightforward design, such as a standard warehouse unit or farm building, then it can be more cost-effective to use a building contractor who has already completed similar projects rather than to opt for a separate design service. Cost reimbursement appears to be a fair and reasonable way of dealing with construction costs in a fair world. Society is not fair, and such an arrangement is often too open-ended for many of the industry's employers. Some clients may choose to use it out of necessity, setting objectives other than cost more highly. The majority of employers, however, will need to know, as a minimum, the approximate cost of the scheme before they begin to build, and there is nothing new in this (Luke 14:28).

Where a firm price is required before the contract is signed, then one of several procurement arrangements can be used (see Chapter 8). A firm-price arrangement relies to a large extent on a relatively complete design being available. Where more price flexibility is possible, then one of the more advanced forms of procurement might be used whilst at the same time achieving a measure of cost control.

Provisions exist under most forms of contractual arrangement for a fixed- or fluctuating-price agreement. The choice is influenced by the length of the contract period and the current and forecasted rates of inflation. Where the inflation rate is small in percentage terms and falling, then a fixed-price arrangement is preferable. When the rate of inflation is small and stable, then it is common to find projects of up to thirty-six months' duration set up on this basis. When the rate of inflation is high, and particularly when it is rising, then contractors will be reluctant to submit fixed-price tenders for more than about twelve months' duration. Although a fixed price is attractive to the employer, it cannot be assumed that this will necessarily be less expensive than a fluctuating type arrangement. Some contract conditions limit fluctuation reclaims to increases caused directly by changes in government legislation.

It is difficult to make cost–procurement comparisons, even where similar projects are being constructed under different contractual arrangements. It can be argued that competition reduces price, price certainty in the case of premeasurement or fixed price might not be the most economic, and that using serial tendering, bulk purchase agreements and similar concepts are good ways of reducing building costs. Cash flows and projected life cycle costs on larger projects should be considered in any overall cost evaluation.

There have been suggestions in recent years that the lowest price might not be to the employer's best advantage. It is difficult to argue this case where the specification has been correctly prepared and contractors are simply competing on cost alone. However, it is important to realise that the least expensive design is necessarily always in the employer's best interests. It is frequently a sensible approach to spend more at first on the basis of recouping savings later during the project's life. Sometimes, access to funds initially prohibits this option. Here are the cost factors to consider:

- Price competition/negotiation.
- Fixed-price arrangements.
- Price certainty.
- Price forecasting.
- Contract sum.
- Bulk purchase agreements (economies of scale).
- Payments and cash flows.
- Whole life costs.
- Cost penalties.
- Variations.
- Final cost.

9.2.3 Time

The majority of employers, once they have made the decision to build, want the project to be completed as quickly as possible. The design and construction phases in the UK are known to be lengthy and protracted. Some of the apparent delays are linked to the drawn-out planning processes rather than the actual design or the construction phases. It is difficult to make comparisons on a global basis since there are a wide variety of influences to be considered, such as methods of construction, safety, organisation of labour and quality assurance. This is evident from studies comparing UK, Australia and US performances. Construction techniques also vary, and these in turn produce different qualities and costs, and the time available will influence the type of construction techniques which might be used. A need for rapid completion may force the employer to consider using an *off-the-peg* or package deal type building that can be constructed quickly.

In order to secure the early completion of the project, several different methods of procurement have been devised with this objective in mind. Such approaches have implications for the other factors under consideration, such as design, quality or cost. There is an optimum time solution depending upon the importance which is attached to these other considerations. For example, in terms of cost, shorter or longer periods of time on-site tend to increase building costs. The former is due in part to overtime costs and the latter to extended on-site costs. Some clients are prepared to pay extra costs in order to achieve earlier occupation. Different types of project also have different time concepts. A shopping centre redevelopment may require rapid completion, since earlier revenue receipts may easily cover the extra costs. Different techniques of design and construction may also need to be used to achieve early completion. New educational building completions are linked to school term times, whereas housing starts and completions are at the rate at which they can be either let or sold.

Research has shown that construction work should not commence on-site until the project is fully designed and ready for construction, with only the minimal involvement thereafter by the designer. Variations will still be allowed, but are not encouraged. Such an approach helps to eliminate a large amount of construction uncertainty that is common on many supposedly designed projects today. This approach allows the contractor to plan the organisation and management of the project better and to spend less time awaiting drawings, details and other information. Overall, the design and construction period is shortened and earlier completion is achieved. The contractor is on-site for a shorter period of time, which saves on building costs. The mitigating difficulty of adopting this approach is that the client wants to see work beginning on-site as soon as possible.

Many of the newer methods of contract procurement have been devised specifically to find a quicker route through the design and construction processes. In some ways, they have been assisted by changes in the techniques used for the construction and assembly of the building products and materials. Early selection was developed to allow construction work to start on-site whilst some of the scheme was still at the design stage. It also allowed the contractor an early involvement with the project. Some forms of cost forecasting and control can be used, but they will be much less precise than with some of the more conventional methods of procurement. Critical path analysis can be used to find the quickest way through the construction programme, and the American fast-track system was imported for the sole purpose of securing early completion of a project.

Inaccurate design information, coupled with a need for speed in completion of the works, has often resulted in abortive parts of a project, poor quality control and higher costs to both the contractor and the employer. The later forms of management contracting offer some solution to the time delay problem. Project management contracts, where an independent organisation controls both the design and the construction teams, have provided some good examples in terms of project coordination.

Faster Building for Commerce (NEDC 1998) was a study aimed at helping the industry achieve earlier completions of its projects. The study claimed that the major influences on projects' time performance were customer participation, design quality and information, contractor's control over site operations and the integration of the subcontractors with design and construction. The study showed that fast times were 20–25 per cent shorter than the average, and the slowest times could take twice as long as the average. Overall, one-third of commercial projects finish on time, the rest overrun by a month or more, often because of extensive design alteration. These were claimed as the largest single factor affecting delays. Common ingredients in fast projects are organisation to promote unity of purpose, competent management at all stages of the project and working practices which interlock smoothly and leave no gaps between activities and responsibilities. Many of the delays result from a lack of information, underestimating the supervision of subcontractors and late changes in requirements by employers. In some cases, quality control was a problem, where work needed to be redone and hence caused delays. Slow reactions by the utilities companies can cause extensive delays, especially on sites with high services contents. Here are the time factors to consider:

- Completion dates.
- Delays and extensions of time.
- Phased completions.
- Early commencement.
- Optimum time.
- Complete information.
- Fast tracks.
- Coordination.

9.2.4 Accountability

According to the dictionary definition, *accountability* is the 'responsibility for giving reasons why a particular course of action has been taken'. In essence, it is not simply having to do the right things, but having to explain why a particular choice was made in preference to others that were available. It is of greater significance when dealing with public employers where it is necessary to justify why a particular course of action was taken. It has also become increasingly important with all types of employers where an emphasis is placed on achieving value for money on capital works projects. The documentation used for construction works is often complex and the technical and financial implications are considerable.

Employers need the assurance that they have obtained the best possible procurement method against their list of objectives. The possible trade-offs between competing proposals will need to be evaluated. Where tenders are sought in the absence of any form of competition, it is difficult to satisfy the accountability criteria in respect of price. There is also the difficulty of justifying subjective judgments where these appear to be in conflict with common practice, but the process of selection will never solely be a mechanistic process. The elimination of procedural loopholes should be such as to provide the employer with as much peace of mind as is possible.

Accountability is interlinked with finance and an emphasis on paying the smallest price for the completed project. It may be easy to demonstrate to some employers that to pay more for a perceived higher-quality or earlier completion is worthwhile. Other employers may need more convincing and some will feel doubtful about non-monetary gains.

The procedure for the selection, award and administration of contracts must be as precise as possible. Auditing plays a useful role in the tightening up of the procedures used, with ad hoc arrangements that breach these procedures being discouraged. Systems that require huge amounts of documentation and the subsequent checking and cross-checking of invoices, time sheets, etc., are not favoured because of loopholes that can occur. Prime cost contracts are an example of this. Open-ended arrangements which are unable to provide a realistic estimate of cost are fraught with difficulties in terms of accountability. They provide difficulties in demonstrating value for money at the tender stage, since any forecast of cost will be too imprecise for reliability purposes. Here are the accountability factors to consider:

- Contractor selection.
- Ad hoc arrangements.
- Contractual procedures.
- Loopholes.
- Simplicity.
- Value for money.

9.2.5 Design

There is a good argument that the best design will be obtained from someone who is a professional designer. The design will then not be limited by the capabilities of the constructor or restrained to those designs which might be the most profitable in terms of their construction to such a firm. However, a design and build contractor is more likely to be able to achieve a solution which takes into account buildability, and produces a design which is sound in terms of its construction. There are examples where the employer, having been provided with an unsatisfactory building, has to wait impatiently whilst the consultant and constructor argue about their respective liability. There are also examples where design and build projects, some using industrialised components, have had to be demolished after only a few years of life because of their poor design concept, impossible and costly maintenance problems and an unacceptable user environment which they helped to create.

Contractors, on the whole, have been better at marketing their services and this has reaped benefits in the growing increase in design and build schemes (see Chapter 7). Designers have to some extent been thwarted in their response to this upsurge in activity because of the restrictions imposed on advertising by their profession and a failure to respond adequately to the changes that have taken place in the industry and throughout many aspects of society. Only in recent years has the design and manage approach to the procurement of buildings been introduced.

The traditional methods of contract procurement fail in many aspects of building design, not least because of the absence of any constructor input. This is not a common feature in other comparable industries. However, they continue to remain a common method of procurement despite their publicised drawbacks. Some of the forms of management contracting, which still largely retain the independence of the designer, did gain in popularity, although bad publicity has curtailed their development. Where the employer has been encouraged to form a contractual relationship with a single organisation, then, on the whole, this appears to have been beneficial.

Designs which evolve and develop only marginally ahead of the construction works on-site must be of questionable worth in design solution terms, unless they are working within the constraints of either previously completed schemes or the confines of an existing structure. Here are the design factors to consider:

- Aesthetics.
- Function.
- Maintenance.
- Buildability.
- Contractor involvement.
- Standard design.
- Design before build.
- Design prototypes.
- Innovation.

9.2.6 Quality assurance

Open tendering can result in a lower standard of work than might have been achieved by using a building contractor who submitted a higher price. You only get what you pay for is certainly true of construction quality. Where a building contractor has had to submit an uneconomical price at cost, then quality may suffer unless improved supervision is provided. Consistent and good quality control procedures are frequently lacking in the construction industry and are often relegated in favour of other criteria in the list of objectives. The quality of buildings has not improved in line with the quality of, for example, motor vehicles.

The quality of buildings depends upon a whole range of inputs: the soundness of the design, a correct choice of specification, efficient working details, adequate supervision and the ability of the builder. The skills of the operatives are also important, perhaps even more so today than a decade ago, now that the designs are tending to become more complex in their detailing and a higher level of skill is expected. The choice of a contractor who has a good reputation for the type and quality of work envisaged is important in achieving this objective. The use of labour-only operatives and the general subcontracting phenomena have sometimes contributed towards a deterioration in quality and performance, due to poor site coordination and supervision by site management.

The quality of design and build schemes depends largely upon the reputation of the building firm selected, particularly where the employer chooses not to involve any professional advisers. The quality of the materials and the work will be regulated entirely by the contractor alone. Speculative building schemes where the quality assurance is determined solely by the builder are not necessarily renowned for their high-quality work, although they are market driven, relying upon customers to purchase their speculative units. Certainly, in the past, speculative house building was deficient in terms of quality. This has now improved to meet house buyers' expectations. Government regulations to improve building structures have been monitored for some time by external agencies, requiring defects to be corrected for periods of up to ten years.

Fast-track procurement methods, which may involve a number of contractors on the same project, can, without adequate supervision, result in widely varying standards of quality and work. There is also the added difficulty of coordinating different contractors on the site, such as occurs with work packages on management contracts. The use of selective tendering, properly managed, continues to offer a good solution in terms of quality control.

Quality standards cannot be judged at the building's completion alone, but need to be considered in the longer term. A virtually complete design prior to the commencement of work on-site is likely to be beneficial in improving the qualitative aspects of the project rather than the more ad hoc design approach to problems as they occur on-site.

The turnkey method, where the designer-contractor has a contractual responsibility for the long-term repair and maintenance of the project, does offer advantages, particularly in terms of quality assurance. Under this method of procurement, there is the incentive that the designer-contractor will wish to reduce the likelihood of future defects arising by a more careful design and effective site management during construction. This may result in less innovation, but it is preferable to inconvenience and costly failures in the future. Progress in construction is necessary, but not at the expense of prototype designs which result in poor quality assurance.

Although serial and continuation contracts, using the same design and detailing, should improve the quality aspects, in practice this has not always been achieved. The learning curves of the operatives in these cases have been an aid towards improved quality. However, poor design, detailing and construction methods have in too many cases been unfortunately repeated. The following are quality assurance factors to consider:

- Quality control.
- Independent inspection.
- Teamworking.
- Coordination.
- Subcontracting.
- Buildability.

- Future maintenance.
- Design and detailing.
- Reputation of craftspeople.
- Quality standards.

9.2.7 Organisation

Allowing the contractor total control of the building project, as in design and build or management contracting, removes a layer of organisation and eliminates dual responsibility. This should result in fewer things being overlooked or forgotten, work left undone or subcontractors being unable to complete their work on time owing to a lack of information. An additional tier of organisation also has the disadvantage of the parties blaming each other when disputes arise. The traditional methods of contract procurement appear to set the lines of demarcation between designer and constructor clearly, although the large number of disputes in practice might suggest that this is not the case. In practice, the designer probably relies too heavily on the constructor. To cope with the organisation of construction work, elaborate conditions of contract have had to be drawn up to anticipate most of the eventualities that might arise.

The employment of a single firm, such as a design and build contractor, allows for quick response management, the ability to deal with problems as they occur and more freedom in the execution of the works. The extent of such freedom will vary with the conditions of contract being used. Where a separate designer is used, the response time is often much longer and this can result in delays to the contract. Where complex or difficult contract arrangements are employed, they can have the effect of removing the initiative from the contractor.

The more parties that are involved with a building project, the more complex will be the organisation and the contractual arrangements. The employment of a group design practice should therefore result in fewer organisational difficulties than where individual firms are used. Management contracting is based upon awarding individual work packages to a range of specialist and general subcontractors. This can create problems of organisation and coordination. There is much less control over such firms than when directly employed operatives of the main contractor are used. However, general contracting is now unusual since about 90 per cent of construction work is subcontracted regardless of the method of procurement being used. These individual firms need to be programmed for precise periods of time, and a delay in allowing them to proceed with their work or a failure by them to complete on time can have a knock-on effect for the whole project. This presents even greater difficulties with tightly scheduled construction programmes. Here are the organisational factors to consider:

- Complexity of arrangements.
- Single-point responsibility.
- Levels of responsibility.
- Number of individual firms involved.
- Lines of management.

9.2.8 Complexity

Projects which are complex in design or construction require more precise and more comprehensive contractual arrangements. Complexity may be the result of an innovative design, the utilisation of new construction methods, the phasing of the site operations or the necessity for highly specialised work. It can also be the result of employing several contractors on the same site at one time in order to achieve rapid progress, or the complicated refurbishment of an existing building whilst still in use by its occupants. It is often necessary in circumstances like these to devise new contractual arrangements and to apply different types of procedures to the various parts of the construction work. Where work can be reasonably well defined and forecasted, then traditional estimating processes can be used and the work paid for on the usual basis. Where the work is indeterminate, of an experimental type or requiring a solution from the contractor, then a lump sum or cost reimbursement approach with contractor design may need to be employed. In the latter case, the contractor is given the opportunity to offer an acceptable solution to the problem as a part of the contract.

Where the project is very complex, then the employer is likely to choose a separate designer with the skills required to produce the right solution. However, it is important to involve the constructor as soon as possible with the project, particularly where this might influence the sequencing of site operations. A form of two-stage tendering might therefore be appropriate against this eventuality. Here are the organisational points to consider:

- Nature of complexity.
- Capabilities of parties.
- Main objectives of employer.

9.2.9 Risk

Risk is inherent in the design and construction of a building. The employer's intentions will be to transfer as much of this as possible to either the consultants or the contractor. Risk may be defined as possible loss resulting from the difference between what was anticipated and what actually occurred. Risk is not entirely monetary. An unsatisfactory design, although completed successfully, can result in a weakening of the designer's reputation, with a consequent loss of future commissions. Risk can be reduced, but it is often difficult to eliminate it entirely. For example, the risk associated with a particularly specialised form of construction can be reduced by selecting a contractor with the appropriate experience and expertise.

The transfer of risk from the employer to others involved with the project may appear to satisfy the accountability criteria. It may be argued as an appropriate course to follow for the employer, but it may not be a fair and reasonable approach. It may also not be the best route for the employer to follow, since the risk needs to be evaluated. All contractors' tenders contain a premium to cover contractual risk. Where the risk does not materialise, this then becomes a part of the contractor's profit. The employer may thus be better advised to assess and accept some of the risk involved and thereby reduce the contractor's tender sum and also costs accordingly. This is a more common way of dealing with risk on construction projects.

The lump sum contract with a single price which is not subject to any variation is at one extreme, and at the other is cost reimbursement, where risk and financial predictability are uncertain. In the former, the employer is paying for eventualities which might not occur. In the latter, the client is accepting the risk, but only pays for events that happen. A balance has to be drawn. Risk should always be placed with the party to the contract who is in the best position to control it. When this is not possible, then it should at least be shared, although it may be difficult to convince the employer that this course of action is the most financially appropriate. Some projects involve a large amount of risk in their execution. In some cases, the risk might be so high that it is impossible to get a contractor even to consider tendering under conventional arrangements. Some form of risk sharing may then become essential in order for the project to proceed. Here are some risk factors to consider:

- Risk analysis.
- Risk sharing.
- Risk transfer.
- Risk control.

9.2.10 Market

The selection of a method of procurement will be influenced by the state of a country's economy and the industry's workloads. An appropriate recommendation for today may have different implications for some time in the future. When there is an ample amount of work available, contractors are able to choose those schemes which are the most financially lucrative. Under these circumstances, employers will be unable to insist upon onerous contractual arrangements and conditions. When the risk involved is high, it will be even more difficult to persuade contractors to tender for the work. Employers may need to be advised to delay their building projects at such times, and wait until the economy is more favourable. Many employers will, however, be unable or unwilling to follow this advice.

When construction prices are low, then a form of cost reimbursement or management fee approach can be expensive. During times like this, contractors are sometimes prepared to do work at cost, and take a gamble with the risk factors that nothing of financial significance will go wrong. In times of full order books, the opposite is true and paying contractors their actual costs plus an agreed amount for profit can be a better proposition. When work is plentiful, contractors often have difficulty in recruiting a competent workforce of skilled operatives, and this, coupled with similar restrictions in the availability of good supervision, can result in a deterioration in quality standards. When the available amount of construction work is restricted, the standards of work, coupled with more intensive inspection, are likely to enhance the overall quality of the project. Here are the market factors to consider:

- Availability of work.
- Availability of contractors.
- General economic condition of the country.
- Global market (fuel, steel and other building materials).

9.2.11 Finance

The usual way of paying the contractor for the building work is through monthly or stage payments. These payments help the contractor to offset the financial borrowing that is required to pay for wages and salaries, goods and materials. Two alternatives to this can be used.

The first is a delayed payment system similar to that used on speculative developments. The employer effectively pays for the work using a single payment upon completion of the project. The employer has to accept the design as it is built, but acquires immediate occupation of the project. The financial borrowing requirements of the contractor are higher, but the employer makes savings by paying for the work at the end of the project.

The other alternative is for the employer to fund the work in advance and thereby reduce the contractor's interest charges that are otherwise included in the tender. In this situation, the employer needs to be beware. The industry is notorious for insolvencies, and the employer needs to ensure the financial soundness of the appointed contractor. Contractors also tend to be less interested in a project once they have received payment for work.

The employer can devise remedies to deal with these factors. With the former, a performance bond can be adopted; with the latter, liquidated damages can be applied. Here are some financial factors to consider:

- Payment systems.
- Financial soundness of parties.
- Financial remedies.
- Contract funding.

9.3 Conclusions

The choice of a particular method of contract procurement for a construction project involves identifying the employer's objectives, balancing them with the procurement methods available and taking into account the considerations outlined in this chapter. Figure 9.2 provides a comparison of some of the payment methods. It compares factors such as the information requirements with the control and risk consideration. For example, if an employer's main

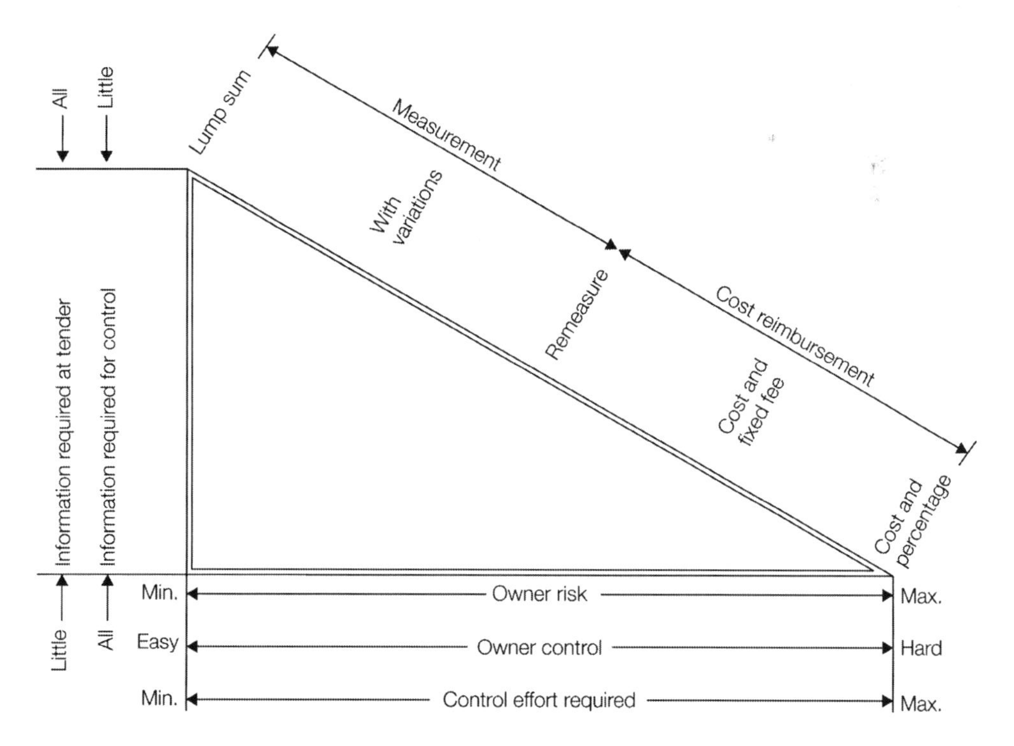

Figure 9.2 Range of contract types.

Source: Adapted from Burgess 1982, p. 149.

Table 9.1 Identifying the client's priorities

			Traditional selective tendering	*Early selection*	*Design and build*	*Construction management*	*Management fee*	*Design and manage*
TIMING	Is early completion important to the success of the project?	Yes:						
		Average:		**Average**	**Average**	**Average**	**Average**	**Average**
		No:	No	**No**	**No**	**No**	**No**	**No**
VARIATIONS	Are variations to the contract important?	Yes:	Yes	Yes		Yes	Yes	Yes
		No:						
COMPLEXITY	Is the building technically complex or highly serviced?	Yes:	Yes	**Yes**		**Yes**	**Yes**	**Yes**
		Average:		**Average**	**Average**	**Average**	**Average**	**Average**
QUALITY	What level of quality is required?	High:	**High to Average**	**High**		**High**	**High**	**High**
		Average:		**Average**	**Average**	**Average**	**Average**	
		Basic			Basic			
PRICE CERTAINTY	Is a firm price necessary before the contracts are signed?	Yes:	**Yes**		**Yes**	**Yes**		
		No:	No	No			No	No

Table 9.1 (continued)

			Traditional selective tendering	*Early selection*	*Design and build*	*Construction management*	*Management fee*	*Design and manage*
RESPONSIBILITY	Do you wish to deal only with one firm?	Yes:			Yes			Yes
		No:	**No**	**No**		**No**	**No**	
PROFESSIONAL	Do you require direct professional consultant involvement?	Yes:	**Yes**	**Yes**		**Yes**	**Yes**	Yes
		No:			No			
RISK AVOIDANCE	Do you want someone to take the risk from you?	Yes:						
		Shared:	Shared	Shared		Shared		
		No:					No	No

Source: Adapted from National Economic Development Office (NEDC 1985).

priority is for a lump sum price from a contractor, then full information at the tender stage must be provided for an accurate price to be prepared. Risk and control effort throughout the duration of the contract will then be minimal to the employer. On large and complex projects, the ability to provide this detailed information is difficult. The quality and reliability of the design information will determine how precisely the building's costs can be forecasted and controlled. The poorer and more imprecise it is, the greater will be the risks to the employer. The risk, of course, may never materialise and hence there will be no loss to the employer.

The three broad areas of concern for the employer and the expectations that are required from any building project have already been identified in Figure 9.1. The balance of these will vary, but their analysis will help to influence and select the most appropriate procurement method.

Table 9.1 offers a checklist of questions to help determine an appropriate contract strategy. It is based upon an NEDC report entitled *Thinking About Building* (NEDC 1985), but has been adapted accordingly. The table provides examples of some of the more usual contractual arrangements available; however, it does need to be emphasised that the solutions recommended are based largely upon judgment rather than objective analysis. In answering the questions, users can arrive at appropriate solutions. Under differing circumstances, or where other factors need to be considered, the solutions must be adjusted accordingly. The questions themselves are not weighted and users will need to do this in order of importance. Some employers may wish to emphasise only a single aspect, such as quality, and choose a method and contractor that are capable of securing this. The majority of employers are, however, interested in an amalgam view, trading off the various factors against each other. It is inappropriate to use the table in an incremental fashion by adding the various answers together.

The correct choice of a procurement method is a difficult task, owing to the wide variety of options that are available. Some of the changes in methods of procurement are the result of a move away from the craft base to the introduction of off-site manufacture, the use of industrialised components and the wider application of mechanical plant and equipment. Improved knowledge of production techniques, coupled with the way in which the workforce is organised, has enabled the contractor to analyse the resources involved and move towards their greater optimisation. Contractors also have a much greater influence upon the design of the project, and the recognition of buildability has influenced the design and how the work is carried out on-site, hence it has influenced the quality of the finished work. The time available for construction and the subsequent costs involved have been affected by these changes.

10 Contract documents

The contract documents under any building or civil engineering project should include the following information as a minimum:

- *The work to be performed*: this usually requires some form of drawn information. It assists the client's own understanding of the project through the use of schematic layouts and elevations. Even the non-technical employer is usually able to grasp a basic idea of the architect's design intentions. Drawings will also be necessary for planning permission and building regulations approval. Finally, they will be needed by the prospective builder in order that the architect's intentions can be carried out during estimating, planning and construction. On all but the simplest types of project, therefore, some drawings will be necessary, and they will often be supported by additional information such as three-dimensional models and computer-aided design.
- *The quality of work required*: it is not easy on drawings alone to describe fully the quality and performance of the materials and the standards of work expected. The usual way is to describe the quality and standards by a specification or a bill of quantities.
- *The contractual conditions*: in order to avoid possible misunderstandings, it is desirable to have a written agreement between the employer and the builder. For simple or small-scale projects, conditions of contract such as the minor works agreement may be sufficient. On more complex projects, one of the more comprehensive forms of contract should be used (see Chapter 6). It is preferable to adopt the use of a standard form of contract rather than to devise separate conditions of contract for each project.
- *The cost of the finished work*: this should be predetermined, wherever possible, by an estimate of costs from the builder. This is best achieved through the use of some form of measured quantities of the work. On some projects, it may only be feasible to assess the cost once the work has been carried out. In these circumstances, the method of calculating this cost should be clearly agreed.
- *The construction programme*: the length of time available for the construction work on-site will be important to both the employer and the contractor. The employer will need to have some idea of how long the project will take to complete in order to plan arrangements for the handover of the project. The contractor's costs will, to some extent, be affected by the time available for construction. The programme should include progress schedules to assess whether the project is on time, ahead or behind the programme.

The contract documents will normally comprise a combination of the following documents:

- Form of contract, including:
    - Articles of agreement.
    - Appendix.
- Contract drawings.
- Bill of quantities, specifications or measured schedules.
- Programme of work (in some instances).

On a civil engineering project, it is more usual to include both a bill of quantities and a specification. Most building contracts allow only one of these documents, depending on whether quantities have been provided.

10.1 Form of contract

The form of contract is the principal contract document and will generally comprise one of the preprinted forms described in Chapter 6; that is, forms of contract that have been developed and agreed by the various parties in the construction industry. The form of contract will usually take precedence over the other contract documents. The conditions of contract seek to establish the legal framework under which the construction work is to be undertaken. Although the contract clauses aim to be precise and explicit, and to cover any eventuality, disagreement in their interpretation does occur. When disputes arise, in the first instance an attempt is made to resolve the matter, as amicably as possible, by the parties concerned. Where this is not possible, it may be necessary to refer the disagreement to a form of alternative dispute resolution, adjudication or arbitration. The parties to a contract usually agree to take any dispute initially to adjudication rather than to litigation. This can save time, costs and adverse publicity, which may be damaging to both parties. Where the dispute cannot be resolved, it is taken to court to establish a legal opinion. Such opinions, if held, eventually become case law and can be cited should similar disputes arise in the future.

It is always preferable to use one of the standard forms available, rather than to devise one's own personal form, although several private forms have been established by the large industrial corporations for their own use. These corporations often undertake a considerable amount of construction work on a regular basis. The imposition of conditions of contract which are biased in favour of the employer are not to be recommended. Contractors will tend to overprice the work, even in times of shortage, to cover the additional risks involved that have been placed with them. Unless there are very good reasons to the contrary, architects or engineers should attempt to persuade the employer to use one of the published standard forms available. The modification of some of the standard clauses, or the addition of special clauses, should only be made in exceptional cases and with experienced individuals.

The majority of the standard forms of contract comprise, in one way another, articles of agreement, conditions of contract and appendix.

10.1.1 Articles of agreement

The articles of agreement are the part of the contract which the parties sign. Note that the contract is between the employer (the person or organisation who wants the project to be constructed) and the constructor (the contracting firm). The blank spaces in the articles are filled in with:

- the names of the employer, contractor, consultant
- the date of signing the contract
- the location and nature of the work
- the list of the contract drawings
- the amount of the contract sum.

If the parties make any amendments to these articles of agreement, or to any other part of the contract, then the alterations should be initialled by both parties.

In some circumstances it may be necessary or desirable to execute the contract under seal. This is often the case with local authorities and other public bodies. The spaces for the signatures are then left blank and the seals are affixed in the appropriate spaces indicated. After sealing, the contract must be taken to the Department of HM Revenue & Customs, where upon payment of stamp duty, a stamp, will be impressed on the document. Without this, the contract will be unenforceable.

10.1.2 Conditions of contract

In general, the conditions of contract in any of the forms have a large degree of comparability, but are different in their details. They include, for example, the contractor's obligations to carry out and complete the work shown on the drawings and described in the contract bills to the satisfaction of the architect, engineer or supervising officer. They cover matters dealing with the quality of the work, cost, time, specialist firms, insurances, fluctuations and VAT. Their purpose is to attempt to clarify the rights and responsibilities of the various parties in the event of a dispute arising.

10.1.3 Appendix

The appendix to the conditions of contract needs to be completed at the time of signing the contract. The completed appendix includes that part of the contract which is peculiar to the project in question. It includes information on the start and completion dates, the periods of interim payment and the length of the defects liability period for which the contractor is responsible. The appendix includes recommendations for some of the information.

10.2 Contract drawings

The contract drawings should ideally be complete and finalised at the tender stage. Unfortunately, this is seldom the case, and both employers and designers rely too heavily upon the clause in the conditions allowing for variations or change orders. Occasionally, the reason is due to insufficient time being made available for the pre-contract design work or, frequently, because of indecision on the part of the employer and the design team. One of the original intentions of the Standard Method of Measurement (SMM7) was only to allow bills of quantities to be prepared on the basis of complete drawings. To invite contractors to price work that has yet to be designed does not appear to be either a sensible or fair course of action to follow. Tenderers should be given sufficient information to enable them to understand what is required in order that they may submit as accurate and realistic a price as possible. The contract drawings will include the general

Table 10.1 Drawings supplied by the architect

Drawing title	*Suggested scale*
Survey plan	1:1,250
Site plan and draining	1:500
Floor plans	1:200
Roof plan	1:200
Foundation plan	1:200
Elevations	1:200
Sections	1:100
Construction details	1:10, 1:5
Engineering services	1:200

arrangement drawings showing the site location, the position of the building on the site, means of access to the site, floor plans, elevations and sections. When these drawings are not supplied to the contractors with the other tendering information, the contractors should be informed where and when they can be inspected. The inspection of these and other drawings is highly recommended, since it may provide the opportunity for an informal discussion on the project with the designer.

The preparation of the drawings should comply with the appropriate British Standard, BS 1192. Each drawing should include:

- the name and address of the consultant (architect, engineer or surveyor)
- the drawing number for reference and recording purposes
- the scale – if two or more scales are used, they should be sufficiently dissimilar as to be readily distinguishable by sight
- the title, which will indicate the scope of the work covered on the drawing.

The contractor, upon signing the contract, will usually be provided with two further copies of the contract drawings. The contract drawings may include copies of the drawings sent to the contractor with the invitation to tender, together with those drawings that have been used in the preparation of either the bill of quantities or specification. The contract drawings are usually defined in the articles of agreement as those which have been signed by both parties to the contract. It will be necessary during the construction phase for additional drawings and details to be supplied to the contractor. These may either explain and amplify the contract drawings or, because of variations, identify and explain the changes from the original design. An information release schedule should be provided to aid the regular progress of the works. The provision of further drawings and details is the norm and is to be expected.

The contractor must have an adequate filing system for the drawings. Superseded copies should be clearly marked. They should not, however, be discarded or destroyed until the final account has been agreed, since they may contain relevant information used for contractual claims. The drawings should be clear and accurate, but, because paper expands and contracts, only figured dimensions should be used. Scaling dimensions is therefore very much a last resort. Depending upon the size and type of the project, the designer may supply the building contractor with some or all of the drawings in Table 10.1.

10.2.1 Return of drawings

Once the contractor has received the final payment under the terms of the contract, the drawings may be requested to be returned to the designer along with details, schedules and other documents that bear the name of the architect or engineer. The copyright of the design is normally vested in these. If the employer or the contractor wishes to repeat the work, then a further agreed fee will be paid to the designer. None of the documents prepared especially for the project can be used for any other purpose than the contract, by either the employer or the contractor.

10.2.2 Schedules

The preparation and use of schedules is particularly appropriate for items of work such as:

- Windows.
- Doors.
- Access holes.
- Internal finishes.

Schedules provide an improved means of communicating information between the architect and the contractor. They are also invaluable to the quantity surveyor during the preparation of the bills of quantities. They have several advantages over attempting to provide the same information by correspondence or further drawings. Checking for possible errors is simplified, and the schedules can also be used to place orders for materials or components.

During the preparation of schedules, the following questions must be borne in mind:

- Who will use the schedule?
- What information is required to be conveyed?
- What additional information is required?
- How can the information be best portrayed?
- Does it revise information provided elsewhere?

The designer must supply the principal contractor with two copies of the schedules that have been prepared for use in carrying out the works. This should be done soon after the signing of the contract, or as soon as possible thereafter should they not be available at this time. The designer will supply the information release schedule to assist the principal contractor with their work.

10.3 Contract bills

Bills of quantities were described in *The Placing and Management of Building Contracts* (Simon Report 1944) as 'putting into words every obligation or service which will be required in carrying out the building project'. In the 1950s, the National Federation of Building Trades Employers (now the Construction Confederation) recommended that its members should not tender for work exceeding £8,000 in value without the provision of bills of quantities. However, this recommendation fell foul of the Restrictive Practices Act of 1976 as being contrary to public interest.

The provision of bills of quantities removes the duplication of effort by the different principal contractors involved in tendering for a project. It also ensures that each firm is tendering on exactly and precisely the same basis. The use of bills is widespread in Commonwealth of Nations countries and in other developing and industrialised economies. In other countries, such as the USA and mainland Europe, each principal contractor has to calculate the amount of work involved themselves. This places a greater risk on the principal contractors at a time when they are busily trying to meet the tender submission date. This risk has a premium that the use of bills of quantities removes. There are a great many myths surrounding bills of quantities, especially from those who would like such a process to be eliminated. There is a trend at the present time for their increased use by employers and for contractors to individually and directly employ quantity surveying practices to prepare bills for tendering purposes. The costs of such activities are added into tenders.

Some form of bill of quantities or measured schedules should be prepared for all types of construction projects, other than those of only a minor nature. The bill comprises a list or schedule of items of work to be carried out, providing a brief description and the quantities of the finished work in the building. A bill of quantities allows each principal contractor tendering for a project to price the construction work on exactly the same information as other contractors using a minimum of effort. The bill may include firm or approximate quantities, depending upon the completeness of the drawings and other information from which it was prepared. The principal contractor's rates in contract bills must not be divulged to others or used for any purpose other than the contract.

10.3.1 Uses

Although the main use of a bill of quantities is to properly assist the principal contractor during the process of tendering, it can be used for many other purposes, such as:

- Preparation of interim valuations in order that the architect may issue the certificate.
- Valuation of variations that have either been authorised or sanctioned by the architect or engineer.
- Cost control purposes.
- Ordering of materials, if care is properly exercised in the checking of quantities and the implications of future variations.
- Preparation of the final account, the bill is used as a basis for agreement.
- Production of a cost analysis for the building project.
- Determination of the quality of materials and standard of work by reference to preamble clauses.
- Obtaining subcontract quotations for sections of the measured work.
- A form of cost data.
- Taxation and grant purposes.
- Preparing contractual claims, when it then becomes essential.

10.3.2 Preparation

The preparation of bills of quantities should be undertaken by qualified quantity surveyors, adopting best practice manual or computer-based procedures. The items of work should be measured in accordance with the rules of a recognised method of measurement. Building projects carried out in the UK should be measured in accordance with the Standard Method of Measurement of Building Works (SMM7) or the New Rules of Measurement (NRM). However, there are several other methods of measurement available to the construction industry. On mass housing projects, it may be preferable to use a simplified version, such as the Code of Measurement of Building Works in Small Dwellings. Separate methods also exist for civil engineering work, known as the Civil Engineering Standard Method of Measurement (CESMM), and work on petrochemical plants.

Many countries have developed their own methods of measurement, some of which are based on UK methods. An international version has been published by the RICS, Principles of Measurement International (POMI). Descriptions and quantities are derived from the contract drawings and specification notes. These are on the basis of the finished quantities of work in the completed project. ASMM6 is the version currently used in Australia as their standard method of measurement.

10.3.3 Contents

Preliminaries

The preliminaries cover the employer's requirements and the contractor's obligations in carrying out the work. The different SMMs provide a framework for this section of the bill. These include:

- names of parties
- description of the works
- form and type of contract
- general facilities to be provided by the contractor.

In practice, although the preliminaries may comprise more than twenty pages of the bill of quantities, only a small number of items, approximately ten to fifteen, are priced by principal contractors. The remainder of the items are included for information and contractual purposes only. The value of the preliminary items may typically account for 8–15 per cent of the contract sum.

Preambles

The preamble clauses contain descriptions relating to:

- the quality and performance of materials
- the standard of work
- the testing of materials and work
- samples of materials and work.

This section of the bill of quantities has in some cases been replaced by a set of standard and comprehensive preamble clauses, such as the National Building Specification. The contents of the usual preambles bill are generally extracted from a library of standard clauses. Contractors rarely price any of this section of the bill. It is there for information and clarification purposes.

Measured works

The measured works include the items of work to be undertaken by the principal contractor or to be sublet to subcontractors. There are several forms of presentation available for this work. Here are some examples:

- *Trade format*: the items in the bill are grouped under their respective trades. The advantages of this format are that similar items are grouped together, there is a minimum of repetition and it is useful to the principal contractor when subletting.

- *Elemental format*: this groups the items according to their position in the building, on the basis of a recognised elemental subdivision of the project, e.g. external walls, roofs, wall finishes, sanitary appliances. This type of bill should, in theory, help in tendering by locating the work more precisely. In practice, principal contractors tend to dislike it since it involves a considerable amount of repetition. It is, however, useful to the quantity surveyor for cost analysis purposes and during future cost planning of new projects.

A recognised order for the inclusion of the various items is important in order to provide quick and easy reference. This often follows that of the relevant standard method of measurement.

Prime cost and provisional sums

Some parts of the construction project are not measured in detail, but are included in the bill as a lump sum item. These sums of money are intended to cover specialist work not normally undertaken by principal contractors (prime cost sum) or for work which cannot be entirely foreseen, defined or detailed at the time that the tendering documents are issued (provisional sum). In some rules of measurement, they are separately described as for defined or undefined work. Prime cost sums cover work undertaken by named or nominated subcontractors, nominated suppliers and statutory undertakings. They include lump sums that have been based upon quotations for items of work such as electrical installations, lifts, escalators and other similar systems.

Appendices

The appendices include the tender summary, a list of the principal contractors and subcontractors, and a basic price list of materials.

The form of tender is the principal contractor's written offer *to undertake and execute the works in accordance with the contract documents for a contract sum of money*, and will also state the contract period and whether the contract is on the basis of a fixed price. The tenders are submitted to the architect or engineer, who will then make a recommendation regarding the acceptance of a tender to the employer. If the employer decides to go ahead with the project, the successful tenderer is invited to submit a bill of quantities for checking. The form of tender may also state that the employer:

- may not accept any tender
- may not accept the lowest tender
- has no responsibility for the costs incurred in their preparation.

10.4 Contract specification

In certain circumstances, it may be more appropriate to provide documentation by way of a specification rather than a bill of quantities. The types of project where this may be appropriate include:

- Minor building projects.
- Small-scale alteration projects.
- Simple industrial shed type projects.

The specification is similar to a bill of quantities. The difference is that it does not include a measured works section. In place of this are the detailed descriptions of the work to be performed, to assist the contractor in preparing the tender. A specification is used during tendering to help the estimator price the work that is required to be carried out, during construction by the designer in order to determine the requirements of the contract legally, technically and financially, and by the building contractor to determine the work to be carried out on-site.

The National Building Specification (NBS) is the construction industry standard specification in the UK. It is a library of clauses for selection and editing to produce a project specification. Extensive guidance notes assist the specification writer to choose and complete the appropriate clauses for the job.

NBS Plus is the library of building product manufacturer details linked to clause guidance in the NBS, which is available online as a subscription service. NBS Plus is designed to improve proprietary specification practice for both specification writers and manufacturers. It is now fully integrated with the NBS specification software. It is reported that more than 89 per cent of the UK's top architectural practices have access to NBS Plus. NBS Educator has been designed for lecturers and students to understand contract documentation. Both NBS Plus and NBS Educator are available on the NBS website (www.thenbs.com).

10.5 Schedule of rates

A schedule of rates is a compromise between a specification and a bill of quantities. It is essentially a bill of quantities that has omitted to include any quantities for the work to be carried out. Its main purpose is therefore in valuing the items of work once they have been completed and measured. A schedule of rates may be used on:

- Jobbing work.
- Maintenance or repair contracts.
- Projects that cannot be adequately defined at the time of tender.
- Urgent works.
- Painting and decorating.

10.6 Master programme

The principal contractor is responsible for preparing a master programme showing when the works will be carried out during the contract period. Unless otherwise directed, the type of programme and the details to be included are to be at the discretion of the principal contractor. If there is an agreement to a change in the completion date because of an extension of time, then the principal contractor should provide the amendments and revisions to the programme, usually within fourteen days.

10.7 Information release schedule

This schedule informs the principal contractor when information will be made available by the architect or engineer. The schedule is not annexed to the contract; however, the architect or engineer must ensure that the information is released to the principal contractor in accordance with that agreed in the information release schedule. In practice, this is coordinated with the principal contractor's master programme for the works.

10.8 Contractual provisions

10.8.1 Copies of contract documents

All contracts will specify which documents are to be provided to the principal contractor free of charge. Typically, these will include:

- one copy of the form of contract certified by the employer
- two further copies of the contract drawings
- two copies of the unpriced bill of quantities
- two copies of any descriptive schedules or similar documents.

The formal contract drawings and the contract bills will usually remain in the custody of the employer. These are to be available for inspection at all reasonable times. Where there are no bills of quantities, the principal contractor should be provided with copies of a specification or schedule rates as may be appropriate. The principal contractor will normally provide two copies of the master programme, without charge to either the employer or the designer.

Before the date of practical completion of the works, the principal contractor must supply the employer with any drawings or other information for any performance-specified works. This would include, for example, details of any operating schedules for mechanical plant or equipment that has been installed in the project by the principal contractor or a subcontractor.

10.8.2 Availability of documents on-site

The principal contractor must keep one copy of the following documents on-site at all times for reasonable inspection:

- Contract drawings.
- Bills of quantities (unpriced).
- Descriptive schedules.
- Master programme.
- Additional drawings and details.

10.8.3 Discrepancies in documents

The principal contractor should write to the architect or engineer if any discrepancies between the documents are discovered. Only the form of contract takes precedence over the other contract documents. The contractor cannot therefore assume that the drawings are more important than the bills. However, drawings drawn to a larger scale will generally take precedence over drawings that have been prepared to a smaller scale. The principal contractor should adopt a similar course of action should a divergence be found between statutory requirements and the contract documents.

Where discrepancies or differences result in instructions to the contractor requiring variations, these will be dealt with under the respective contract conditions.

11 Design and build

What has become known as the traditional method of constructing building projects requires the separation of the roles of the designer from those of the constructor. They each do their work in isolation from the other and have separate legal responsibilities towards the employer. During the latter half of the last century, numerous attempts were made to find alternative procurement solutions, which to many is a failure of the traditional method of project development.

Design and build is a procurement arrangement where a single entity or consortium is contractually responsible to the employer for both the design and the construction of a project. It is sometimes referred to as design and construct, usually to incorporate works of a civil engineering nature. However, design and build is not a modern-day concept. In centuries past it was the only procurement arrangement that was available. Its roots originate in the ancient master builder concept, where responsibility for both design and construction resided with a single individual.

Design and build can be traced to ancient Mesopotamia, where the Code of Hammurabi (*c* 1800 BC) fixed absolute accountability upon master builders for both the design and construction of projects. In classical Greece, great temples, public buildings and civil works were designed and built by master builders. Enduring structures, such as the Parthenon and the Theatre of Dionysus, are testimony to this master builder process.

During the Renaissance, architecture and construction evolved as separate professions, and the presence of the master builders disappeared. Project complexity, in both design capability and construction methods, increased during this time as the need arose for specialists in design and specialists in construction. As time went by, these specialists subdivided even further, being recognised for different types or classes of building, and different specialists evolved taking on responsibility for specific functions of the project. As statutory and case law evolved during the 1800s, courts determined the extent of liability of architects, and builders also were responsible for the adequacy of their own work.

Design and build came back into prominence during the late 1960s, although it would be a decade or so later when its real impact would be felt. Subsequently, the use of project procurement would consist of a range of variable design and build type arrangements. Package deal arrangements, for example, which are sometimes believed to be another name for design and build (see Chapter 8), are really a special type of arrangement usually involving a contractor's own type of prefabricated or predesigned building system. Turnkey projects are a further example, where a contractor's responsibility far exceeds the designing and initial construction alone, to include fitting out and long-term maintenance agreements with the employer (see Chapter 8).

11.1 Design and build practice

In the more common design and build arrangement, an employer chooses to engage a contractor in preference to a designer. In some countries, for instance France, an architect may complete a design to about 50 or 60 per cent and then invite design and build proposals. The primary motivation of design and build, in preference to other forms of procurement, is to eliminate the inherent conflicts that exist between the architect and the contractor. The construction industry has a poor reputation for its adversarial nature. Other reasons for the choice of design and build include the greater assurance of completion on time, at an agreed and lower cost and improvements in the quality and performance of the building. These further considerations are often contested by design-only specialists and may in practice not always be achieved.

Whilst design and build procurement has many obvious advantages, employers remain wary of its ability to achieve design excellence. This concern is largely due to the importance of the contractor within a design and build organisation.

Contractors' main concerns are with aspects of buildability and profitability and the fact that a design and build project places a much greater responsibility on the design and build contractor. Inevitably, innovative design principles that may never have been properly tested in practice are excluded as far as possible from design and build solutions. Since some of the design ideas are of an innovative nature, they may not always work in practice as expected.

However, design and build contractors are wanting to expand their businesses and must therefore aim to deliver projects that employers require. Such firms are aware of the criticisms that design and build projects lack architectural merit. This view may now be outdated, since there are many examples of design and build projects that have received praise from architectural critics.

11.2 Advantages and disadvantages

There are several reasons why design and build projects offer many advantages in respect of design, price, construction and time:

- The contractor is normally involved from the commencement of the project and is thus aware directly of the employer's requirements and priorities.
- The contractor is able to offer the benefits of specialised construction knowledge and methods.
- By eliminating the traditional tendering procedure, the time from inception through to completion is reduced to a minimum.
- There is a direct contract between the contractor and the employer for the delivery of the whole building/structure.
- A functional building at a reasonable cost should be achieved. A past criticism of design and build schemes was their apparent limitations in designing projects with aesthetic merit. There is no reason why this cannot be achieved today.
- It should be possible to minimise initial tendering and pre-tender costs.
- The costs associated with the construction works, which also include the design fees, can be agreed prior to the commencement of the project on-site.
- The main contractor (design and build contractor) has control over all aspects of the project, including the design and the appointment of subcontractors.
- The need to appoint nominated subcontractors is virtually eliminated in most cases.
- There can be no claims by the contractor for delays due to lack of information (other than direct delays by the employer) since the contractor has full responsibility for the design details and project information.

Besides these advantages, there remain some disadvantages in the use of design and build. Many of these are actively being addressed by design and build contractors. The typical disadvantages are as follows:

- Because only a performance brief is provided to the contractor, alternative design solutions may not always be fully investigated.
- The contractor's solution may be governed to some extent by the capabilities of the firm rather than by the needs of the employer.
- Only a limited number of design and build contractors are able to offer in-house design facilities and this may be restrictive to an employer who initially desires to appoint a firm on the basis of some competition. However, in practice many contractors employ external consultants for this purpose.
- The aesthetic aspects of the design, both internally and externally, may be sacrificed in favour of easier and more cost-effective solutions.
- By designing and building to minimum performance requirements, the employer's long-term interests may be compromised.
- The normal external supervision, typical with traditional procurement arrangements, is only available where the employer retains a separate representative. This increases the costs involved, although external objectivity is desirable. A quantity surveyor may be required to authorise interim payments that become due to the contractor.
- If the contractor becomes bankrupt, the employer's legal and financial interests in the project can be considerable, even where independent consultants have been retained.

11.3 Choice of design and build

The recommendation to adopt design and build for a construction project will depend upon a number of different factors. These may include:

- user familiarity with design and build procurement arrangements
- preferences of an employer or consultant
- the desire for single-point responsibility for the employer
- greater certainty of outcomes, initially and in the longer term.

Design and build arrangements are useful in the following circumstances:

- Relatively simple and standard forms of construction, involving little architectural design such as warehouses, small factory units and farm buildings.
- Buildings using a proprietary building system, especially those described as closed systems of construction. These arrangements are more correctly described as package deals (see Chapter 8).
- Construction projects where contractors may have become specialist firms within a region or locality.
- Where the cumulative knowledge and skills of an experienced general contractor, and their ability to utilise the attributes of a range of subcontractors, can be used to advantage.
- Design and build, or the use of a contractor's design, is a common component on many traditional contractual arrangements for the design and construction of specialist elements.

However, design and build arrangements far exceed these four circumstances. Employers may choose to employ a contractor in preference to a designer for a whole range of reasons and types of project. The trio of design, quality and cost will need to be considered by the employer prior to appointment. Where an employer appoints a contractor, often an architectural practice will be employed to look after the design and therefore no project is exempt from this consideration.

Design and build contractors are also sometimes known to make extravagant claims, which cannot be substantiated, about the benefits of using this form of procurement in preference to other alternatives. These have been considered separately in Chapter 8. However, the single-point responsibility will reduce administration, contractual claims and litigation as conflicts between a designer and constructor are eliminated.

In many cases, design and build is well suited to the task of delivering standard types of building using tested construction technology. These can include building projects of any size. Where the project involves relatively few construction trades, it will be particularly suited to design and build. Conversely, design and build may be ill-suited to more complex solutions for building projects.

11.4 Contractual arrangements

The contractual arrangements that are used for design and build can vary considerably. In some cases, the contractor is invited to design and then build the project. In other circumstances, the employer may have already instructed an architect to provide sketch plans and outline designs which, after planning consent has been received, are then passed over to the contractor to develop into working drawings. The employer may choose to retain the services of an architect or quantity surveyor to oversee compliance with aspects of quality and to authorise payments due to the contractor.

The payment of the contractor may be based on a lump sum with arrangements for interim or stage payments. Alternatively, detailed quantities may be requested from the contractor in order to more accurately value the works for interim payment purposes and as a basis for agreeing variations and final accounts.

Some element of competition may be required, although the evaluation of such schemes as design and build is by no means straightforward. Design and build contracts are frequently awarded on a fixed-price basis that will not be subject to adjustment except in exceptional circumstances. This aspect is viewed by employers as an advantage, but it may also be partially seen as a disadvantage. Many design and build contracts exclude the possibility of employer-directed variations.

11.5 Procurement: three main methods

11.5.1 Competition

It is not unusual at the outset of the project for employers or their advisers to discuss outline proposals and ideas with a selected group of suitable building firms. On the basis of these initial contacts, firms may be requested to submit tenders and outline design solutions together with other criteria that the employer may have requested, such as time for completion, standards of work, etc. However, it is an extremely difficult task to judge such solutions since the best one may frequently cost the most, and building firms, or even employers, may not have considered the associated implications.

Competition is a useful method where the project is of a simple design, for example, a rectangular warehouse building of known size and requirements. Competition is also appropriate where a package deal solution is to be sought. The differences between design and build and package deals are described in Chapter 8.

11.5.2 Negotiation

Negotiation is the common approach to procurement, although it may be preceded by some form of competition, as described above. Negotiation will be sought where a relationship already exists between the employer and the contractor or in those circumstances where the particular specialist skills of the contractor are sought by the employer. Negotiation is used where the firm's interests, qualifications and experience meet the needs of the project.

11.5.3 Design competitions

In these circumstances, the employer awards the contract to the firm that wins the competition held between a number of design and build contractors. The successful proposals are usually highly developed and include significant detail in their preparation. Where the project is not complex, then standard details and layouts may be used, but this is more common where the project is a package deal representing a contractor's own patented system. Whilst unsuccessful designs do not usually receive any reimbursement, some employers are prepared to offer some minimum level of compensation.

11.6 Selection factors

Design and build contractual arrangements offer a range of advantages to employers; the most important ones are given in Table 11.1. Employers will, of course, consider the various attributes differently, setting their own priorities and goals. Whilst design and build organisations are vociferous about the apparent advantages of this method of procurement, the evidence in practice remains inconclusive. Some of the claims appear to be obvious, but there is a lack of empirical data and unbiased research to support them. For example, even identical projects, carried out across the road from each other, frequently display different outcomes. The evaluation therefore depends not solely upon a range of factors that are directly related to the project, but also on other factors that may have little to do with design or construction methods adopted.

Table 11.1 Selection factors for design and build procurement arrangements

Selection factor	*Description*
Cost certainty	Determine the overall project cost before awarding the contract.
Cost reduction	Increase the project's value by reducing costs in comparison to other procurement methods.
Contract period	Agree the time for completion and handover to the employer.
Reduced contract period	Reduce the length of time from inception to completion.
Quality and standards	Enhance initial quality, reduce defects and long-term maintenance.
Constructability	Introduce this knowledge into the process, during the design phase.
Litigation	Reduce the possibility of contractual claims.

The question of whether design and build is just a fashionable solution needs to be considered. The traditional method is flawed, but design and build may be shown to be no better when an improved method is discovered. Design and build may be the best arrangement for the current times, but may not be the best alternative in the future. Design and build is no doubt aptly suited to certain types of project, but its uniform application for all types of construction project is not necessarily the best advice.

11.6.1 Detailed review of factors

Cost certainty

Some owners will choose to use design and build because a fixed price can be guaranteed. There are usually fewer opportunities for variations or changes to the design, because it is recognised that such changes often have a bad effect on the project as a whole. The single organisation should be better able to control the costs that are involved.

Cost reduction

Although little empirical data exists to indicate that design and build offers better value for money than other procurement methods, there is sound reasoning to believe this is so. The cost reduction stems from three factors:

- The shortening of the contract period and the overall time required to completion.
- The introduction of the contractor's knowledge and expertise into the construction process and its influence upon the design formulation.
- The *hidden* design fees that are included within the overall tender sum. Whilst design fees are charged, they are set at a more realistic level. The designer is able to work closely with the constructor to achieve a solution without the possibility of future litigation between the two roles as they come under the one single entity.

Contract period

The design and build contractor is able to exercise much greater control over the project's time management. This is achievable because the single entity is able to control not just the contract period, but also the time available for the design phase.

Reduced contract period

Design and construction are able to be carried out in parallel, so reducing the project's overall completion date. Through integrating design and construction, methods of construction can be included that will further reduce the amount of time available on-site.

Quality and standards

Design and build should result in a project that has inherent quality attributes, in terms of the performance of the building in respect of the structure, materials, services and finishes. The holistic development of the design along with the construction capability should result in fewer defects that need to be remedied at completion and there should be improved performance in the longer term.

Constructability

The early involvement of the constructor is a good practice with any type of procurement. Designers know what needs to be built and constructors know how to build. Constructability is an important component in ensuring that a design concept can be easily developed into an efficient and effective building programme.

Litigation

One of the really attractive features of design and build procurement is the potential to reduce contractual claims arising out of disputes between designers and constructors. Design errors or omissions, for example, are solely the responsibility of the design and build company. Claims for delays due to a lack of information from the designer are entirely removed. Design and build will not remove litigation entirely from the process, but it does have the potential to remove a considerable amount of conflict in construction.

11.7 Use worldwide

The use of design and build projects is common in most countries around the world, as indicated in Table 11.2. It is the most common in Europe and least common in Africa and the Middle East. The increased globalisation of the construction industry will help to propagate different construction methods. Debates still exist in some countries about the role of design and build in the work of the architect. In some countries, design and build is seen as contrary to the role of the architect as defined in their Architects Act.

Where design and build is commonly used, the frequency of use varies between the public sector and the private sector. Typically, it is much more common in the private sector. The public sector has for years been anxious about accountability, and issues relating to design and build projects appear at times to run contrary to this. For example, aspects of competition are more difficult to justify with design and build arrangements than with more traditional procurement methods. A number of countries are an exception to this rule (see Table 11.2).

Table 11.3 indicates some typical projects likely to use design and build. The obvious responses included warehouses and industrial buildings, but the range of projects using design and build in Norway and Australia is interesting. Where there is a strong design and build culture in a country, contractors will permanently employ a range of design professionals, such as architects, engineers and surveyors. A more common approach is for the contractor to form a joint venture with consultants. In these circumstances, the contractor will frequently lead the joint venture company.

11.8 Design and manage

Design and manage is the consultants' response to design and build. With the growth in subcontracting, it provides the opportunity for designer-led design and build. It is discussed more fully in Chapter 8.

11.9 Partnering

In some parts of the world, design and build appears to be the old-fashioned word for partnering. Partnering is a very different concept and it is described in Chapter 12.

Table 11.2 Use of design and build

Country	*Use is common*	*Frequency of use*		*Standard contract*
		Private %	*Public %*	
Europe				
Austria	No	3	2	No
Belgium	No	5	5	No
Denmark	Yes	30–50	10–30	Yes
Finland	Yes	10–15	0–5	No
France	Yes	85–95	85–95	Yes
Great Britain	Yes	40–50	25–35	Yes
Greece	Yes	50	70	No
Ireland	No	2–3	2–3	No
Italy	No	0–10	Very little	No
Norway	Yes	85	15	Yes
Russia	Yes	30–40	10–15	Yes
Spain	No	30	10	No
Sweden	Yes	40–50	20–35	Yes
North America				
Brazil	Yes	30	2	No
Canada	Yes	–	–	Yes
Mexico	No	60	40	Yes
USA	Yes	25–35	5	Yes
Pacific				
Australia	Yes	30–40	20–30	Yes
China	Yes	5	1	Yes
Japan	Yes	30–35	0–3	No
New Zealand	No	5–10	2	No
Thailand	Yes	50	30	No

11.10 Public private partnerships

Further interest in design and build arrangements have been encouraged through the private finance initiative (PFI) (see Chapter 12) and public private partnership (PPP) arrangements.

Different arrangements exist for PFI projects, and combinations of arrangements are not uncommon. As the process evolves, new arrangements are being developed, which inevitably go under a range of different acronyms including:

- Design, build and finance (DBF).
- Design, build, finance and operate (DBOF).
- Build, own and operate (BOO).
- Design, build, operate and maintain (DBOM).
- Build, own, operate and transfer (BOOT).
- Build, lease and transfer (BLT).

These descriptions are largely self-explanatory, each with their own comparable advantages and disadvantages – between themselves and when compared with other methods available. Depending on the agreements reached, the private sector employer and the public sector organisation will determine the part and extent each party will play. It is unusual with such arrangements for any payments to be made prior to the completion of the project. In the case of

Table 11.3 Typical projects using design and build arrangements

Country	*Types of project*
Europe	
Belgium	Industrial, some non-prestige offices
Denmark	Office, industrial, warehouses, hotels
Finland	Warehouses, projects with tight schedules
France	Any type of project
Germany	Manufacturing, residential, commercial
Great Britain	Warehouses, retail parks, educational
Greece	Any type of project
Ireland	Industrial, warehouses, decentralised public offices
Italy	Small simple projects
Netherlands	Offices, industrial, infrastructure
Norway	Most types of project
Russia	Standard offices, industrial
Spain	Industrial, warehouses
Sweden	Industrial, offices
North America	
Canada	Public sector jails, administration, industrial
Mexico	Industrial
USA	Warehouses, offices, rental
Pacific	
Australia	Schools, stadiums, universities, most large projects
Japan	Industrial, warehouses
New Zealand	Industrial, warehouses, suburban offices

income generating projects (see below), the public sector may make no payment whatsoever. The arguments that are used for these arrangements include the ability to fund the project more cheaply, and the contractor's greater ability to decide how the project should be carried out to meet the agreed specification. In some cases, a performance specification is used:

- *Joint ventures*: these are where both the public and the private sector contribute towards a project and then jointly share the responsibility and rewards of the project.
- *Income generating projects*: in this case, the private sector may be encouraged to take full responsibility for the project on the basis that it will regain its costs and profits through charging the users of the project. This concept has been used on toll bridges and toll roads where the responsibility for repair and maintenance remains with the contractor for a fixed number of years. During this time, the contractor receives the income that such projects generate.
- *Private sector projects*: on these projects, a scheme is designed and constructed by a company in the private sector to a brief supplied by a government department. The government department has no legal interest in the project on completion, but agrees to make a leasehold arrangement with the private sector company.

11.11 Form of contract

The Joint Contracts Tribunal has published a standard form of building contract with contractor's design, based on the standard form of building contract. This was introduced with a suite of forms in 1998, replacing the original of 1981. It superseded the design and build form of contract originally published by the National Federation of Building Trades Employers

(NFBTE), which had been in use since 1970. The principal contractor's duties are necessarily wider than usual, since greater responsibility has to be taken for a much wider remit that combines design and construction. There are no references to the architect.

11.11.1 Articles of agreement

Employer's requirements

In the first recital, the employer, in lieu of supplying the contractor with drawings and bills of quantities, has issued the contractor with basic requirements. These requirements will broadly be the same as the information that would have been provided to the architect.

Contractor's proposals

In the second recital, the contractor's proposals are identified. These include details of the contract sum analysis that will be required for the execution of these proposals.

Employer's acceptance

The third recital states that the employer has accepted these proposals and the contract sum analysis and that the employer is satisfied they meet the requirements for the project. The employer's requirements, the contractor's proposals and the contract sum analysis are described in detail in Appendix 3 to the conditions. Article 3 names the employer's agent. Since there is no independent designer or supervisor for the works envisaged, provision is made for someone to act on behalf of the employer. The duties of this person may include receiving or issuing applications, consents, instructions, notices, requests and statements as well as acting on the employer's behalf.

11.11.2 The conditions

Clause 2

Clause 2 includes the provisions of the counterpart clause of the standard form, but additionally covers matters appropriate to the contractor's design and their liability for it. The contractor in clause 2.1 shall carry out and complete the works, including the selection of any specifications of materials and work standards in order to meet the employer's requirements. Clause 2.5 indicates that the contractor must carry out the design work in an equal manner to that of an independent architect or professional designer. The contractor's responsibility to the employer for any inadequacy of the design work includes the exclusion of defects or any insufficiency in the design work. Where the work includes the design and construction of dwellings, reference is made to liability under the Defective Premises Act 1972. If the Act does not apply, the contractor's design liability for loss of use of profit or consequential loss is limited to the amount, if any, set out in Appendix 1. The contractor's design under this clause includes whatever may have been prepared by others on behalf of the contractor.

Clause 4

The contractor must comply with all instructions issued by the employer, as long as the employer has the contractual power to issue such instructions. Variations are referred to as change instructions in clause 12, and where the contractor makes a reasonable objection in writing to a change instruction there is no need to comply with it. The provisions relating to an employer's instructions are generally similar to those dealing with an architect's instructions under the standard form.

Clause 5

This clause requires that both the employer's requirements and the contractor's proposals are to remain in the custody of the employer. The contractor is to be allowed access to them at all reasonable times. When the contract has been signed by the parties, the employer will provide the contractor, free of charge, with:

- a copy of the certified articles of agreement, conditions and appendices
- a copy of the employer's requirements
- a copy of the contractor's proposals, which includes the contract sum analysis.

The contractor must then provide the employer with two copies of the drawings, specifications, details, levels and setting out dimensions proposed to be used on the works. All this information largely represents the equivalent of the contract documents of the standard form, but is not described as such in this form. A copy of all the above information is to be kept on-site and available to the employer's agent at all reasonable times.

Prior to the commencement of the defects liability period, the employer is to be provided with copies of the 'as-built' drawings and other relevant information. This information may include details of the maintenance and operation of the building works, including any installations contained in the works. The information provided by either party is for the use of this contract only. It would appear, therefore, that copyright of the documents is vested in the party who actually prepared them.

Clause 7

The employer is required to define the boundaries of the site, and this information is to be written into the conditions of contract at this point.

Clause 12

Variations are referred to as changes in the employer's requirements. The alteration in wording from the standard form may lead one to suspect that changes are intended to be of a minor nature only, the employer's requirements in the first instance being comprehensive and complete. A change in the contractor's proposals is not envisaged, but, where the contractor considers them to be necessary, the employer's permission will need to be obtained; the contractor will possibly incur extra cost where this occurs. In other respects, this clause is very similar to clause 13 of the standard form, which deals with variations and provisional sums.

Clause 16

Upon practical completion of the works, the contractor will receive a written statement from the employer to that effect. The defects liability will operate from that date, and the employer must provide the contractor with a schedule of defects within fourteen days. Practical completion marks the end of the contract, and no further new instructions from the employer are permissible. A notice of completion of making good any defects is issued when the work has been properly rectified.

Clause 26

Clause 26.2 includes the normal grounds for claiming loss and expense. It also includes the provision for dealing with a delay in receipt of any permission or approval for the purposes of development control requirements necessary for the works to proceed. The contractor must have taken all practicable steps to avoid or reduce this delay. This can also give rise to an extension of time under clause 25.4.7 as a relevant event. It is also considered to be an important factor that could result in the determination by the contractor under clause 28.1.2.8.

Clause 27

The employer may terminate the employment of the contractor for one of the reasons suggested under clause 27.1. In the event of this occurring, the contractor must provide the employer with two copies of all the current drawings, details, schedules, etc., in order that another contractor may be engaged to complete the works. A similar condition will still apply where the contractor has terminated the contract for one of the reasons listed in clause 28.1.

Clause 30

The processes to be used for interim payments follow those of normal practice. Two alternatives are, however, specifically suggested, and these are described in Appendix 2. Alternative A is on a stage payment basis. The stages are predetermined and the appropriate amounts set against them, the total of which adds up to the contract sum. Any adjustments to the employer's requirements or for fluctuations and claims must be properly documented. Alternative B describes the periodic payment basis that is the more usual method with the standard form. Payment on this basis must be supported by the appropriate information. The other matters, of retention and when the payments are due to be paid by the employer, are in accordance with the standard form.

The final account must be presented to the employer within three months of practical completion of the works (clause 30.5.1). The employer must then agree to this within a maximum period of four months from the time of submission. Thereafter, the account is conclusive evidence of the amount due. It is also conclusive evidence that the employer is fully satisfied that the project is in accordance with the terms of the contract and the employer's requirements.

11.12 Conclusion

The role of the building contractor in the design of any project is valid. Criticism has often been levelled at the absence of any contractor input, and the fragmentation of the design and construction process. There is no doubt, therefore, that this method of contracting plays a vital role and function within the industry today. It does not include all the advantages over the more traditional methods, since, if it did, these methods would now be moving nearer towards extinction. The more one can involve the contractor in both the constructional detailing and the method, the more satisfactory buildings are likely to be. However, an employer considering this contractual option for a proposed project is well advised to retain the professional advice of an architect and quantity surveyor in some form. The normal building employer is unlikely to be familiar with building contracts or the processes involved. An independent adviser is therefore likely to be able to offer both constructive comments on the contractor's proposals and assistance during the building's erection.

12 Procurement in the twenty-first century

The procurement process is one of the components that brings construction projects into existence. The others are design and construction. Each project possesses a number of different variables that will determine the choice of procurement method in providing the most advantageous route for employers. The selection of the procurement route may have repercussions on the operation of the building throughout its life. This chapter explores the issues surrounding an employer's main priorities and the methods that are available to produce the building on time, within budget and to the specified quality. Reference will be made to the interface between employers and the industry, how the industry organises the development process and an evaluation of the different procurement methods.

The construction industry is unique. Its characteristics separate it from all other industries. These include:

- The physical nature of the product.
- Manufacture normally takes place on the employer's own construction site.
- Projects often represent a bespoke design.
- Design separate from manufacture is not mirrored in other industries.
- The organisation of the construction process.
- The methods and manner of price determination.

Traditionally, employers who wished to have projects constructed would invariably commission a designer, normally an architect for building projects or a civil engineer for civil engineering projects. These would prepare drawings for the proposed scheme and, where the project was of a sufficient size, a quantity surveyor would prepare budget estimates and contract documentation on which contractors could then prepare their tenders. This was the procedure used during the early part of this century. Even up to thirty years ago there was only limited variation from this method. The term *procurement* was not used for tendering practices until the 1980s. Since the 1960s there have been several catalysts for change in the way that projects are procured. These are shown in Box 12.1.

Procedures will continue to evolve in order to meet new circumstances, situations and fashion. Procurement is similar to quality in that improvements to present procedures within current practices can always be achieved. Procurement methods of a hybrid nature are being developed to help utilise best practices from the various competing alternatives. Each of the different methods has been used in the industry, and some have been used more than others largely owing to:

- User familiarity.
- Ease of application.

- Employer insistence.
- Recognition.
- Reliability.

As suggested in Box 12.2, many studies concerned with the organisation and management of the construction industry have been undertaken since the twentieth century. Whilst the industry has responded in some measure to change, this has been more attributed to pragmatic courses of action and commercial pressures rather than the advice offered by government, industry or researchers.

The construction industry comprises many different parties, organisations and professions. It is seen as fragmented and does not speak with one voice. This is both an advantage and a disadvantage. Views are held by many with vested interests and traditions that represent their own power and authority. Such bodies are clearly loath to relinquish these positions freely. Some of the recommendations have been implemented. Other practices continue to persist, even though the structure of the industry and its employers have changed dramatically.

Employers are at the core of the process and their needs must be met by industry. Implementation, after all, begins with them. Employers are also dispersed and vary greatly. In the past, the government used to act as a monolithic employer. The privatisation of many government departments and activities has changed this perception, resulting in the fragmentation of this important employer base. Existing government departments now operate different procurement strategies and practices, and this became more pronounced after the demise of the Property Services Agency.

12.1 Relevant published reports

Several reports, many of which have been UK government-sponsored, have been issued over the past seventy-five years. Their overall themes have been on improving the way the industry is organised and the way construction work is procured. These various reports are listed in Box 12.2.

Box 12.1 Catalysts for change in procurement

- Government intervention through committees, such as the Banwell Report of the 1960s and, more recently, through the Department of the Environment, Transport and the Regions (DETR) and the Latham Report (1994).
- Pressure groups formed to encourage change for their members, most notably the British Property Federation.
- International comparisons, particularly with the USA and Japan and influence of the Single European Market in 1992.
- The apparent failure of the construction industry to satisfy the perceived needs of its employers, especially in the way in which projects are organised and executed.
- The influence of educational developments and research.
- Trends throughout society towards greater efficiency, effectiveness and economy.
- Rapid changes in information technology both in respect of office practice and manufacturing processes.
- The attitudes amongst the professions.
- The overriding wish of employers for single-point responsibility.

Box 12.2 Reports influencing procurement

- *The Placing and Management of Building Contracts* (Simon 1944), the Simon Report.
- *A Code of Procedure for Selective Tendering* (NJCC 1959).
- *Survey of Problems before the Construction Industry* (Emmerson 1962), the Emmerson Report.
- *The Placing and Management of Contracts for Building and Civil Engineering Works* (Banwell 1964), the Banwell Report.
- *Action on the Banwell Report* (Banwell 1967).
- *Faster Building for Industry* (NEDC 1983).
- *Manual of the British Property Federation System* (BPF 1983).
- *Construction Contract Arrangements in European Union (EU) Countries* (Bruges Group 1983).
- *Thinking about Building* (NEDC 1985).
- *Faster Building for Commerce* (NEDC 1998).
- *Building Towards 2001* (University of Reading 1991).
- *Trust and Money* (Latham 1993).
- *Constructing the Team* (Latham 1994), the Latham Report.
- *A Statement on the Construction Industry* (Barlow 1996), the Barlow Report.
- *Rethinking Construction* (Egan 1998), the Egan Report.
- *Modernising Construction* (National Audit Office 2001).
- *Rethinking Construction Achievements* (Rethinking Construction 2002).
- *Accelerating Change* (Strategic Forum for Construction 2002), report by the SFC chaired by Sir John Egan.
- *Rethinking Construction Innovation and Research: A Review of Government R&D Policies and Practices* (Fairclough 2002), the Fairclough Report.
- *Low Carbon Construction* (Department of Business, Innovation and Skills 2010).
- *Government Construction Strategy 2011–2015* (Cabinet Office 2011).
- *Government Construction Strategy: Final Report to Government by the Procurement/Lean Client Task Group* (Cabinet Office 2012).
- *Construction 2025: industrial strategy for construction – government and industry in partnership* (Department of Business, Innovation and Skills 2013a).
- *Government Construction Strategy 2016–2020* (Cabinet Office and Infrastructure and Projects Authority 2016).
- *The Farmer Review of the UK Construction Labour Model: Modernise or Die, Time to Decide the Industry's Future* (Farmer 2016).
- *Fixing our Broken Housing Market*, Housing White Paper (Department for Communities and Local Government 2017).

12.1.1 Government Construction Strategy 2016–2020

The *Government Construction Strategy (GCS) 2016–2020* follows from the *GCS 2011–2015* aiming to deliver £1.7 billion over the course of the next Parliament. The report was published by the Cabinet Office and the Infrastructure and Projects Authority (IPA) in 2016. It forms part of a series of documents prepared by the IPA about improving the performance of the built environment in the UK. The main objectives of the *GCS 2016–2020* are fourfold:

- Improving central government's capabilities as a construction employer.
- Improved digital technology and BIM adoption.
- Deploying collaborative procurement techniques.
- Driving whole life approaches to cost and carbon reduction across construction, operation and maintenance.

In particular, the strategy focuses on achieving overall procurement process improvements through early contractor supply chain involvement, developing capacity and capability through 20,000 apprenticeships and promoting fairer payment, in compliance with the Construction Act (Local Democracy, Economic Development and Construction Act 2009). The IPA is working closely with the Fair Payment Working Group to develop the scheme for project bank accounts as a means of improving payments to supply chains.

There are several other reports that generate influence on improving or regulating construction procurement, and some of them are explored in the appropriate sections through this book.

12.2 Employers' main requirements

Employers' main requirements have already been considered in Chapter 9. The three main requirements are further examined below. They are subject to change to meet current and perceived future needs, and different employer groups will each have differing requirements and priorities.

12.2.1 Performance

The designer prepares plans and details for the proposed project. When completed, it should offer aesthetic appeal and add to the environment in which the project is to be located. The completed project must also meet needs and requirements in terms of the spatial layout, the structure's function and the environmental controls. The specification will define the quality of the materials to be used and the standards of work to be expected.

An inadequate design concept, poor constructional detailing, an incorrect choice of materials and a wrong choice of constructor are problems that will create obstacles to achieving overall good performance. Such problems may create difficulties throughout the project's life, so that the project will never operate effectively.

It is important to consider the needs of future maintenance and refurbishment. The better-informed employers will consider the project in terms of the longer-term needs, rather than solely hoping to satisfy present-day requirements. Their evaluation will therefore represent a holistic view, taking into account the provision of an immediate design solution coupled with the elimination of possible future problems.

12.2.2 Cost

Before employers commit themselves to a detailed design that they cannot afford, some information on costs and prices needs to be provided by the quantity surveyor. In some circumstances, perhaps where a project has to be carried out as a matter of urgency, cost may be of less importance. But cost can never be ignored, and competing proposals must be evaluated as fully as possible. In those rare circumstances where cost does not matter, some cost advice will nevertheless prove to be beneficial.

A budget price is usually prepared on the basis of scant outline information. This may be based upon the units of accommodation or the floor space that the employer is considering. This figure will be imprecise because it is based upon imprecise design data. It cannot be too high, otherwise the employer may decide not to build and a possible commission will be lost. But it cannot be too low, since as the design evolves it may become apparent that what the employer desires in a design cannot be afforded. The project should be cost planned as the design develops. This should result in tenders being received that remain within budget.

Employers will not evaluate a project solely on the initial cost alone, but rather on the basis of a life cycle cost. They will want to consider the future recurring costs associated with owning or using the project. It may be possible to reduce overall costs by introducing construction that requires less maintenance or repair.

12.2.3 Time

Once employers decide to build, they are frequently in a hurry to see the project completed. Some research has shown that if more time could be allocated to the design stage, then the overall project would be completed more quickly, and this might also improve the project's quality. A shorter time on the construction site would be likely to reduce initial construction costs. However, prior to the design getting under way, a large amount of time is spent by employers in deciding whether to build.

The design of the project will have some influence on the time required for construction. The use of an *off-the-peg* type building will greatly reduce construction time on-site. The methods adopted by the contractor on-site for construction purposes will also influence the length of the contract period. The involvement of a contractor as part of the design process will help to influence the methods used for construction and may ensure buildability in the design.

One way of measuring the success of a project is whether the building is available for commissioning by the date suggested in the contract documents. Where time is considered to be very important by the employer, then it may be necessary to consider adopting fast-track construction methods that have the effect of reducing the length of the contract period.

12.3 Major considerations

This section supplements the discussion provided in Chapter 7.

12.3.1 Consultants versus constructors

The arguments for engaging either a consultant or a constructor as the employer's main adviser or representative are linked with tradition, fashion, loyalty and satisfaction or disappointment with a previous project. There is also the belief (sometimes mistaken) that had an alternative approach to procurement been used, some of the difficulties would not have occurred or problems would have been more easily solved. The emphasis on a single point of responsibility for the employer is an attractive proposition. However, this does not automatically mean design and build by the constructor, but a re-evaluation of existing arrangements.

12.3.2 Competition versus negotiation

Businesses, designers or constructors, are able to secure their work or commissions in a variety of ways. These can include invitation, recommendation, reputation and speculation. They are usually appointed to a project by negotiation or through competition with other firms. Where some form of competition on price, quality or time exists, then employers supposedly obtain a better deal. However, there are circumstances when negotiation with a single firm or organisation may offer direct benefits to an employer.

12.3.3 Measurement versus reimbursement

The methods of calculating the costs of construction work are either on the basis of paying for the work against some predefined criteria or rules of measurement, or through the reimbursement of the contractor's actual costs. Measurement contracts distribute more risk and incentive to the contractor to complete the works efficiently. Reimbursement contracts result in the contractor receiving only what is spent plus an agreed amount to cover profits.

12.3.4 Traditional versus alternatives

Until recently, the majority of the major building projects were constructed using a procurement system known as *single-stage selective tendering*. This system had evolved within the parameters of good practice and procedures. However, largely due to better understanding, the availability of other procurement procedures and an improved knowledge of practices in other countries, new methods have now become available for consideration.

Various procurement procedures have been devised and developed in an attempt to address some or all of these issues. However, it is difficult to select only a single procurement method to solve all problems. There is no universal solution. None of the newer contractual procedures address all the past criticisms. In fact, some of the newer procedures have been abandoned in favour of existing practices that have been shown to perform better against a range of different criteria. Procurement practice is, in reality, trading off the employer's clear objectives, if known, against the different methods that are available. The employer's needs and wants have to be differentiated in an attempt to achieve a best possible overall solution.

12.4 Procurement selection

There are a wide variety of procurement options aimed at addressing criticisms of poor quality, lengthy construction periods and high costs. The methods vary from traditional single-stage selective tendering (where an employer uses a designer to prepare drawings and documentation on which contractors are invited to submit competitive prices), to schemes where a single construction firm will provide the truly all-in service, the turnkey project. There are methods devised to get the contractor on-site as quickly as possible, such as two-stage tendering and fast tracking, and these anticipate that the project will be available sooner than by using other arrangements. Other methods recognise the contractor's improved management skills or have evolved to meet changes that have occurred in the industry, such as the proliferation of subcontracting, the increase in litigation and the need for single-point responsibility. The Latham Report identified partnering between employers, consultants and contractors as a useful arrangement for the procuring of buildings, emphasising teamwork. The *GCS 2016–2020* encourages the use of collaborative forms of contract and procurement, moving further away from adversarial forms.

12.5 Current considerations

12.5.1 Fair construction contracts

It is recognised that the existing arrangements used in the construction industry mitigate against cooperation and teamwork and against the employer's own requirements. These arrangements also contribute towards helping to perpetuate the poor image of the industry. A 1995 report from the Construction Sponsorship Directorate summarised the fundamental principles of a modern construction contract:

- Dealing fairly with each other and an atmosphere of mutual cooperation.
- Firm duties of teamwork, with shared financial motivation to pursue those objectives.
- An interrelated package of documents, clearly identifying roles and duties.
- Comprehensible language with guidance notes.
- Separation of the roles of contract administrator, project or lead manager and adjudicator.
- Allocating risks to the party best able to control them.
- Avoiding the need, wherever possible, of changes to pre-tender information.
- Assessing interim payments through milestones or activity schedules.
- Clearly setting out the periods for interim payments and automatically adding interest where this is not complied with.
- Provision of secure trust funds.
- Provision of speedy dispute resolution.
- Provision of incentives for exceptional performance.
- Provision for advance payment to contractors and subcontractors for prefabricated off-site materials and components.

The construction payment practices have been further streamlined with the introduction of the Local Democracy, Economic Development and Construction Act 2009 and the consequent amendments to the Scheme for Construction Contracts.

12.5.2 Trust funds

It is fundamental to trust, within the construction industry, that those involved should be paid the correct amounts at the right time for the work they have carried out. It may be argued that a problem does not exist and that:

- employers only award work to firms with integrity
- contractors are at liberty to decline work from dubious employers
- subcontractors can adopt similar business practices
- bonds and indemnities are already available
- bad debts are not singularly a problem in the construction industry.

However diligently employers, contractors and subcontractors verify each other, the realities of the construction industry and its markets continue to exist. In circumstances such as a recession, contractors and subcontractors are often prepared to undertake work for any employer. This is frequently done at a minimal profit margin. Bad debt insurance is available, but this adds extra costs at times when firms are seeking to reduce overheads. In times of prosperity, employers are prepared to undertake work with almost any firm who is available in order to get an important project constructed.

The contractor's goods and services become part of the land ownership once incorporated within the project. Any *retention of title* clause that might be incorporated by suppliers or contractors in their trading agreements does not protect them once the materials are incorporated within the works. The building contractor is also likely to be far down the queue when an employer is unable to make payment within the terms of the contract. In some countries, legislation has been provided to deal with the potential injustice that might be suffered. The most comprehensive is the Ontario Construction Lien Act 1993.

An effective way of dealing with this problem is to set up a trust fund for interim payments and retention monies. An employer, for example, could be requested to pay into such a fund at the start of the payment period, e.g. at the beginning of the month. The authorised payment would then be paid to the contractor at the appropriate time. Where a form of stage payment is used, then the amount of the particular programme stage would be deposited in the trust fund at the commencement of the work in this stage. The amount authorised should correspond with the contractor's approved contract programme. The main contractor and the subcontractors would be informed of the amounts deposited. If a party considered that the sums were inadequate then the adjudicator would be consulted. There may also be some rationale of making payments to the subcontractors directly from this fund rather than through the main contractor's account. Any monetary interest accrued in the trust fund belongs to the employer. Where the fund necessitates bank charges, then these charges would need to be determined at the time of tender.

Trust funds are not really required for public works projects, since it is unlikely they will become insolvent. However, trust funds would be a source of reassurance for subcontractors where a main contractor becomes insolvent during the course of the work.

12.5.3 Project bank accounts

In 2009, the Government Construction Board introduced project bank accounts (PBA) as the primary vehicle for payments for public sector construction projects. The PBA documentation (*Briefing* and *Guide to Implementation*) were published in 2012 (Cabinet Office and Infrastructure and Projects Authority 2012). This provides an efficient and fairer payment system for long and complex supply chains where all payments are made through the PBA and received within five days or before the due date. The accounts are usually set up in the name of the contractor or in joint names with the employer/employer. The account is set up as a trust with members of the supply chains, such as subcontractors and suppliers, as beneficiaries. As a trust, it serves as insurance/protection against the insolvency of the contractor.

In Australia, the government of New South Wales is proposing to use a scheme of retention money trust accounts for their projects, though it does not offer the fully-fledged payment benefits offered by PBA. However, the situation differs from state to state and Western Australia have trialled PBAs.

12.5.4 Compulsory competitive tendering

The philosophy behind compulsory competitive tendering (CCT) is that if market forces are allowed to operate, then services can be provided with greater efficiency and at a lower cost. Government accounting and purchasing policies have made it clear that value for money and not the lowest price should be the aim. CCT was intended to lead towards better managed, more innovative and more responsive services. However, some argue that if the Transfer of Undertakings Regulations (TUPE) apply, then many of the opportunities for cost savings could be lost. The provision of publicly funded services through CCT has been growing in recent years around the world. It remains highly controversial in Australia and in the industrialised areas of Europe and the USA. It raises fundamental questions about competition and ownership in the provision of such services. Some of the more problematic issues with policy implementation are:

- fair and effective competition
- incentive compatibility
- performance monitoring
- whether CCT provides the best value for money.

The preliminary assessments of contracting suggest generally successful outcomes. Empirical evidence at the present time suggests that efficiency gains have been made, and effectiveness and quality of service have been maintained, if not enhanced. Few professional consultants, who come within this directive, are likely to admit openly that they have reduced their services because of professional fees. However, a survey by the Association of Consulting Engineers found that less time, resources and consideration were given to projects where fee competition was used.

12.5.5 Reverse auction tendering

Under reverse auction tendering (aptly abbreviated as RAT), contractors bid against each other in a live telephone auction to offer the lowest possible price or the best value for money. Bidders remain anonymous, with their bids relayed through an auction assistant to the auctioneer, who acts for the employer. The aim is to do away with the, often unfair, practice of one-off, sealed bids, and to offer contractors a chance to lower their bids against their competitors. Whilst a number of employer groups are considering piloting this idea, contractors are understandably much less enthusiastic. Contractors are raising issues of confidentiality, cartels, inequitable pricing and intellectual property rights. It is important that employers obtain the best possible price. However, if such a system resulted in contractors bidding too low to obtain work, this might have repercussions in terms of disputes arising or even more liquidation amongst construction firms and their suppliers, which would benefit no one.

12.5.6 Appointment of specialist firms

The traditional arrangement of appointing specialist firms on a construction project is to use one of the nomination procedures. Whilst many specialist firms would like this procedure to be extended, it has been estimated that as few as 11 per cent of specialist engineering contractors are appointed in this way. Alternative methods are available:

- *Joint venture*: this is a particularly helpful approach where there is a large engineering services input to the project. The joint venture arrangement is between the main contractor, who may typically carry out the role of a project manager, and the specialist contractor. The companies work together as a joint company. It is therefore suitable for design and build arrangements.
- *Separate contracts*: the employer in this case lets individual contracts to different firms, i.e. the main contractor and the specialist contractor. In the past, this has not been easy to administer, particularly where problems have arisen.
- *Management and construction management*: this is believed to be the most effective way of dealing with such firms. The different trade and specialist firms are appointed and a

contractual arrangement formed with each company. This arrangement allows for full participation by the firms in design and commercial decisions at an early date.
- *Appointing a specialist firm as the main contractor*: where the specialist work represents the largest portion of the project, then employers may choose to reverse the arrangements and appoint a specialist firm as the main contractor. The more usual construction trades would then be employed by this firm.

12.5.7 ***Quality assurance***

Every employer in the construction industry has the right to assume a standard of quality that has been specified for the project. The Building Research Establishment has reported that defects or failures in design and construction cost the industry and its employers more than £1 billion per year. This represents 2 per cent of total turnover. The construction industry is an industry in which:

- there has never been a requirement for the workforce to be formally qualified, and skills are generally developed through time serving
- much of the work is carried out by subcontractors in a climate in which some fifty firms come into existence every day and a similar number go into liquidation or bankruptcy
- there is a paucity of research and development involving new materials, designs and techniques
- there is often poor management and supervision – studies in the UK have indicated that about:
    - 50 per cent of faults originate in the design office
    - 30 per cent on-site
    - 20 per cent in the manufacture of materials and components.

An investment in quality assurance methods can therefore reap substantial long-term benefits by helping to reduce such faults, the inevitable delays, the costs of repairs and the all too frequent legal costs that often follow.

ISO 9001 certification has been increasingly taken up within the construction industry by consultants and contractors. Quality assurance is therefore seen as a good thing for the industry. Procurement methods that fail to address this issue adequately are not doing the industry or its employers any favours. The use of quality-certified firms, who have been independently assessed and registered, therefore offers some protection within the context of getting it right first time. Work that has to be rectified is rarely as good as work that was right in the first place.

ISO 9001 is seen by some as an additional expense and an unnecessary overhead. Also, whilst it should ensure that quality standards are achieved, it does not ensure that the appropriate quality has already been set in the first place. However, some employers are now refusing to employ consultants, contractors or suppliers who do not have the relevant Kitemark. Within the total quality management scenario, quality remains an ongoing process of *continuous quality improvement*. Quality must be appropriate to the work being performed. It should only be insisted upon where it adds value to the finished construction project.

The system being employed by the main contractor for the management of quality must incorporate the integrated quality management activities of the various members of the supply chain. This is especially important in respect of the quality assurance of construction projects, as frequently the main contractor often outsources work to the various members of the supply chain.

12.5.8 Latent defects liability and BUILD insurance

The Construction Sponsorship Directorate of the DETR has suggested compulsory latent defects insurance, or Building Users' Insurance against Latent Defects (BUILD). Employers frequently look for some degree of protection against:

- the risk of latent defects
- the costs of remedying any defects
- the cost of any damage caused by defects, including:
  - loss of rents – loss of profit
  - other consequential losses.

The physical complexity of the construction process and the integration of its constituent parts, the number of (temporary) teams involved, the sums of money involved and the complex tangle of potential legal and financial responsibilities of the different parties involved all require greater clarity. The issues arising from the consideration on latent defects liability appertain to:

- joint and several liability
- limitation periods and prescription of actions
- transfer of employers' rights.

Difficulties exist where more than one party is involved, hence the desire by some employers for a single point of responsibility for the project. The proposal in the consultative document is that the period of liability should commence from the date of practical completion of the works. This might appear unfair to those specialist firms who have completed their portion of the project, perhaps several years earlier on a large project.

The provisions of the Limitation Act 1980 and the Latent Damage Act 1986 should be brought into line with the provisions of the Consumer Protection Act 1987, which provides for a ten-year limitation period. This should apply for all future new commercial, retail and industrial building works in both the public and private sectors.

The doctrine of privity of contract means that, as a general rule, a contract cannot confer rights or impose obligations arising under it on any person except the parties to it. The tenants and the subsequent owners have no privity of contract with contractors or consultants. In July 1991, a House of Lords judgment (*Murphy* v. *Brentwood*) closed down the law of negligence as a route for recovery of economic loss, except in exceptional cases. Since then, owners or tenants without privity of contract have used other contractual techniques, such as collateral warranties, to create contractual rights where none would otherwise have existed.

The BUILD insurance would be financed by the employer at, currently, about 1 per cent of the contract cost. Such policies normally cover the structure, foundations and the weather shield envelope. They often exclude engineering services, which is seen as a weakness. The possibility of mechanical or electrical failure could be provided for an additional premium. Policies could also cover loss of rental or extra rental expenditure.

12.5.9 Added value

The Latham Report, published in 1994, recommended that initial construction costs should be able to be reduced by 30 per cent by the early part of the twenty-first century. Such cost reductions should not reduce quality, but at least maintain it and preferably improve it overall. The

Box 12.3 Adding value in construction

- Reduce the amount of changes to the design.
- Optimise specifications.
- Improve design cost-effectiveness.
- Apportion risk efficiently.
- Improve productivity.
- Reduce waste.
- Examine cost-efficient procurement arrangements.
- Improve the use of high technology for both design and construction.
- Reduce government stop-go policies.
- Develop more off-site activities.
- Standardise components.
- Consider construction as a manufacturing process.
- Get it right first time, i.e. avoid defects.
- Make better use of mechanisation.
- Improve the education and training of operatives.
- Reduce staff, noting their high costs and their reduction in manufacturing industry.

implication is one of adding value, a principle of doing more with less. It is essential that such reductions in cost do not refocus the industry backwards fifty years, towards the emphasis on initial costs alone. The importance of ensuring that whole life costs are given their rightful importance in the overall building process must be maintained. The principle is one of changing cultures and attitudes, and benefiting from changes that have already taken place in other industries. Box 12.3 shows some of the areas of possible investigation in attempting to meet this aim.

The benchmarking of good practices in the procurement of construction projects is one way of achieving added value. The use of benchmarking techniques has gathered a wide interest in the UK (Construction Task Force), the USA (Construction Industry Institute) and Australia. Evidence for the use of benchmarking has arisen from its use in the manufacture of motor cars. Toyota were identified as the best after several years of benchmarking car production activity. This was done by the International Motor Vehicle Programme based at the Massachusetts Institute of Technology.

12.5.10 Private finance initiative (PFI)

The purpose of the PFI was to encourage partnerships between the public and private sectors in the provision of public services. The scheme is outlined in a DETR report that was published in 1993. In 1992 the then Chancellor of the Exchequer announced a new initiative to find ways of mobilising the private sector to meet needs that had traditionally been met by the public sector. Achieving an increase in private sector investment would mean that more projects could be undertaken. This also takes into account that public spending should decline in the medium term. The broad aims of such a partnership will be to:

- achieve objectives and deliver outputs effectively
- use public money to best effect
- respond positively to private sector ideas.

In exploring the possibilities for the use of private finance, including proposals from the private sector, consider these questions:

- Can the project be financially free-standing?
- Is it suitable for a joint venture?
- Is there potential for leasing agreements?
- Is there potential for government to buy a service from the private sector?
- Can two or more of these elements be brought together in combination in any particular instance to form innovative solutions?

Concessionary contracting falls neatly into such an arrangement, whereby the private sector is encouraged to construct public projects such as roads and then charge a levy on this provision for a fixed period of time specified in the contract. Throughout the entire period, the contractor is responsible for the maintenance of the works. Upon handover to the public sector, the contract will also specify the required condition of the asset (see Chapter 14).

12.5.11 Competitive advantage

The key contribution for shaping and reshaping the thinking in the current context of world economies is described by Porter (1980). In competitive advantage, the rationale is not directed towards organisational structures or change, but has profitability as the strategic driver. Porter argues there are five competitive forces that determine profitability:

- Potential of new entrants into the industry.
- Bargaining power of customers.
- Threat of substitute products.
- Bargaining powers of suppliers.
- Activities of existing competitors.

One of the most important concepts established by Porter is the *value chain*. This is a systematic way of examining all the activities a firm performs and how they interact. Primary activities are inbound logistics, outbound logistics, marketing, sales and service. Support activities include procurement, technology development, human resource management and infrastructure. The way in which one activity in the chain interacts with another can be crucial. This can occur within the organisation, or externally with suppliers. Porter argues that a firm gains competitive advantage by performing these activities, alone or linked, more economically or in a better way than its competitors.

12.5.12 Business process re-engineering

The idea of business process re-engineering is to learn as much as possible from other industries who have had to respond to massive cultural changes. Improvement or added value can be achieved through construction re-engineering. No other concept has recently received more interest and criticism than re-engineering. This is because it is a concept that is easy to understand, but difficult to put into practice. The successful re-engineering projects are founded on six basic principles:

- Organise around outcomes not tasks.
- Have those who produce the output of the process perform the process.

- Subsume information processing into the real work that produces the information.
- Treat geographically dispersed resources as though they were centralised.
- Link parallel activities instead of integrating their tasks.
- Put the decision point where work is performed, and build control into the process.

In spite of a few well-known successes, there is much evidence to suggest that re-engineering fails, or at best produces only marginal results, in the majority of organisations where it is implemented. Sometimes, this is because the programmes are not sufficiently radical and only tinker with the most easily accessible processes.

12.6 Carbon pricing

It is now becoming very important to establish the carbon footprint of a construction project. This is aimed at minimising the carbon emissions involved. The carbon footprints of buildings may be invisible, but they need to be considered when designing for sustainability, which is currently a hot topic internationally. Achieving international agreement is therefore vital.

The establishment of a carbon price is one of the most powerful mechanisms available to reduce greenhouse gas emissions. Much of these emissions comes from the mining and manufacture of building materials and components, construction work on-site and the use of buildings and other structures by their owners. The European Union has set the goal of achieving a 20 per cent reduction in carbon emissions by 2020. It hopes to lead the way through the development of green products across all industries, including the construction industry, using carbon pricing to encourage a reduction in CO_2 emissions.

A global benchmark of CO_2 emissions is an urgent challenge for international collective action by governments around the world. A global approach can, in theory, be created through internationally harmonised taxation or inter-governmental collaboration, but neither is straightforward in practice. It may be necessary to set caps across countries; should the cap be exceeded, then carbon monitoring will take place and the government responsible will be financially penalised.

12.6.1 What is a carbon price?

This is an administrative approach imposing a cost on the emission of greenhouse gases that contribute to climate change. Paying a price for carbon spewed into the atmosphere is a way of motivating countries, businesses and individuals to reduce their carbon emissions. It also provides an incentive to invest, and deploy technology that does not emit carbon into the atmosphere. Such a pricing mechanism also acts as a disincentive for electricity generating companies to use relatively more-polluting coal-, gas- and oil-fired power stations.

For a carbon pricing policy to be effective, it is important to have significant government support for the concept and practice. It is also important to encourage the innovation of new technologies that will reduce CO_2 outputs.

A carbon price is needed to stabilise global greenhouse gas concentrations at levels that limit the risk of severe future climate change damage. Annual global emissions will need to be reduced substantially in the coming decades. Paying a price for carbon emissions will slow the output.

To reduce greenhouse gas emissions substantially, a combination of measures needs to be introduced, including carbon pricing, directed government funding of alternatives, regulations and standards, organisational change, education and information. Emissions trading schemes should not be seen as either a licence to pollute or a means of unnecessarily prolonging the world's dependence on fossil fuels. To have a significant impact, climate policies will need to fundamentally change the basis of our fossil-fuelled economies.

Climate change policy that will take us away from fossil-based fuel will cost us more, but in the long term benefit us more, and will require more changes in behaviour by firms and individuals than any other environmental policy. The magnitude of this challenge has drawn attention to the potential use of market-based or economic-incentive instruments to ensure that polluters face direct cost incentives to mitigate emissions at the lowest possible cost.

The imposition of a carbon tax is expected to increase the cost of electricity and is very likely to reduce the competitiveness of industries that are energy intensive. This reduction in competitiveness can result in negative economic and environmental outcomes. For instance, companies may transfer their facilities to places without climate change policies, which will increase emissions at these new locations and, in the process, bypass many of the benefits of the cap and trade or tax policy.

12.7 Risk analysis and management

Construction projects are full of risks, including those that may relate to external commercial factors, design, construction and operation. The construction industry will always be a risk business, owing to the intrinsic nature of its activities. Some of the reasons for this include:

- incomplete design and investigation at tender stage
- high levels of contract disputes and disagreement
- requirement to complete the works on the employer's premises
- low profit margins
- need to complete works within short timescales
- bespoke nature of projects
- competition from other firms and increasingly within an international dimension
- budgets that often result in cost overruns.

It is necessary to distinguish between risk and uncertainty. Risk arises when the assessment of the probability of a particular event is statistically measurable. It differs from uncertainty, which cannot be mathematically predicted. Risk relies upon the availability of previous, known events. Some risks represent a feature of most construction projects. Other risks may only occur on some projects or may arise during construction operations. Uncertainty occurs where there is no data on previous performance. Uncertainty arises in one of two ways. In the first case it can arise because it can be imagined or anticipated. Activities, for example, that involve the use of new materials or construction techniques may result in uncertain consequences. Some of these can be contemplated. However, because they are untried and tested in practice, they may have outcomes that have not been fully investigated beforehand. In other examples, the risks may be due to events whose cause and effect cannot be imagined or predicted. In dealing with the management of risk, an overemphasis on risk avoidance may lead to over caution.

The management of risk is a practice that all of us use on a regular and routine basis. However, the complexity and scale of most construction projects is such that good risk management in the construction industry requires more than pure common sense and instinct. An increasing number of companies advocate the promotion and benefits of risk management. The success of risk management is supported by the growing list of employers using it within the UK. Risk management is now understood to be an important factor, critical to the success of projects by providing a method with which to improve value for money in construction.

In principle, risk management is a straightforward process that requires the evaluation of risk and the execution of a risk management strategy. The assessment of risk first entails *risk identification*, followed by the *analysis of risks* identified. This imparts a level of understanding required to facilitate the adoption of a suitable *risk management response*.

12.7.1 Risk identification

The customary method of carrying out a risk analysis is by utilising a workshop at which participants 'brainstorm' risks that they consider could have an effect upon a project. The workshop forum, which brings together specialists from a variety of relevant disciplines, promotes a wide project viewpoint, which, if managed well, will lead to meaningful debate and communication. This should be considered as an exercise that is beneficial in itself. Brainstorming activity is not the only approach to risk identification. Historical data may be used, possibly using the experience of the participants' records, formal or otherwise.

The use of checklists may assist in providing structure to the thought processes used. An example of such a checklist is provided in Table 12.1. The examples of categories given in the table incorporate a large range of risks; some categories are particularly wide in their potential scope. There is a danger with checklists that their use may limit deliberation to those categories contained in the list, and it should be borne in mind that this could result in ruling out some major and possibly significant items.

The success of the risk identification process will be reliant upon several factors, including: the level of experience and ability of the personnel concerned with the workshop; the amount of data readily available; the skill and experience of the analyst or facilitator; the time available; and the timing of the workshop. It is important to realise that the process of risk identification is not likely to result in the discovery of all possible risks.

Table 12.1 Checklist of risks categories

Risk category	*Indicative examples*
Physical	The collapse of the sides of a trench excavation, resulting in delays, additional costs and injury.
Disputes	The disruption to a third party's business due to noise or construction traffic, resulting in financial loss or litigation.
Price	Increases in inflation, causing excessive financial loss. It might be easy to ignore low inflation in the UK, but not the rates in some countries around the world.
Payment	Delayed payments by the main contractor to nominated subcontractors, causing programme delays.
Supervision	Delays in issuing drawings or instructions to a contractor, resulting in abortive work and claims from the contractor.
Materials	Delays in the dispatch of unique flooring materials, resulting in delays to the contract programme.
Labour	The shortage of labour due to the construction of another nearby project.
Design	Anomalies in the design due to poor communication between engineering consultants, resulting in abortive work.

Source: Ashworth and Hogg 2000.

12.7.2 Risk analysis

It is important to be aware that problems in construction do not necessarily restrict themselves to cost, although in due course, all problems may have a cost effect. In numerous situations, time or schedule risk is of more significance than pure cost and, in some cases, quality may be the most important priority. Therefore, it is essential that risk analysis addresses the needs of a given situation and centres upon applicable areas of concern.

There are a range of risk analysis tools that may be used to evaluate the identified risks. The choice of the most appropriate approach will depend upon project size, type and opportunity, and they may be categorised as qualitative, semi-quantitative and quantitative.

12.7.3 Risk management

Following the identification and evaluation of the risk, the way in which the risks should be managed needs to be determined. The successful management of risk requires:

- focus upon the most significant risks
- consideration of the various risk management options
- understanding effective risk allocation
- appreciating the factors that may have an impact upon a party's willingness to accept risk
- appreciating the response of a party if and when a risk eventuates.

Whilst there are doubts about the need to artificially restrict the number of risks to be actively managed, it will be clearly advantageous to give attention to those risks which are considered to be high impact and/or high probability.

There are only a small number of risk management options available for consideration, which is helpful in simplifying the process. These may be categorised as follows:

- A risk can be *shrunk* or reduced by, for example, establishing more and better information about an unknown situation.
- A risk can be *accepted* by a party as unavoidable, and any alternative strategy may be considered as being inefficient or impossible to adopt.
- A risk may be *distributed* to another party; for example, contractors usually distribute construction risk by selecting reputable subcontractors to carry out the work.
- A risk may be *eliminated* by the rejection of a project or by the rejection of a particular part of the proposed works.

It is important to recognise that when risk management action is taken, in each case, including that of the elimination option, secondary risks should also be considered. When taking into consideration the allocation of risk to another party, the following factors should be considered:

- The ability of the party to manage the risk.
- The ability of the party to bear the risk if it eventuates.
- The effect that the risk allocation will have upon the motivation of the recipient.
- The cost of the risk transfer.

There are many examples of inappropriate risk allocation within the construction industry that occur from the strong desire to minimise risk exposure at all costs.

12.7.4 Willingness of a party to accept risk

The readiness with which a party may be prepared to accept a risk will depend upon several key factors:

- *Attitude to risk*: a party who is risk averse is someone who is less willing to accept risk than someone risk seeking.
- *Perception of risk*: a party who has recently experienced a serious injury on a construction site is quite likely to perceive the probability of a similar occurrence on a new project more highly than someone without the experience.
- *Ability to manage risk*: a party unable to manage a risk due to lack of resources or experience should be less willing to accept a risk than someone with the necessary expertise. However, in practice this may not always be the case.
- *Ability to bear risk*: a party unable to bear a risk due to the lack of necessary financial back up should be unwilling to accept risk.
- *The need to obtain work*: a party in need of work is more willing to accept risk as a necessary means of business survival. Risk acceptance is therefore market sensitive.

12.8 Other industry comparisons

It is always relevant, when examining a subject like procurement, to see how it is done elsewhere. This comparison may be made against similar or competing firms, perhaps in the form of a benchmarking study. Alternatively, comparisons can be made with firms or organisations overseas, in countries that mirror UK practices and in countries where different traditions are employed. It is also important to consider other industry comparisons – as illustrated in the Latham Report, which compared the performance of the construction industry with that of the motor car industry. Table 19.1, in Chapter 19, is an adaptation of that comparison. Other comparisons have been made with the aerospace industry (Flanagan 1999). The outcome of such studies acts as a guide to good practices found elsewhere that might have been overlooked. Current comparisons do not place the construction industry in a good light, but do act as motivators to help change the culture of the construction industry.

The vehicle manufacture and aerospace sectors include the following attributes that are generally absent from the construction industry:

- The recognition of a manufacturing culture.
- The integration of design with production.
- The importance of the supply chain network.
- A focus on innovation; that this will only be secured through adequate research and development.
- An acceptance of standardisation in design, components and assembly across the product range.

13 Lean construction

The lean construction process is a derivative of the lean manufacturing process. This has been a concept popularised since the early 1980s in the manufacturing sector. The original thinking was developed from Japan, although it is now being considered and introduced worldwide. It is concerned with the elimination of waste activities and processes that create no added value. It is about doing more for less.

Lean production is the generic version of the Toyota Production System. This system is recognised as the most efficient in the world today. (Incidentally, Toyota's activities in the construction industry are larger than those of its more well-known automobile business!) It needs to be acknowledged that construction production is different from that of making motor cars, although it is possible to learn from and adapt successful methodologies from other industries. In the automobile manufacturing industry, spectacular advances in productivity, quality and cost reduction have been achieved in the past ten to fifteen years. Construction, by comparison, has not yet made these advances. It also remains the most fragmented of all industries, although this can be seen both as a strength and a weakness.

The application of lean production techniques in automobile manufacturing has been a huge success and is associated with three important factors:

- The simplification of manufacturing dies.
- The development of long-term relationships with a small number of suppliers, to allow just-in-time management.
- Changes in work practices, i.e. the culture and ethos of practices, most notably the introduction of teamworking and quality circles.

Lean thinking is aimed at delivering what clients want, on time and with zero defects. Lean construction has identified poor design information, which results in a large amount of redesign work. Several organisations around the world have established themselves as centres for lean construction development. The aim, for example, of the Lean Construction Institute in the USA is a dedication towards eliminating waste and increasing value.

Few products or services are provided by a single organisation alone. The elimination of waste therefore has to be pursued throughout the whole value stream, including all who make any contribution to the process. Removing wasted effort represents the biggest opportunity for performance improvement.

Several companies around the world are attempting to introduce lean construction methods into their core businesses. *Rethinking Construction* (Egan 1998) provides two examples and these are briefly discussed later in this chapter. One of the firms is based in Colorado and the other in San Francisco. One of them has already reduced project times and costs by 30 per cent through developments such as:

- improving the flow of work on-site
- using dedicated design teams
- innovation in design and assembly
- supporting subcontractors in developing tools for improving processes.

This suggests that perhaps the most useful way of achieving cost reductions, whilst still maintaining value, is to consider profitable ways of reducing the time spent on construction work on-site. Design readiness is the same principle. This suggests that to fully complete the design prior to starting work on-site will save both construction time and the respective costs that are involved (see Figure 8.5).

Lean construction is a philosophy that is about managing and improving the construction process to profitably deliver what the customer, the construction employer, requires. Engineering, in all its different kinds, has had to develop a wide range of strategies to remain at the competitive edge and improve its products. Comparisons have been made on several occasions between the high technology engineering approach and the low technology that is adopted generally throughout the construction industry sector.

Because it is a philosophy, lean construction can be pursued through a number of different approaches. The lean principles have been identified as follows:

- Eliminate all kinds of waste; this includes not just the waste of materials on-site, but all aspects, functions or activities that do not add value to the project.
- Precisely specify value from the perspective of the ultimate customer.
- Clearly identify the process that delivers what the customer values; sometimes referred to as the value stream.
- Eliminate all non-added-value steps or stages in the process.
- Make the remaining added-value steps flow without interruption, through managing the interfaces between the different steps.
- Let the customer pull, do not make anything until it is needed, then make it quickly. Adopt the philosophy of *just-in-time management* to reduce stockpiles and storage costs.
- Pursue perfection through continuous improvement.

13.1 Lean manufacturing

Lean manufacturing was adopted by the large Japanese car manufacturers and has been implemented by a number of Japanese, American and European manufacturers with some considerable success, particularly in the automotive and other engineering industries.

The lean process is about designing, constructing and operating the right systems to deliver the right product first time. Essential to this is the elimination of what the industry describes as *snagging* work, i.e. the remedying of defective work prior to handing over to the employer or client. There have been examples quoted in the construction industry where snagging work has taken up almost as much time as doing the work in the first place. Activities or processes that absorb resources but create no additional value must be eliminated. This waste can include:

- mistakes
- working out of sequence

- redundant activities and movement
- delayed or premature inputs that are the result of bad programming
- products or services that do not meet a customer's needs or requirements
- non-conformance to specified standards or quality.

The primary focus of lean thinking is moving closer and closer to providing a product that customers really desire, and, by understanding the processes involved in construction manufacture, identifying and eliminating the waste that is normally generated.

13.2 Production and management principles

Lean thinking is focused on value rather than on cost. However, whilst cost is not unimportant, the emphasis has been switched to adding value across the whole range of services and processes that are used in the construction of buildings. It seeks to remove all components that do not add value, especially the various processes involved, whilst improving components that do add value.

Its aim is to define value in customer terms, identifying key points in the development and production process, where that value can be added or enhanced. The goal is the seamless, integrated process or value stream wherein products flow from one added-value step to another. This is all driven by the philosophy and the pull of the customer (the employer).

The idea of getting it right first time is fundamental to the process of the lean philosophy. In this context, right means making it so that it does not require any rectification at a later stage. The approach involves an extremely rigorous analysis of every detail of product development and production, and seeks to identify the ultimate source of problems. Only by eliminating the source is it possible to prevent a fault recurring.

Production management techniques have traditionally focused on the need to schedule discrete activities in the building process. This is in contrast to adopting a philosophy of seeing it as a manufacturing process involving the management of resources across a network of firms. This perspective has been increasingly criticised and there is already a growing body of research on supply chain management techniques in the construction industry. These include buildability, just-in-time management and lean construction practices. Production management techniques have also focused too much on the process rather than the eventual product that is produced.

13.3 Design and product development

Extensive product development work has been undertaken in the engineering industry in recent years, in order to maintain its viability and feasibility wherever possible. This industry has responded positively to the demands placed upon it by adding value to its products. Lean manufacturers, regardless of the industry, have developed systems for product development which first identify the right product to be made, in terms of customer needs and expectations. The product is then designed correctly in order that it can be manufactured efficiently. For example, where an analysis of architectural detailing is carried out, simpler and less costly solutions can often be found without any detrimental effect to the design itself.

Design in manufacturing terms is concerned with the development and integration of systems and components into coherent, efficient and manufacturable products. The construction industry has had a considerable interest in buildability solutions for more than twenty-five years. There is some evidence to suggest that this aspect is now considered more frequently than previously, although the separation between the design and the construction professions remains, and probably remains at an overall disadvantage. In design education, there is too little emphasis given to manufacturing problems, and an insufficient consideration of design methodologies on courses concerned with building production management.

Tools have been developed to capture and analyse customer perceptions and requirements for product quality and performance. These tools enable product development and manufacturing performance targets to be established. Design development targets include reductions in design changes and process iterations. The critical success factors can be summarised as follows:

- Design is informed by extensive data on the performance of products, systems and components.
- Carry-over to new models of a high proportion of systems and components from previous successful models.
- Front-loading of resources towards design to prevent problems during construction manufacture.
- Concurrent working between manufacturers and suppliers during design development.

13.4 Lean production

Lean manufacturers arrange production in closely located cells, so that work flows continuously, with each step adding more value to the product. The standard time for all activities is known and the objective is to totally eliminate all stoppages or delays throughout the entire production process. Only minimum stocks of materials are kept, as buffers between processing stages. The efficient application of materials handling, through the use of information technology, was pioneered for stock control purposes in the early 1970s.

For the system to be effective, every machine and worker must be completely capable of producing repeatable perfect quality output at the exact time required. Employees are responsible for checking quality as the product is assembled, and in some cases given the authority to halt production should defects arise. In such a system, quality problems are exposed and rectified as soon as they occur.

The workforce is kept informed of progress towards production and cost targets. Information displays are provided in order that everyone can see the status of operations at all times. The work teams in lean manufacturing are highly trained and multi-skilled. In some cases, the supervisory and management functions have been devolved to them. The critical success factors include:

- in-depth understanding of production processes and the resources required
- responsibility and authority placed with the workforce
- real-time feedback on performance
- training and multi-skilling.

This last point is especially important. The value of appropriate training cannot be overemphasised. Without the adequacy of this investment, the whole philosophy has a danger of not becoming a reality. The construction industry is frequently seen as low technology and has not changed to meet aspirations or practices that are seen elsewhere.

The use of lean production techniques must be placed within the context of the construction industry. The comparison between mass production in factories with generally bespoke buildings on a construction site must not be minimised. Furthermore, lean manufacturing appears to achieve the greatest improvements in efficiency and quality when design and manufacture occur in close proximity. With traditional procurement arrangements in the construction industry, this is frequently not the case.

13.5 Supply chain management

Lean manufacturing is based upon the elimination of waste process and practices. This includes the time waiting for others to complete their tasks, the delay caused through the lack of late deliveries, unnecessary storage and the value that is tied up in large stocks or parts awaiting assembly. Just-in-time delivery is an important concept of the process of lean construction. Lean construction firms have therefore had to develop their own reliable network of suppliers and subcontractors. Significant efforts need to be made to encourage these firms to adopt the same principles and systems. The fragmented nature of the construction industry makes this point of even greater importance and significance.

The adoption of lean manufacturing in engineering has had the effect of moving away from traditional relationships with suppliers towards partnering arrangements with a fewer number of firms based upon good communications and open-book accounting. These arrangements work for both parties, sharing the philosophy of continuous improvement. This is especially the case in the area of defect reduction and the cost and timeliness of deliveries. There is a sharing of business and development strategies sufficient for both parties to know enough about each other in order to make forward planning effective. The critical success factors include:

- the lack of reliance on formal contracts
- the use of benchmarking of suppliers' performance against each other on a range of generic criteria.

13.6 Benchmarking

Benchmarking has been described as a method for organisational improvement that involves continuous and systematic evaluation of products, services and processes of organisations that are recognised as representing best practice. It is a system that uses objective comparisons of both processes and products. It may make internal comparisons within a single firm, perhaps by comparing the performances of different building sites. It can be an external system, where comparisons of performance are made against similar and dissimilar firms. The enterprises that are internationally recognised as world leaders are described as having the best practices and are the most efficient. The Construction Industry Board in the UK has published a list of key performance indicators (KPIs) that are updated and can be used by any firm to measure its performance.

Benchmarking is an integral part of lean construction methods, since they seek to identify where improvements in processes or products are possible. There is considerable interest in, and practice of, benchmarking techniques across a wide range of industries. It has been suggested that probably all of the top firms and companies use them in one way or another. It is also worth remembering that benchmarking as a tool for maintaining competitive edge is not just practised by those firms that are lagging behind and need to improve their performance, but also by the international world leaders. The focus of benchmarking is on the need for continuous improvement.

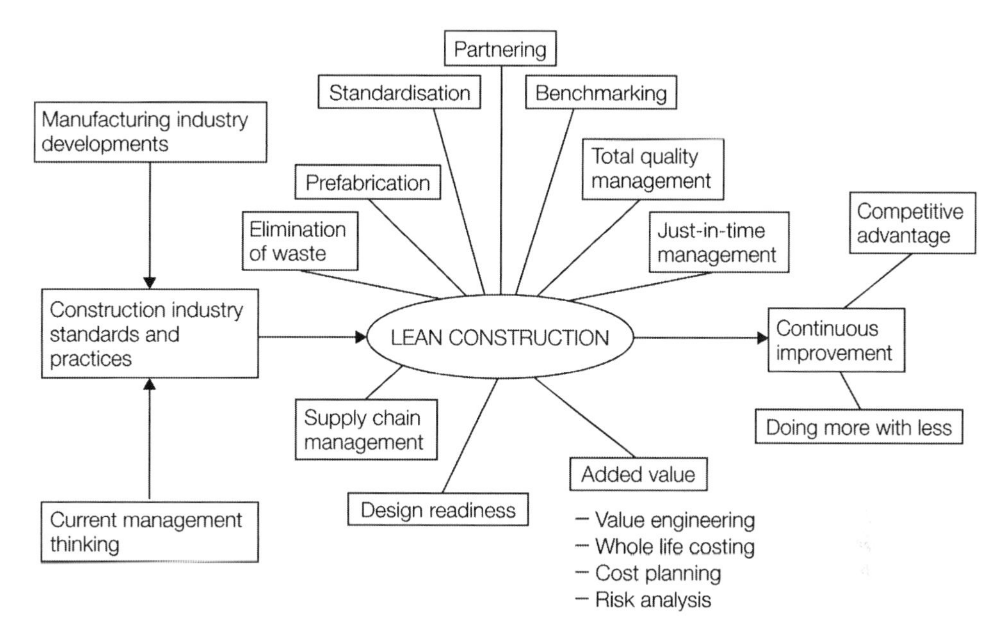

Figure 13.1 Lean construction.

13.7 Lean thinking construction

The lean principles that have been used effectively in engineering manufacturing can only be applied effectively in construction through focus on improving the whole design and construction process. This requires a commitment from all the parties involved, including the employer. The obstacles that may arise through traditional contractual arrangements can then be removed. Figure 13.1 is a schematic description of lean construction. Here are some issues to consider.

13.7.1 Design

- Use visualisation techniques such as three-dimensional computer-aided design and virtual reality. These will help to promote product definition and provide a clearer perspective and understanding on the part of the employer.
- Apply value management to achieve a greater understanding of those aspects that employers value.
- Use integrated design and build arrangements, including partnering.
- Encourage greater cooperation between designers, constructors and specialist suppliers.
- Design for greater standardisation.
- Use prefabrication or pre-assembly of components in order to achieve higher quality, and cost and time savings.

13.7.2 Procurement

- Incorporate a supply chain management process.
- Rationalise and integrate the number of suppliers who are involved.
- Provide a seamless integrated process.

- Eliminate waste in the procurement process.
- Ensure that decisions on customer value can be taken.
- Ensure confidentiality of construction costs and cash flows.
- Introduce the concept of partnering with common goals where the boundaries between companies and firms become less critical.

13.7.3 Production

- Introduce benchmarking practices to establish best-in-class production methods and outputs.
- Establish a stable project programme with a clear identification of a critical path.
- Apply risk management techniques to manage risk throughout the project.

13.7.4 Logistics

- Adopt just-in-time delivery of materials to the point of use to eliminate on-site storage and double-handling.

13.7.5 Construction

- Ensure clear communication of project plans to the whole team.
- Provide relevant and ongoing training.
- Encourage a teamwork approach.
- Develop multi-skilling.
- Provide daily reporting and improvement meetings.
- Develop a well-trained, highly motivated, flexible and fully engaged workforce.

13.8 Case studies

13.8.1 Pacific Contracting of San Francisco

This is a specialist cladding and roofing contractor. It used the principles of lean construction to increase its annual turnover by 20 per cent within an eighteen-month period, using only the same number of staff. The key to its success was improvement of the design and procurement processes in order to facilitate construction on-site. It did this by investing in the front end of projects to reduce costs and construction time. The firm identified two major problems in achieving a flow of the whole construction process:

- Inefficient supply of materials, which prevented site operations from flowing smoothly.
- Poor design information from the prime contractor, which frequently resulted in a large amount of redesign work.

To tackle these problems, Pacific Contracting combined more efficient use of technology with tools for improving construction processes planning. The planning tool, known as Last Planner, improved the flow of work on-site by reducing constraints such as lack of materials and labour. Their computerised three-dimensional system provided a better, faster method of redesign that lead to better construction information. Their design system now provides a range of benefits, including isometric drawings of components and interfaces, fit coordination, planning of construction methods, motivation of work crews through visualisation, first run of tests of construction sequences and virtual walk-throughs of the product. They also used a planning tool, known as last planner, to improve the flow of work on-site through reducing constraints such as lack of materials and labour.

13.8.2 Neenan Company, Colorado

This is a design and build firm and is one of the fastest-growing construction companies in its region. The firm has worked to understand the principles of lean thinking and to look for applications in its business. It has used *study action teams* of employees to rethink the way they work. This firm has reduced project times and costs by up to 30 per cent, through developments such as:

- Improving the flow of work on-site by defining units of production and using tools such as visual control of processes.
- Using dedicated design teams working exclusively on one design from beginning to end.
- Developing a tool known as *schematic design in a day* to dramatically speed up the design process.
- Innovating in design and assembly, for example, through the use of prefabricated brick infill panels manufactured off-site and pre-assembled atrium roofs lifted into position.
- Supporting subcontractors in developing tools for improving processes.

13.8.3 BAA Airports Limited

BAA Airports Limited is one of the world's major airport operators. It is also a key client of the construction industry in the UK and, increasingly, within an international context. Typically, at any one time it has more than 1,000 construction projects in operation and spends more than £1 million per day. Its core business is to operate airports. It has been highly innovative in applying the principles of lean construction to all of its activities, which involved: developing, documenting and implementing a standard construction process (now known as the BAA process); selecting partner firms of designers and constructors for its long-term relationships; implementing lean project processes; and adopting information technology to support, integrate and improve the process and the product.

One of the first objectives was to achieve consistent best practice. This has been achieved through the BAA project process. It is a broad framework used to control all BAA projects. It also allows further development to take place. Lean construction cannot be achieved without enterprise integration across the supply chain. Emphasis is therefore placed on its partnering programme. A framework has been established with suppliers, construction managers and designers to meet this objective. BAA places emphasis on off-site fabrication, using the construction site as an assembly process. The standardisation of components is a key feature. Driving out waste means specifying correctly, i.e. not under- or over-specifying the work that needs to be carried out.

13.9 Conclusion

The vision of lean construction stretches across many traditional boundaries. It challenges our current practices, and has required changes in work practices and in understanding new roles and different responsibilities. It recognises that technology has an important role to play. It has required a change in attitudes and culture, and expects that those involved in a construction project will share a common purpose. It is setting new standards that can be measured to indicate improvement. It is a vision for the future.

The response to lean construction from the construction industry is not unanimous. With many new ideas, some do not always live up to the theoretical expectations when put in to practice, but lean construction is not just a theory, since its techniques have been successfully applied in other industries. However, there must also be the desire for success, so a belief system is very important. Lean construction is not a phenomenon that is likely to disappear, since there is a groundswell of opinion and worldwide interest in its principles.

14 Public private partnerships (PPP) and the private finance initiative (PFI)

14.1 Reasons for change

The construction industry is continually evolving its processes and practices. Sometimes new methods of procurement are very successful. In other cases, although they are initially welcomed by the industry and its clients, they may be found to be flawed and eventually abandoned as an idea. The introduction of PPP was initially in response to the growing levels of public debt in the 1970s and 1980s. New buildings and structures were required, but new approaches to procurement were also essential if such projects were to come to fruition. New ideas were therefore developed. Other countries had similar difficulties, and their solutions were not dissimilar to those employed in the UK. The existing arrangements for the procurement of public sector projects also raised questions of value for money and these questions dated back to the rebuilding of the UK after the Second World War. Whilst PPP was considered to be a solution, as shall be explained later, this would prove not to be a panacea for public sector procurement in general. The protagonists for PPP were often those who had the most to gain. Some also argued that, even with its faults, PPP offered a better solution than other methods that had previously been employed. The construction industry was also changing in shape and direction and this, too, had a bearing on the ways projects were procured. However, all things considered, PPP was considered to be a method that was worthy of use for the replacement of the existing building and infrastructure stock that through use was in decline and in need of updating. Changes in work and leisure practices also meant that many existing buildings were no longer fit for purpose and needed to be replaced.

Wherever possible, governments of different political persuasions around the world have sought to involve the private sector in public sector projects. In some cases, they have sought to do this by not clearly separating recurrent and capital expenditures, and this has enabled them to disguise the true costs of PPP. At one time, some politicians, supported by government officials, suggested that infrastructure could be provided at zero cost to the Treasury. Unfortunately, this concept has proved to be false accounting and flawed, and is now abandoned as a means of improving the built environment. Even so, this has not reduced the interest in this method of procurement for certain types of project. Also, some lessons have been learned so that the use of the private sector for such projects has not been completely abandoned. The private sector continues to accept responsibility for a project, including an allocation of risk, whilst the public sector accepts a role of public accountability for such schemes. It must be accepted that many of the projects that have been constructed under PPP would never have occurred without this method of procurement. It may have been at a high price, but benefits nevertheless have accrued.

Initially, the use of PPP was in connection with single projects. These were negotiated as single one-off arrangements. Prior to PPP, a large proportion of the public sector had already been disposed of through the sell-off of most of the utilities companies, including rail services. As with all actions, varying degrees of success have been achieved. The principles of PPP were a second stage. Whilst the private sector would be integrally involved, the projects completed would still largely be in public ownership and were not privatised in the same way as, for example, electricity or gas. Government felt that, through this approach, the public sector would get a better deal for the money that it spent. Public bodies such as hospitals, prisons and schools could then be given greater freedom to procure new buildings. The initiative was about shifting expenditure from capital to revenue and this would, in terms of the government's own balance sheet, be more palatable than would otherwise be the case. The major driver was that private sector capital could be used when such monies were not available from the public purse. This was a form of creative accounting that suited both public and private sectors.

In 1997, a Labour government came to power in the UK. It persisted with the use of PPP, but attempted to shift the emphasis towards a better achievement of value for money, largely through a change in the way the risks involved were allocated. Value for money is a concept with which we are all familiar, but it is not always easy to properly measure since individuals or organisations all have a different interpretation of what this might mean (see section 14.7); for example, some people fly business class and believe that this represents good value for money; for others it is construed as a waste of money. This change in practice was accompanied by the concept of *best value* in the public sector (for a better understanding of this concept, see Ashworth and Perera 2015).

It should be understood that all developed countries around the world were wrestling with the same problems of how to improve public facilities when the public purse was empty. Australia, for example, developed a similar model to the UK, and there was widespread use of these methodologies throughout Europe and the USA.

Shortly after the Labour government was elected in the UK in 1997, a PPP unit was appointed to provide a central knowledge base for the use of PPP in an attempt to reduce some of the pitfalls that had already been experienced under the previous regime. It was based in the Treasury and became known as the Treasury Task Force (TTF). This unit included civil servants and private sector executives. Procurement methods would be standardised to capture best practices from different government departments, and government staff would be better trained in the best ways of implementing PPP.

In recent years, PPP has run out of favour and had many criticisms levelled upon it through mass media journalistic evaluation, so that political will has deteriorated as a consequence. It has been seen as using a credit card for infrastructure investment, with hugely disadvantageous PPP deals being highlighted, which has subsequently led to the re-evaluation of the PPP landscape in the UK.

14.2 Public private partnerships

A public private partnership (PPP, or P3) is a method of procuring different kinds of government service through a partnership with the private business sector. It has been used extensively in the procurement of construction projects. Some have argued that we would not now have all the new buildings and infrastructure had government relied only on its funds acquired through taxation. Others will point to the fact that, whilst the construction industry has typically suffered from booms and slumps, PPP has helped in smoothing out workloads.

PPP provides the basis for a contract between a public sector organisation and a private party (or parties). The most common form of PPP arrangement in the UK is the PFI (or PF2, see section 14.3). HM Treasury (2008) defines PPP as:

> arrangements typified by joint working between the public and private sectors. In their broadest sense they can cover all types of collaboration across the private–public sector interface involving collaborative working together and risk sharing to deliver policies, services and infrastructure.

Historically, government borrowed its own finance, often at preferential rates, to carry out and complete the replacement of its existing building and infrastructure stock. PPP differs in that the private party is responsible for this funding. In the best examples, the project is delivered free of charge to government and even its maintenance and repair come within this remit. The second River Severn crossing, constructed in 1996, is a good example. The contractor and consultants financed and developed this scheme and then recouped their money through a toll charge on its users. Eventually, the bridge will revert to public ownership. A business consortium called River Crossing plc, led by John Laing plc, was formed to build the new river crossing and this was also used in the management and maintenance of it.

The PPP consortium assumes a considerable financial, technical and operational risk in the development and management of the project. In the River Crossing example, the costs of providing and maintaining the structure are not borne directly by the taxpayer. Charges are recouped directly only from those who choose to use the service. In other examples, the capital investment is provided on the strength of a contract with a government body, or is partially shared by the different parties who are involved. Sometimes, a government contribution to such a scheme may involve the transfer of existing assets, and these may represent the government's part or contribution to the project. In other cases, a government department may provide grants and subsidies for projects that it considers worthy of development but where funds from the more usual sources are not available. This will have the effect of making such a project more financially attractive to potential private investors. In other circumstances, specified tax relief may be made available. Government has a number of options and incentives available to encourage developments that it considers desirable, but cannot afford at the time (see Ashworth and Perera 2015).

14.2.1 Types of PPP

There are many types of PPP arrangements in operation. These can be classified based on the level of risk (responsibility) to the private and public sectors. The more private involvement on projects there is, the more risk allocated to the private sector, and vice versa (Figure 14.1).

The terminology used to describe these arrangements will vary depending upon the service that is provided by the consortium. The consortium will establish a new trading company to deliver the initiation, development, construction, maintenance and operation over a specified period of time. As in all contractual arrangements, these can be changed at any time if the different parties agree. The most common arrangements may be typically referred to as design, build and operate (DBO). Alternatives to this include: build, operate and transfer (BOT); build, own and operate (BOO); design, build, finance and operate (DBOF); build, own, operate and transfer (BOOT). The popularity of these different methods depends to some extent on the fashion that prevails, although changes have been made in light of previous arrangements that have not always worked out as expected. The consortium involved typically comprises a building or civil engineering contractor, a consultancy group, a maintenance company and a finance organisation such as a commercial bank. Professional advisers, such as architects, engineers and quantity surveyors, will also be involved and, in some cases particularly, look after the client's interest. Some arrangements may be for the initial design and construction alone, although for many it includes maintenance and management over a time period. Turnkey

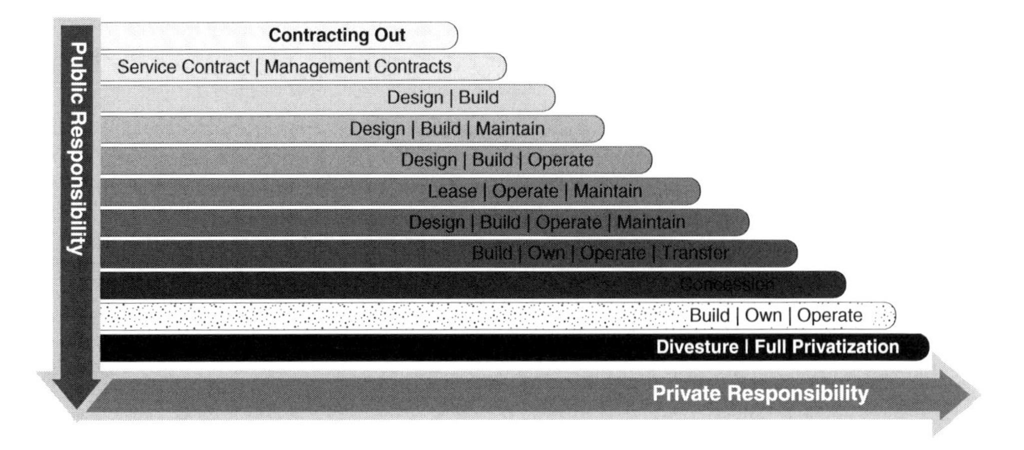

Figure 14.1 PPP models: private and public sector risk.

Source: Colverson and Perera 2012.

contracts (see Chapter 8) have similar characteristics, although these do not usually involve the public sector client as a partner in the process.

PPP has been employed on a wide range of government-sponsored projects, such as hospitals, schools and highways. Universities have also employed these techniques for the development of their own building stock. For example, a hospital wants to extend its facilities by building a new ward block. The development and finance may be provided by a PPP consortium and the project, when completed, then leased to the hospital authorities. The consortium or private developer then acts as a landlord for the building and its associated works and the hospital provides the medical services that were intended. In this way, each of the parties involved uses their own particular expertise. In other examples, a new motorway may be constructed that is needed, but government does not have the necessary funds available to undertake the works. This was used for the M6 toll road constructed in 2002–03 to help relieve congestion on the existing road network in the area. A competition for this project took place in 1989 and was won by Midland Expressway Limited (MEL) in 1991. The contract was for a fifty-three-year concession (three years construction, fifty years operation) to build and operate the road. MEL paid for the construction and maintenance and recoups these costs through toll charges. At the end of the concession period, the project reverts to government. In this example, there is no cap on the tolls that can be charged. With some arrangements, charges and increases in charges are fixed at the outset and in line with, or linked to, projected inflation.

14.2.2 The PPP arrangement

Projects that use PPP models involve many stakeholders that influence the process and the outcomes. The primary stakeholders are indicated in Figure 14.2. The concession agreement that relates to PPP is between the 'special purpose vehicle' (SPV) and the public body responsible for the project. The SPV is the consortium formed to deliver the project and typically involves building consultants such as architects, surveyors and engineers and builders, specialist suppliers and facility operators or managers, amongst others. The lenders, insurers and other equity shareholders or sponsors are the other main stakeholders. There are hosts of

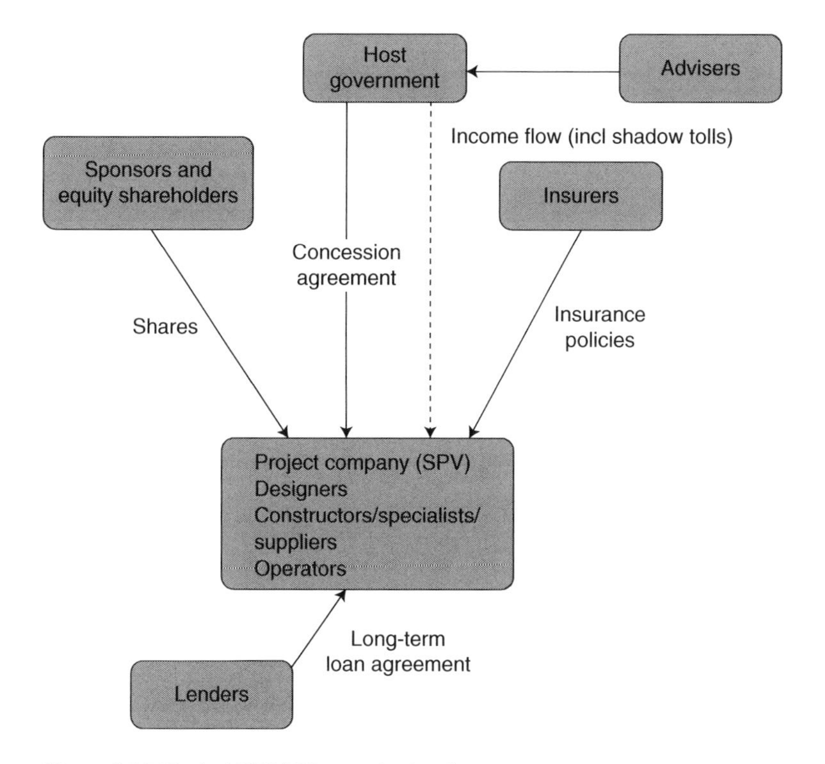

Figure 14.2 Typical PPP/PFI organisational structure.

peripheral advisers that may be recruited by these stakeholders for specialist advice, such as financial, legal and technical advisers. They provide a crucial service in making the PPP agreement work effectively and efficiently. The consumers/users of PPP-built facilities are important stakeholders as well. They communicate the ability and willingness to pay for the services generated. They also act as a useful feedback mechanism in identifying the strengths and weaknesses of existing services in order to design new services of higher quality as well as to identify priorities for quality and level of service.

The payment structures for PPP can take the form of a unitary charge paid by the public authority commissioning PPP to the SPV (See Figure 14.3). In the cases where direct user income is generated, this may happen in terms of a toll payment by the users. For example, a road or bridge constructed under a PPP agreement may result in a toll charge paid by the users. Both unitary payments and toll payments can co-exist simultaneously, depending on the nature of agreement and service provided.

The most attractive element of PPP (or PFI) to the government is the fact that it is not required to make capital investments in delivering public sector products and services, or only minimal investments. This is explained in Figure 14.4.

In the case of traditional procurement, the public sector has to make an upfront capital investment to make it possible to deliver the service from the facility. For example, if it is a public sector school and the new school building costs £1 million, this would be the initial capital investment along with the land and other infrastructure required. If PFI procurement is to be used, then the Education Department is not required to make this initial investment, but only provide the land and necessary infrastructure to build it.

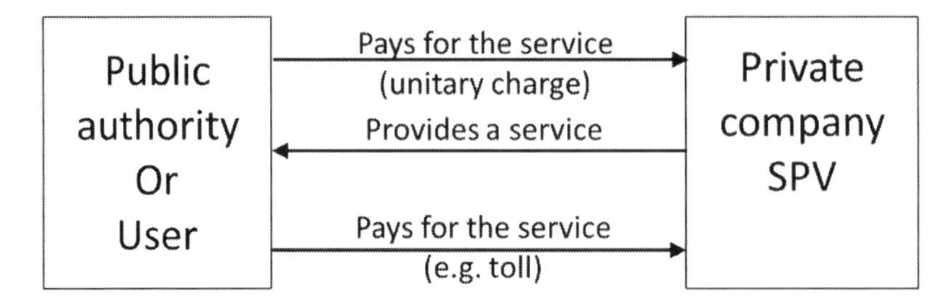

Figure 14.3 PPP payment system.

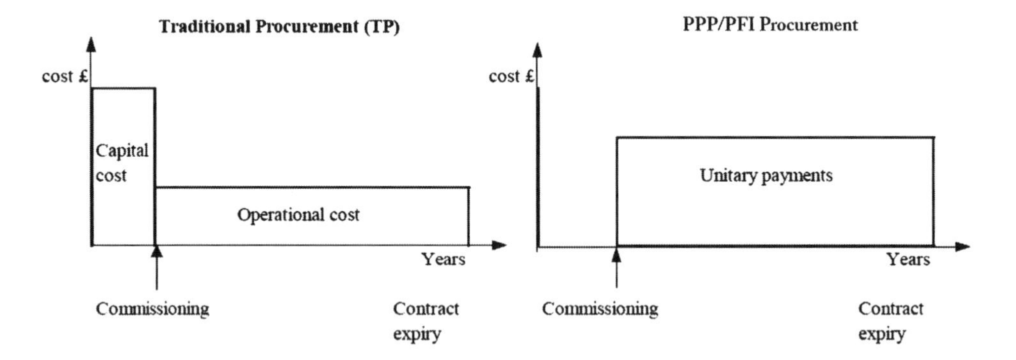

Figure 14.4 Traditional vs PFI cost to client.

So, in the case of PPP (or PFI) procurement, the public sector will not make an initial capital investment, but will make that commitment through the PPP agreement the SPV. The public authority will then make regular unitary payments to the SPV for a longer period (could typically vary from fifteen to thirty years). The unitary payment will cover the capital investment, the cost of finance (interests), costs of operating and maintaining the facility, including relevant service charges, and other costs.

14.2.3 The importance of PPPs

It needs to be recognised that the use of PPP has generated a large number of projects in the construction industry and that this in turn has improved the quality of the built environment in towns and cities around the UK. In the last twenty years, more than 1,500 PPP deals have also been signed for new projects across the European Union. These represented an estimated €260 billion. However, since the start of the worldwide economic crisis in 2008, the total number of such projects has fallen by about a third. Governments, who previously relied extensively on this form of procurement for public sector works projects, suddenly found that infrastructure and building projects were either abandoned or postponed. This decline in PPP occurred at a time when such schemes were becoming even more important as a means of maintaining economic activity during the financial crisis across most countries of the world. It is believed by some that investment in construction projects is vital at a time of slowdown in economic activity. The industry employs a large workforce, and a large number of other industries and activities rely extensively on a buoyant construction industry. In addition, a pleasing built environment provides a feel-good factor that can generate optimism and reverse an economic decline.

14.2.4 Issues with PPP

Government has always been able to borrow money at better rates than can be obtained by the private sector because it is able to guarantee any borrowing. Some will argue that, had government acted as a surety for the funding of PPP projects, this would have reduced the overall costs and hence improved the value for money. It is sometimes very difficult to compare contrasting methodologies, since the solutions that are achieved are not always comparable. Constraints on public borrowing are largely determined by the amount of receipts to the Treasury, and this is normally through taxation rather than through the sale of government assets, which is what occurred in the case of privatisation.

A number of studies into the early use of PPP have concluded that, in many cases, the proposals were of inferior quality to those that adopted the standard model of public procurement based on competitive tendering for public works schemes. A response to such negative findings was the development of more formal procedures to assess PPP in terms of value for money. It has been previously noted that, whilst this idea is good in concept, it can result in many varying outcomes depending on the 'price' that is put on value. Value for money is a prerogative term with widely different interpretations in meaning. The new framework thus chose to focus on an appropriate allocation of risk for all of the parties involved.

A further new model of evaluation is currently being proposed, called the public private community partnership (PPCP) model. This model, which was being considered under the previous Labour government, attempts to focus both government departments and private providers on a wider agenda for PPP, and includes social welfare rather than relying only on profit as the major driver for the private sector partners.

14.2.5 Health services PPP

A health services PPP will typically extend over a duration of fifteen to thirty years from its initial inception. A contract is formed by a health trust and one or more private sector companies. This then becomes a legal entity. The government provides an outline for its health system and then empowers the legal entity to meet these objectives through building, maintaining and managing the project for a defined period of time. Private sector firms involved in the project receive payment for their services and assume the financial, technical and operational risk involved in the project. They also share any benefits of potential cost savings throughout the duration of the contract. The model appears to be a good one, but research has not yet shown whether it is the most cost-effective.

There is an opportunity for multi-sector market participants, such as hospital providers, physician groups, pharmaceutical companies and insurance companies, as well as building firms and consultants, such as architects and surveyors, to develop PPP/PFI schemes in this sector. Different models of PPP have been used for health services projects over the last two decades, providing projects from small medical practices to large and complex general hospitals. Whilst it is recognised that good health care is a major policy of any government, nowhere in the world is there a total reliance on the public sector for such services. PPP is therefore a useful option in providing aspects of health care that government funding alone is unable to address. Spending on health care amongst the OECD (Organisation for Economic

Cooperation and Development) nations is expected to grow by more than 50 per cent between now and 2020. This partially represents a growing and ageing world population, but also an expectation that solutions to health problems will be found. A considerable proportion of the increased expenditure will be spent on buildings and associated works over the coming decade.

14.3 Private finance initiative (PFI)

This section discusses the PFI as the most common form of PPP implementation in the UK.

14.3.1 History and development of PFI

PFI was first developed in Australia in the late 1980s. Originally it was used for infrastructure projects such as roads and railways. A Conservative government introduced it into the UK in 1992, where it was opposed by the trade unions and condemned as a form of privatisation by the Labour Party. Although some critics argued that savings in expenditure today would need to be paid for, perhaps many times over, at some date in the future, HM Treasury considered it to offer many advantages. In 1997, after winning the General Election, the new Labour government was encouraged to retain its use. They recognised that public sector capital was limited and so public sector projects would either be constructed on a PFI basis or not at all. Because of this situation, PFI continued to expand in both the number of projects and their overall value. The justification of PFI was built largely around the issue that the public sector was poor at delivery and management and that the private sector was better at providing such services.

By 2014 the total capital value of PFI contracts in the UK had risen to almost £57 billion from 728 current and 671 operational projects. This committed the British taxpayer to paying in excess of £222 billion over the lifetime of these contracts. The current state of the PFI project portfolio in the UK is indicated in Figure 14.5. The financial crisis, which began in 2007, presented PFI with many challenges, not least because the capital used to finance them had

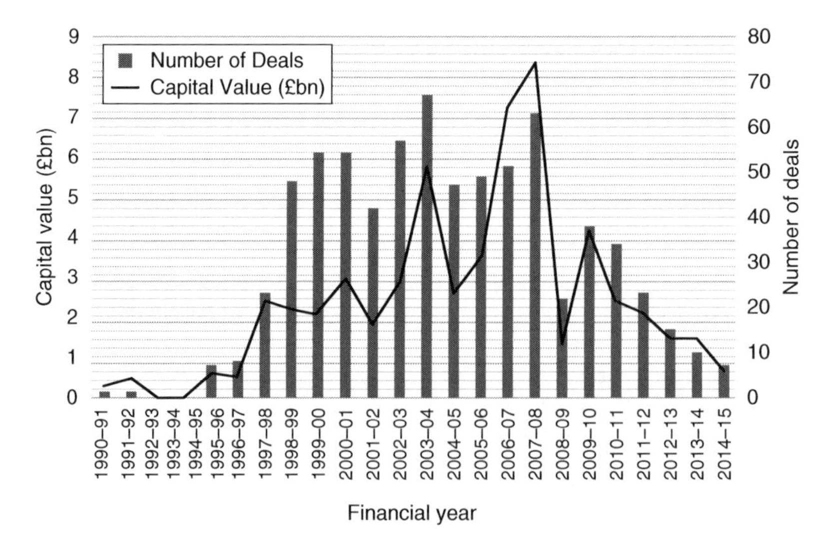

Figure 14.5 PFI project portfolio in the UK.

dried up. PFI still remains the UK government's preferred method for public procurement of major projects, and now pursued through PF2. The NHS has stated that PFI is their 'Plan A' for the construction of new hospital projects, and that there is no 'Plan B'. The shortage of finance after the financial crisis, coupled with uncertainties arising due to Brexit, means the government has now found itself in a situation where there are few options other than improving the PFI process. This resulted in a curtailment of projects during the period of the Coalition government from 2010, some postponed and others abandoned, at least for the foreseeable future. They cancelled several schemes that were far advanced in terms of design and planning, and revised proposals for PFI to consider more carefully the proper risk transfer to the private sector along with more transparency in accounting methods. That government also wanted to renegotiate a number of existing PFI deals in the face of reduced public sector budgets, which led to the initiation of the consultation process to take stock and redevelop PFI in the UK. The UK is considered as the most advanced country for these arrangements, with a mature PFI system and the development of a more streamlined PF2 system, that now continues successfully as further evidence for its advancement and acceptance as one the main forms of public procurement.

14.3.2 The PFI method of procurement

The PFI is a method of funding deemed to be public sector projects with private capital. Although the idea was initially developed by the Australian and UK governments, variants have been developed in many countries around the world as governments have sought to reduce the public estate. The ideas have received approval from the World Bank and the International Monetary Fund (IMF) as ways of funding hospitals, schools, roads, and so on. Universities have developed parts of their own estates through the good use of PFI. Until the introduction of the PF2, there was no standard model for PFI and the methods employed varied depending upon the circumstances involved. PFI is an operational framework which transfers responsibility, but not accountability, for the delivery of public sector services to private companies. PFI usually has a long time horizon covering not just the initial construction but its eventual maintenance as well.

According to an article in the *OECD Journal on Budgeting* (Sarmento 2010), PFI is an arrangement whereby the public sector contracts to purchase services, usually derived from an investment in assets, from the private sector on a long-term basis (often between fifteen to thirty years). This includes concessions and franchises, where a private sector partner takes on the responsibility for providing a public service including maintaining, enhancing or constructing the necessary infrastructure usually performed by the SPV.

PFI attracted a high level of criticism from both the public, encouraged by mass media, and politicians, resulting in a detailed review of the whole PFI system used in the UK. The main issues were about the value for money factor of these projects. It was compared to using a mega credit card for purchase of essential infrastructure for the society, and the cost of borrowing and impact of unitary payments were unjustifiable in certain projects.

14.3.3 A new approach to PFI: PF2

Towards the end of 2011, the UK government announced the review of PFI consultation. The result was the birth of a more refined and accountable PFI system known as PF2. This second generation of PFI was announced as part of the Autumn Statement by the Treasury in December 2012. PF2 retained the basic structure of the PFI model, but has recognised the need to address the widespread concerns. The main changes of the PF2 procurement system are identified below:

- Government will take a minority equity share, between 30 and 49 per cent, to reduce the impact of public sector investment. The investment will be managed directly by the Treasury (not by the procuring authority) to avoid conflict of interest.
- Accelerate delivery of projects: eighteen months procurement standard (PFI was sixty months).
- Improve value for money (public sector share risks and profits).
- Ensure the future of affordable debt finance.
- Institutional funds, such as pension funds, will be encouraged to take a stake in projects.
- An open-book approach will be introduced, with a gain–share mechanism for life cycle funds to facilitate sharing of surplus funds.
- Actual and forecast equity return information will be required for publication.
- All provision of services will be periodically evaluated.
- A control total will be introduced for all commitments arising from off-balance sheet PF2 contracts signed.
- Private sector funding will be sourced through competition, enabling greater transparency and economy.
- A business case tracking system will be introduced on the Treasury website.

The new PF2 system comes equipped with a full suite of documentation, maintained with technical updates:

- A new approach to PFI.
- PF2: A User Guide.
- Standardisation of PF2 contracts.
- Change Protocol Principles.

14.4 Advantages of PPP/PFI procurement

Using the skills and abilities of private sector companies is a sound idea, and there is evidence to support this viewpoint. PPP/PFI has resulted in a number of important and useful projects that would otherwise never have been built using traditional procurement routes alone. As noted earlier, another advantage is the provision of a pleasing built environment, which generates a feel-good factor that has other spin-off benefits. In addition, over the past decade the use of these practices has helped to provide a constant workload for the construction industry, which has otherwise become accustomed to booms and slumps.

Where contracts have been well written, with the correct apportionment of risk and reward, this too has been a beneficial aspect of PPP/PFI contracts. In the early days of PPP/PFI, it was difficult for any party to properly judge the risk that was involved, because arrangements were very different from existing traditional practices. Learning from best practices, and incorporating these ideas into later contracts, has been beneficial to all parties involved. Looking at some existing public sector projects built in the past fifty years, there is evidence that the public sector obtained poor value for money against a number of criteria, including design, construction detailing and building longevity. There were examples of projects not being completed on time, with severe delays for potential users, and projects that overran their initial building tender costs. It has long been argued that contractual arrangements have become too complex and too adversarial. PPP/PFI has gone some way towards simplifying practices and removing the 'us and them' mentality that so frequently appears in the industry. There are other advantages:

- The private sector operator bears the major risks involved in the design, building, financing and operation of the asset. This makes both overruns and additional costs less likely and means that the client (central or local government) does not bear any of the costs involved.
- There may be greater efficiencies in both the building and running costs, e.g. in respect of greater standardisation in components, energy management, etc.
- Government pays on a performance-related basis for the use of the asset and for its continuing management. This provides incentives for the consortium to build the asset to a high standard and maintain it in a good condition and to fully incorporate whole life costing ideas and practices.
- It gives access to an additional source of capital financing through the PPP/PFI credit arrangements and bidding process.

14.5 Disadvantages of PPP/PFI procurement

With any innovation of new practices there is often a trade-off with disadvantages to consider. All forms of procurement have advantages and disadvantages and a one-size-fits-all approach is likely to be only a panacea as far as construction procurement is concerned. However, PPP/PFI has identified a number of disadvantages, particularly to the client, and examples where these types of projects have resulted in less than satisfactory solutions:

- PPP/PFI can distort the primary purpose of a building or structure. Such buildings are not constructed for their own sake but for the functions that are carried out in them. In practice, however, the priority for the use of any funds is first to pay off the PFI loan. In other, more usual circumstances, when funds are limited they are directed towards the core business, such as health or education, and the building itself takes second place. Maintenance work is then delayed. On some PFI projects the opposite is true.
- PFI costs remain fixed as a part of the contract and there are high penalties for terminating contracts. Renegotiation is difficult even though contracts may typically last for thirty-five years or more.
- PFI payments last many years, in order to repay the initial construction costs and the ongoing maintenance and repairs costs. The sums involved are substantially higher than the initial costs alone in the same context as a mortgage. Some commentators have suggested that this represents a millstone around the necks of our children or even grandchildren.
- Some of the initial PFI schemes were very expensive in the context of today's practices. The costs of the risks involved were overestimated.
- These arrangements normally involve maintenance and management of the asset, usually to a minimum standard of repair over its lifetime. This is to ensure that the building is repaired to an adequate level when it is eventually handed over to the client. These costs are fixed and difficult to reduce even though the function of the building's activities may require productivity gains to be made over the lifetime of the PPP/PFI project.
- The annual payments to PPP/PFI consortiums peaked in 2017, at about £10 billion. These are likely to stretch public sector budgets and could well result in refinancing agreements. This will have a negative effect in circumstances where an economy is in downturn. It may also have the effect of reducing core activities to help pay for such agreements.

- If international standards of accounting were applied to these kinds of project, then the PPP/PFI debts would be shown on the government's balance sheet. The UK has adopted the European Accounting System (ESA 95) to avoid this happening.
- Some have argued that the specifications of many public sector projects have been distorted to increase profitability to PPP/PFI consortiums. There are accusations that such schemes have been designed to help maximise corporate profits.
- Some critics suggest that the processes involved are overly bureaucratic and insufficiently transparent to properly allow users to make an evaluation of alternative proposals. There is a misguided notion that only these kinds of procurement arrangements are effective.
- There are examples where a PPP/PFI project has been constructed only to find that the project is no longer required for its original purpose, but the long-term maintenance contract has to be kept in place for the next twenty-five years.

14.6 The Priority School Building Programme (PSBP)

The PSBP is concerned with the rebuilding and refurbishment of schools across England, covering 537 schools in two phases. The whole programme is expected to reach £4.4 billion and was launched in 2011. The PSBP is centrally managed and procured by the Education Funding Agency (EFA), an executive agency sponsored by the Department for Education (DFE). The first phase of 260 schools are being built, the majority (214) through capital grants and forty-six using private finance through PF2 in five batches, and most are expected to be completed by the end of 2017.

The EFA has developed a financing model known as 'the Aggregator', which combines the funding requirements of the five batches. In this way, it has managed to access cheaper finance and streamline procurement by using standard documentation.

The improvement of the PF2 is directly put to the test with the PSBP and has already proven to be more effective in comparison to the previous Building Schools for the Future (BSF) programme. Under the BSF, it took around three years for construction work to commence, whereas, under PSBP, this period has been effectively reduced to one year. The early indications are that PSBP school projects are around 30 per cent cheaper than the BSF school projects.

The second phase of the PSBP, involving 277 schools, has begun with the initial allocation of £2 billion, which was further boosted to £6 billion in 2015.

14.7 Value for money (VfM)

PPP and PFI projects have been criticised for whether there is VfM in the investments made. Since large sums of money are borrowed on long-term loans from the private sector, and the loans are paid back over time from tax collections, it is important to determine whether there is adequate VfM in the investments made using PFI.

The UK Treasury defines VfM as the optimum combination of whole life costs and quality (or fitness for purpose) of the good or service to meet the user's requirement. VfM is not the choice of goods and services based on the lowest cost bid. To undertake a well-managed procurement, it is necessary to consider beforehand, and at the earliest stage of procurement, what the key drivers of VfM will be in the procurement process.

The key drivers for VfM are:

- risk allocation
- focus on outputs rather than inputs

- competition and innovation
- contract duration and scope
- borrowing costs
- private sector management skills
- performance measurement and the use of an output specification.

A VfM analysis often uses a public sector comparator (PSC). A PSC is hypothetical estimate of costs for the public sector if it was to deliver the service by itself. The PSC is used as a tool to determine if the whole life costs for a PFI investment approach is greater or lesser than the PSC. If PFI costs are lesser, then the project can be procured using the PFI method. This is indicated in Figure 14.6.

If $PSC_{NPV} > PFI_{NPV}$, then the PFI procurement option can be recommended. This indicates that if the public authority was to deliver the service concerned, in whole life value terms it will be more expensive to do so, thus the PFI option is more economical.

There are limitations in the VfM approach as well:

- In order for the PSC to be accurate, there should be good set of data on traditionally procured public sector facilities. However, this is often unavailable.
- The whole life cost approach is technically and logically excellent. However, in practice getting whole life cost data, especially for public sector traditionally procured facilities, is often difficult.
- The discount rates are a prediction. Volatilities in markets often can bring about economic turbulence. Therefore, the discount rate assumed might not be accurate. However, sensitivity analysis using variations in discount rates can be used to minimise the impact and the level of risk.

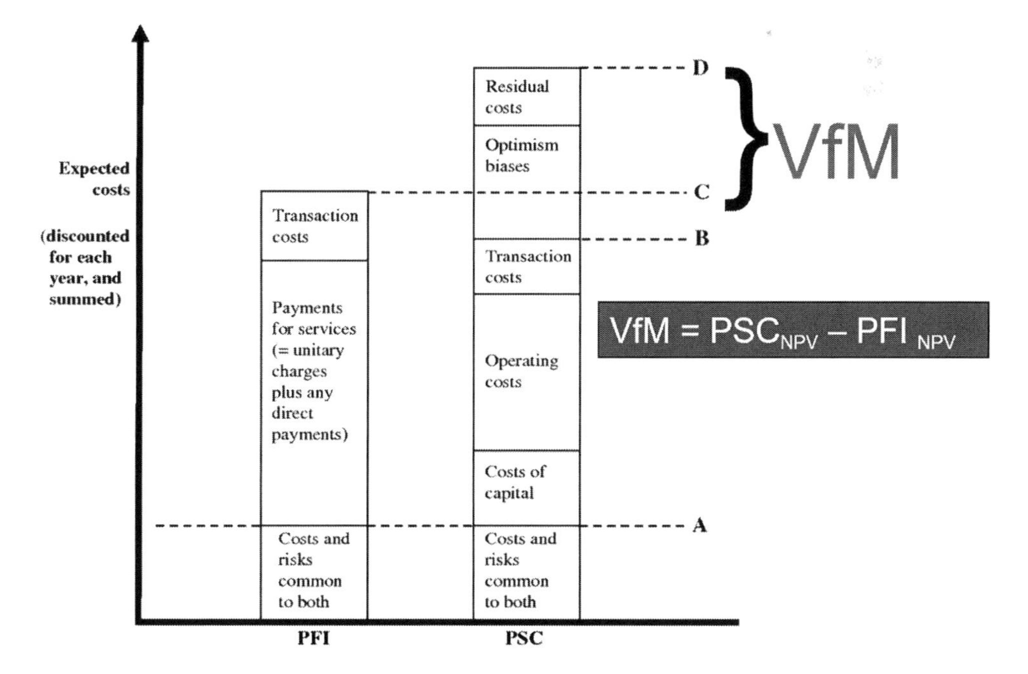

Figure 14.6 VfM in PPP/PFI projects.

- The uncertainty of the future cannot be fully disregarded. These estimates, particularly for longer durations of fifteen to thirty years, are sometimes unrealistic. The world is changing rapidly, with the fourth industrial revolution unfolding; as such, long-term predictions can soon become obsolete because of technological advancements.
- There could be man-made or natural emergencies or disasters. The effects of such events may distort the assumptions made.

Nevertheless, use of PSC as a means of benchmarking against an alternative business-as-usual scenario is a very popular methodology right across the globe, particularly in countries such as the UK, Australia, Hong Kong and Canada, amongst others, where the PPP/PFI procurement method is heavily used.

Part 3

Process and parties

15 Partnering and supply chain management

15.1 Partnering

Partnering is a structured management approach to facilitate teamworking across contractual boundaries. According to the Construction Best Practice Programme (now Constructing Excellence, see Chapter 19) its fundamental components are:

- formalised mutual objectives
- agreed problem resolution methods
- an active search for continuous measurable improvements.

Partnering should not be confused with other good project or construction management practices, or with long-standing relationships, negotiated contracts or preferred supplier arrangements. Whilst these may contain many good features, they lack both the structure and objective measures that are required to properly support a partnering arrangement.

The critical success for partnering is the commitment of all partners at all levels to want to make the project successful. This might sound unusual, but the key word is commitment. The result is that the partnering agreement rather than the contract documentation drives the relationship between the parties.

Partnering is *not*:

- a buzz word
- a new form of contract (it is largely a procedure for improving working relationships)
- a soft option
- a quick fix for a weak business; it is based on the assumption that strong players make each other stronger, weak ones destroy each other
- only about systems and methods; it is essentially about people, enabling them to operate more efficiently, effectively and economically.

15.1.1 Shared values

True partnering can only be carried out by firms and organisations that share similar aims and objectives. It relies on openness, transparency and honesty. Such firms should also have a desire to change and improve their practices and to add value to their business operations. Within this approach, such organisations are able to set up mutually advantageous commercial arrangements. Wherever possible, they should seek to develop long-term strategic relationships that assist all of those involved to work more effectively.

15.1.2 Teamworking

Partnering can be used for one-off projects, but it becomes more effective amongst those clients who have considerable ongoing construction programmes. There are now a number of clients in the UK who already spend in excess of £1 million per day on construction work. Partnering is also appropriate for those organisations who have ongoing maintenance and repair programmes. The benefits of partnering are generally cumulative, in that strategic alliances produce significantly more advantages than single project arrangements. The benefits of teamworking are considerable and are in contrast to traditional arrangements whereby teams are often dispersed on completion of a project and the knowledge gained from understanding each individual's strengths and weaknesses are lost. Firms are only as good as the people they employ, and this resource is the key commodity.

These benefits are also significantly improved where partnering is applied throughout the supply chain rather than solely between the client, consultants and the main contractor. Traditionally, many main contractors have formed working and business relationships with some of their subcontractors and suppliers. Partnering is an extension of this practice. Whilst the concept and practice of supply chain management is still emerging, there are good examples to demonstrate that the theory is effective in practice. The challenge facing participants is to understand the real and long-term benefits that can be achieved and to adopt the principles on projects that are appropriate.

15.1.3 Use

Whilst partnering has a great many attributes, it is not appropriate for every type of construction project. A key consideration is the allocation of risk amongst the different parties involved. It is now generally recognised that risk should be allocated to the party that is best able to assess and control it. Partnering includes a combination and commitment to the following aspects.

Mutual objectives

- Commitment to the project approach from the outset.
- Regular reviews through meetings and effective communication.
- Sustainable long-term goals rather than relying on quick-fix solutions.
- Transparency of practices.
- Confidentiality between the partners.
- Objectives are more easily achieved between businesses with similar cultures and styles.

Problem resolution

- Systematic approach towards problem resolution.
- Seeking solutions to problems rather than a blame culture.
- A win–win-based philosophy for all partners.
- Equality of rights and responsibilities between parties.
- Acceptance that adversarial attitudes waste scarce resources.

Continuous improvement

- Setting measurable targets by which to review performance.
- Adopting best value solutions.
- Focusing on the customer.
- Eliminating waste and then adding value.

15.2 Categories of partnering

The nature of partnering may take several forms depending upon the situations and objectives of the various parties involved. Partnering can be broadly classified as either project partnering or strategic partnering. Whilst the differences between these two classifications, which relate to scale and level of relationship, are significant, the essence of the partnering concept is the same in both.

15.2.1 Project partnering

Project partnering relates to a specific project for which mutual objectives are established and the principles involved are generally restricted to the specified project only. The great majority of partnering opportunities are of this type because:

- *It can be relatively easily applied in situations where legislation relating to free trade is strictly imposed*: in the UK a post-award project-specific partnering methodology has been proposed and was published by the European Construction Institute in 1997 (ECI 1997). This approach is seen as particularly appropriate to the public sector since it allows an openly competitive process of selecting contractors to be adopted. Thus, the European Union's public procurement requirements are respected. It is also fully compatible with the requirements of compulsory competitive tendering as imposed under the various local government legislation.
- *Clients seeking to build on an occasional basis may use it*: the vast majority of clients who, due to the size and timing of their development programme, are not in a position to enter into a long-term partnering relationship.

It should be readily understood that the predicted rewards of project partnering are much less than where longer-term strategic partnering arrangements exist. However, evidence from the USA, where approximately 90 per cent of partnering is of the single project type, as a result of difficulties relating to the competition law, nevertheless shows that benefits may be obtained.

15.2.2 Strategic partnering

Strategic partnering takes the concept of partnering beyond that outlined for project partnering to incorporate the consideration of longer-term issues. This is an important aspect that should be considered before assuming the existence of strategic partnering. In many cases in practice, often what may be identified and described as strategic partnering is effectively a project-based approach used on each of a series of projects. This is done without the added requirement of long-term strategic considerations.

The additional benefits of strategic partnering are a consequence of the opportunity that a long-term relationship may bring, and could include:

- *Establishing common facilities and systems*: the use of shared office accommodation and communication access and storage systems promotes openness, efficiency and innovation.
- *Learning through repeated projects*: construction processes can be developed to reduce defects and improve efficiency, leading to additional cost savings, reductions in the time required for construction and an improvement in quality.
- *The development of an understanding and empathy for the partners' longer-term business objectives*.

15.3 Evaluating partnering

As with all practices and procedures, one method is unlikely to be a panacea for all situations and for all types of construction projects and clients. There are advantages and disadvantages to consider. Most of the following were considered in Ashworth and Hogg (2000).

15.3.1 Advantages

The adoption of partnering, at a strategic level or for a specific project, is considered to bring major improvements to the construction process, resulting in significant benefits to each partner. These include:

- *Reduction in disputes*: adversarial relationships are a common feature of the construction industry. These occur between clients and contractors, contractors and subcontractors, subcontractors and suppliers, consultants and clients, consultants and contractors or subcontractors. In settling disputes, the traditional solution can sometimes result in expensive litigious cases that may impact throughout the supply chain. Where a number of parties are involved, cases may result in simultaneous disputes on a single project. Because of the principles of cooperation that form a foundation of partnering, it is argued that this should be able to reduce the number of disputes.
- *Reduction in time and expense in the settlement of disputes*: it is accepted that partnering will not necessarily eliminate every dispute. However, frameworks have been developed that result in faster and more efficient methods of dispute resolution that may not be available with other forms of contractual arrangement.
- *Reduction in costs*: the key attributes of partnering include the beneficial effects of repetition, improvements in communications, innovation in both design and construction and the search for continued improvement. This approach can result in significant savings in construction cost, especially where partnering is used over a wide range of projects. In addition to these savings associated with the actual construction of the project, the costs associated with procurement are also likely to be reduced.
- *Improved quality and safety*: the existence of mutual objectives and the desire to continually improve design and construction processes are cornerstones of the partnering philosophy that should result in improvements in design and construction quality and safety during construction. Where strategic partnering is functioning, further benefits arise, including those derived from the additional learning and feedback mechanisms. These exist as a consequence of risk sharing and security of work.
- *Improvement in design and construction times and certainty of completion*: the efficiencies arising from the partnering arrangement are able to lead to a large reduction in the time that has been made available for design and construction. The attributes of partnering which include the benefits of teamwork and risk and benefit sharing, lead to an increased certainty in respect of project delivery dates.
- *More stable workloads and income*: a major difficulty faced by contractors is the lack of continuity in respect of future workloads. This can have a negative impact on design and development, including the lack of willingness of contractors to invest in resources such as staff development, plant and equipment and new methods of working. The use of strategic partnering provides an opportunity for forming long-term opportunities and developments.

- *An improved working environment*: several surveys have shown that a less adversarial atmosphere and shared commitment to projects results in perceived improvements in the working environment.
- *Improved trust*: where partnering is used appropriately, it can improve the trust and transparency between the parties concerned. Contractors know that partnering based upon trust brings about a moral obligation more powerful than anything that may be contained in a traditional contract. Clients are able to use periodic competition and benchmarking to ensure that they are receiving the service expected.

15.3.2 Disadvantages

It is important to acknowledge the existence of some important disadvantages when partnering is used:

- *Initial costs*: it is generally accepted that there are some additional costs associated with the partnering process. These may be relatively minor in nature, being calculated at less than 1 per cent of the project costs.
- *Complacency*: it has been suggested that a long-term partnering relationship that has the benefit of improved job security may lead to some complacency. Human nature is such that this could occur. However, there is an equally strong counter argument that job security is highly valued and improved job satisfaction would therefore be safeguarded by continued effort and improvement.
- *Single-source employment*: strategic partnering could also result in a contracting organisation becoming singularly dependent on one client, and thus becoming extremely vulnerable should this source of work be terminated. Clients may be similarly exposed, for example, where a single supplier or contractor has developed a unique product or service for them.
- *Confidentiality*: openness is a feature of partnering and this may cause some clients to be concerned about matters that are of a confidential nature. Disputes may arise if information regarded as commercially sensitive were to be withheld from one or more of the partners. To avoid disputes about disclosure, it is essential to identify as clearly as possible by category what information is to be shared and what is not. This will then encourage disclosure. It will also be important to protect disclosed information by including a confidentiality clause in the partnering agreement.
- *The perceived need for competition*: the accepted route to securing a good value price is through the competitive tendering process. The additional costs associated with an absence of competition are well understood and are used as a powerful argument against, for example, the use of negotiation. Accountability is also a factor that is of concern for many clients. This may also be difficult to establish where there is a lack of competition. The traditional, and easiest, way of demonstrating value is to show the results of a competitive tender selection process. Partnering may be judged to remove the competitive element, irrespective of the measures that may be taken to demonstrate otherwise, e.g. benchmarking.
- *Partnering through the supply chain*: in order to achieve the full benefits of partnering, the same principles need to be applied throughout the supply chain.

Whilst there is a great deal of positive feedback relating to partnering between clients and contractors, there appears to be less enthusiasm for this process amongst subcontracting organisations. There is evidence to suggest that some subcontractors believe that partnering has a negative effect on their already difficult position and that conventional partnering does not work within the normal subcontractor framework.

15.3.3 Legal issues

There is some doubt as to the legal status of the partnering charter, and concern that making explicit statements that a partnering charter does not create a legally binding relationship between partners does not necessarily mean that none exists. Although the construction contract provides a framework of rights and obligations, partnering has the potential to affect the allocation of risk established by the contract and subsidiary contracts. If the partnering arrangement breaks down, a party may find itself in a position where it is necessary, or at least attractive, to assert that the contractual risk allocation has been altered, either by the provisions of the partnering charter or by subsequent conduct or representations in the course of the partnering process.

15.3.4 Payments

Where a lump sum contract is used, the provision for payment under this form will need to be reviewed to ensure consistency with any incentive payment arrangement which may be part of the partnering agreement.

15.3.5 Liquidated damages

The provision for liquidated damages within the underlying contract framework must be dovetailed with any related provision in the partnering agreement.

15.3.6 Quality

The membership, roles and powers of the project management team, if set out or implied in a partnering arrangement, need to be exactly reflected by the underlying construction contract to avoid any confusion in respect of liability issues. Will the project management team be given power to instruct rework, replacement of defective materials and testing of materials and workmanship? Will this power remain with the consultant under the contract?

15.3.7 Disputes

The approach to the settlement of disputes is a central feature of partnering and, again, consistency is required between the partnering agreement and any provision in an underlying contract. To a non-lawyer, there may be some doubt as to the legal effects of partnering, and clearly some caution is required where a partnering agreement is proposed. For this reason, it is suggested that readers undertake further research and seek legal advice whenever the situation demands. The intention of this section is merely to raise awareness.

15.4 Summary of partnering

A majority of those involved in the construction industry recognise the need to move away from the confrontational relationships that cause disputes, problems, delays and, ultimately, unnecessary expense. The definition of partnering is one of a commitment between two or more organisations for the purpose of achieving specific business objectives by maximising the effectiveness of each participant's resources.

As long ago as 1964, the Banwell Report recognised that there was scope for the awarding of contracts without the use of competitive tendering under certain circumstances; for example, where a contractor had established a good working relationship with a client over a period of time, completing projects within the time allowed, at the quality expected and for a reasonable cost. These circumstances may have arisen through serial contracting and particularly in those situations of continuation contracts where projects were awarded on a phase basis. It needs to be recognised that partnering is not the first suggestion aimed at modernising the construction industry (see Chapter 12).

However, in the past there has always been some reluctance on the part of public bodies to adopt such procedures, since they may lack the essential element of public accountability. This was often the case even though it could be shown that a good deal had been obtained for the public sector body concerned. The current UK government is now supportive of partnering in the public sector. Effectively partnering will be possible where:

- it does not create an uncompetitive environment
- it does not create monopoly conditions
- the partnering arrangement is tested competitively
- it is established on clearly defined needs and objectives over a specified period of time
- the construction firm does not become overdependent on this arrangement.

Partnering is now becoming well established in both the USA and the UK and offers a range of benefits, as described above. It has been successfully used in the UK by a number of organisations. The partnering arrangement may last for a specific length of time, perhaps for a single project (project partnering) or for an indefinite period (strategic partnering). The parties agree to work together in a relationship of trust, to achieve specific objectives by maximising the effectiveness of each participant's resources and expertise. It is most effective on large construction projects or projects of a repetitive nature where the expertise developed can be retained and repeated on other projects. The ongoing construction of the McDonald's chain of restaurants provides a good example of the latter. The use of partnering, coupled with innovative construction techniques, has been able to reduce both construction time and costs.

The concept of partnering extends beyond the client, contractor and consultants and includes subcontractors, suppliers and other specialist organisations who are able to add value to the project. The establishment of good supply chain management benefits all those involved.

15.5 Supply chain management

Reviews of the construction industry (see Chapter 12) have consistently criticised the fragmentation, not just between design and construction but also within the industry itself. The industry lacks an integrated approach to its work. About 95 per cent of firms employ fewer than a dozen people, emphasising a reliance upon the performance of a great many small firms. The industry has been equally criticised for its adversarial nature, where problems are resolved in a contractual manner at the different levels suggested in Chapter 5.

This results in confrontation between and amongst clients, designers, main contractors, subcontractors, suppliers and manufacturers. The overall effect of this is demonstrated through:

- inefficient use of labour
- high wastage of site materials
- high costs of construction
- ineffective education and training
- functional inefficiency resulting in high running and maintenance costs
- low profitably throughout the supply chain
- low esteem and image amongst the public.

Clients who construct projects intermittently engage designers and constructors for individual projects with an array of subcontractors and suppliers to meet their short-term needs. Major clients with substantial building programmes have begun to assemble their teams using, wherever possible, a preferred subcontractor or supplier arrangement. In this way, these clients are able to gain from familiarity and team understanding and are thus able to manage their projects in a more integrated way.

The benefits that a carefully nurtured, financially secure and efficient supply chain can bring to improving the overall performance and technological development of the construction sector are largely unrealised. Historically, some local authorities combined to allow for the mass production of building components used in the major school building programme of the 1960s. In this way, through a process of guaranteed bulk purchase agreements, those local authorities were able to procure components at a much lower cost. The manufacturers of these components also gained, since they were able to install mass production technology and initiate improvements that often required considerable investment. Such practices occurred in spite of the process of procurement rather than it being an integral part. Eventually, as school building went into decline, such activities largely came to an end. There is only limited evidence for such collaborative ventures over the longer term.

15.5.1 Other industries

There is ample evidence available from other industries to encourage the construction industry to change some of its practices. This is most notably illustrated in the manufacturing industry. The technical and commercial effectiveness of supply chain management techniques through strategic supply chain partnerships have helped to replace short-term individual project relationships. These supply chains focus on delivering value as defined and required by their clients.

Evidence suggests that long-term supply chain alliances can incorporate continuous improvement to reduce construction costs, enhance quality and deliver the project in a more timely way. The practice adopts the principle of continuous improvement of both good practices and the elimination of weaknesses. The focus has also shifted towards improvement through a whole life project approach rather than to consider the initial construction implications alone. Supply chain integration not only brings together the different organisations involved, but also aims to keep them together over a period of time and from project to project. The long-term relationships are of vital importance if ongoing improvements in practice are to be achieved.

15.5.2 History

Historically, clients used preferred consultant and contractor lists. Some firms not on these lists often attempted, perhaps unsuccessfully, to be invited to tender. Contractors also used a preferred list of suppliers and subcontractors, to a similar end. However, some of these lists were too extensive, sometimes resulting in infrequent working relationships. Also, there was no attempt to integrate these firms with each other through mutual trust and cooperation. Relations sometimes soured, perhaps because of perceived unfairness, but that was, and is, human nature.

Supply chain management challenges some current practices: for example, when specialist firms are asked to price work where they have not been involved or consulted in developing the design; designers might spend a considerable time gathering information from particular firms, but then find themselves working with other organisations that were outside of this consultation process – such firms may have conflicting opinions and judgments, which then create tensions and ineffective communication and working relationships.

15.6 The underlying principles of supply chain management

The supply chain is a system of suppliers and customers that are involved in manufacturing a product or providing a service to the market. It is a vital part of a business's operations. A poor supply chain will result in products or services that do not meet the requirements of the client in respect of quality, time or cost.

The Building Down Barriers (BDB) approach has identified seven underlying principles developed from a number of pilot projects (The Tavistock Institute 2000):

1. Compete through superior underlying value.
2. Define client value.
3. Establish supplier relationships.
4. Integrate project activities.
5. Manage costs collaboratively.
6. Develop continuous improvement.
7. Mobilise and develop people.

15.6.1 Compete through superior underlying value

Mobilisation of key members of the supply chain by the prime contractor achieves mutual benefit to all of the parties involved, and the client should receive improved added value. Where the prime contractor and key suppliers work together towards improvements in practice, this may have the effect of increasing their market share of the work that is available. Supply chain integration is about encouraging the different firms involved to improve their own practices and thus become more efficient in their own work. Being a member of such a supply chain offers incentives in this, and advice and assistance to help achieve the common goals. Supply chain management has nothing whatsoever to do with squeezing the profits of the weakest members or those at the end of the chain.

15.6.2 Define client value

It is sometimes assumed that traditional contracting only uses one measure by which to select the firms involved: price. Most clients will have a range of criteria that they use, but find that by setting the other parameters, such as design, quality and time, it is easier to make choices based against the one variable. Competing contractors in other circumstances do not know how a client may value time when compared with cost. In fact, most clients would not make such decisions until after the tendering procedure was completed, especially in the case of multiple criteria tendering.

At the outset, clients need to brief their teams carefully on what they require, what is considered to be of great importance and what might be omitted as being of only peripheral interest. Failures have arisen in the past where clients have been insufficiently precise on their particular needs and requirements.

Design and engineering issues, wherever possible, are best judged by a combination of members of the supply chain.

15.6.3 Establish supplier relationships

The performance of the whole supply chain affects not only contract profitability for all of the parties involved, but also the client's own expectations of the project. It is especially important that long-term relations are formed for those organisations and suppliers who have a major influence on the project. Such strategic partnerships will enhance the project and the understanding by everyone involved. Such long-term relationships allow the gradual establishment of better and more collaborative ways of working together. The variety of different skills can be harnessed and integrated to help achieve the long-term goals of the client. These supplier relationships will encourage investment and innovation, which will then enhance performance and develop collective expertise.

15.6.4 Integrate project activities

The principles of supply chain management establish the importance of long-term relationships. These will build benefits for all participants in the process. In reality, however, some suppliers and subcontractors only will be involved because of the nature of the work required, and some of these will only be involved intermittently for their own specialisms. Some suppliers and subcontractors, along with clients, consultants and main contractors, will be more able to develop long-term strategic partnerships. In those circumstances where the parties may be different, some form of project partnering remains valuable. In these situations, developing a set of shared and common values will add immediate benefit to the project. Once an overall design strategy has been agreed, the details of the design, construction methods and associated cost implications can be developed. Utilising the skills of all parties in the supply chain is important in achieving an appropriate and effective solution.

15.6.5 Manage costs collaboratively

The forecasting and control of construction costs are important aspects in any project. There is always the danger of reinventing the wheel, and none more so than where costs are concerned. Someone in the team needs to have overall responsibility for costs, and that person should be able to demonstrate a clear understanding of the cost implications of the design and construction process. Cost management seeks to provide a threshold of advice and practice and this might be further extended to provide some aspects of value management. There is little point in clients paying for expensive procedures that result in only half the cost reduction to the inevitable costs of the advice. It is important to involve all of the supply chain partners as soon as possible, as has already been suggested in the context of the design. Different practices can be employed to either cost the various elements of construction or to limit the overall price. In practice, a combination of these practices is usually employed, which has traditionally been described as comparative and elemental cost planning (for further information, refer to Ashworth and Perera 2015). These days, clients must seek to consider not just the initial design costs, but also the whole life costs involved in building and owning an asset.

15.6.6 Develop continuous improvement

This concept is not, of course, unique to a supply chain management approach, but the philosophy underlying supply chain management is believed to make this easier to achieve. Continuous improvement should be a theme running through all of our working practices. Sometimes, this is gradual, but on other occasions it represents a step change in improvements.

Whilst the concept and practice are now well established in manufacturing and some other industries, it is less familiar in the construction industry. It would, of course, be unfair to assume that change has not been taking place. Even before the advent of the Latham Report (1994) the industry was continuing to evolve, but such evolution was often in isolation from other firms and organisations working in the same areas of activity. Trade secrets were assumed to be the barrier to sharing good practices between firms and organisations. Since 1994, the construction industry has seen a more concerted effort aimed at rethinking, making improvements, modernisation and business process re-engineering. The government, research organisations, professional bodies and contractor groups have all sought to bring about modernisation.

Firms that see continuous improvement as one of their hallmarks for success adopt a particular style and approach to management. They attempt to prevent things going wrong in the first place rather than seeking to correct problems retrospectively. They also seek to make the fullest use of their available human resources and use contributions from everyone in the business to find better ways of doing things.

15.6.7 Mobilise and develop people

Developing long-term supply relations in an environment of continuous improvement is a challenging undertaking.

Considering these seven principles means questioning much of what has been seen as good practice in the light of the knowledge and understanding that was available. What is involved in achieving this kind of change? A number of mechanisms can be applied, but key aspects are considered to be:

- commitment from senior managers
- facilitation for project teams
- training in new skills
- economic incentives.

Such programmes of change are utterly dependent on the leadership that is provided. Leaders need to demonstrate and reinforce the principles involved and ensure that others in their supply chains also adopt the same philosophy.

16 The construction process

The statutory definition of development has been defined in the Town and Country Planning Act (1971) and is described as 'the means of carrying out of building, engineering, mining or other operations in, on, over or under land, or, making of any material change in the use of any buildings or other land'. Development can therefore be broadly classified as undertaking construction works, such as building and engineering, or making a material change of use to the land or property.

Property and construction development must first identify a need. This need may be to satisfy the placing of investment funds or to address the requirements in society for housing, health, education or other purposes. These together also require the relevant infrastructure if the project is to function and operate effectively. Factors to consider also include population and age trends, changes in patterns of lifestyles, the implications for new technology and the importance of fashion. The selection of an appropriate site, the choice of appropriate consultants and finding sources of finance are three of the early decisions that the client or developer will need to make.

16.1 Demand for development

Throughout history there has been a demand for buildings for different use and purpose, such as housing, industry, commercial, religious, entertainment and leisure. During the twentieth century, this demand was sustained in the UK as a result of changes in the social, economic and technological aspects of society. Box 16.1 identifies some of the reasons for development during the second half of the twentieth century and Box 16.2 suggestions for development at the beginning of the twenty-first century.

16.2 Marketing

When the speculative development project is to be disposed of after completion, then its promotion and marketing become an integral part of the development process. It is essential to have a clear understanding of what is being provided and a knowledge and understanding of the potential market for the project. The time lag between inception and completion can be considerable. Many projects have been quickly started on-site in a time of relative prosperity and could have been disposed of many times. But sometimes by the time of completion, perhaps five years later, the market has deteriorated or the needs have been switched to other types of property. It is then that those responsible for its disposal may have to work very hard to secure potential purchasers. A clear view, based on a clear analysis of the past and projections into the future, is essential at the outset and needs to be monitored during

Box 16.1 Reasons for development during the second half of the twentieth century

- General rebuilding after damage from two world wars.
- Improvements in housing standards and quality.
- Schools for increasing numbers of pupils.
- Hospitals and health buildings for the National Health Service created after the Second World War.
- Universities to satisfy demand by the increasing numbers going into higher education.
- Redevelopment of commercial and retail town and city centres.
- Factories and warehouses required for increased automation and new technologies.
- Office buildings for the service sector and to accommodate new technologies.
- Roads and motorways for the motor car.
- Multi-storey car parking.
- Power stations to meet increased demand from all kinds of users.
- Airports to meet the needs for increased air travel.
- Out-of-town shopping malls.
- Change of use of existing premises, refurbishment and conservation.
- Decay, obsolescence and redevelopment resulting in demolition.
- Tourism and leisure developments.

Box 16.2 Developments during the early part of the twenty-first century

- Millennium projects.
- Decommissioning and replacement of nuclear power stations.
- Toll roads.
- Domestic housing required for changing patterns of households and increased life expectancies.
- Tourism and leisure facilities.
- Anti-terrorism initiatives within the infrastructure, existing buildings and new projects.
- Sports stadiums.
- Urban redevelopment including inner city areas.
- Airport enlargement, although not on the scale of Terminal 5 at Heathrow Airport.
- International sports arenas where the UK is successful in bidding for events.
- Building Schools for the Future programme.

development. It may be possible to switch a design from, say, offices to flats, when the demand for offices declines and the demand for flats increases. However, this is not generally easy due to planning constraints in terms of a change in use. Alterations in respect of design and construction implications are also difficult unless the design is flexible, perhaps just a shell for fitting out later.

16.2.1 What to build

The first stage that a developer will need to consider is just what type of project should be envisaged. Where a scheme can be designed that might usefully serve a range of client types, then this clearly has some advantages, if planning permission can be obtained. The developer will need to consider the other types of property in the location where the development is envisaged, particularly those that are in direct competition with the proposal. The decision involves an assessment of costs and income. It will be necessary to establish through market research if there is an unsatisfied demand in an area for a particular type of project. The relative shortage of types of property can generally be identified by agents.

16.2.2 Where to build

Developers need to construct projects in areas where there is the expected demand. Some developers work on a national basis, others confine themselves to a close geographical area with which they are familiar. These developers understand the regional planning framework and have built up close working relationships with designers and contractors. The where-to-build scenario depends upon sites looking for projects and projects looking for sites. A developer will have an interest in both of these options.

16.2.3 How to build

In essence, the principle today must be adaptability and durability. The principle is of long life, loose fit and low energy, sometimes referred to as the three Ls concept. A building's life expectancy is sixty years or more. During that time there are likely to be many changes in ownership or tenancy and even more technological developments. The property may also undergo a material change of use during its lifetime. The managing agents for a development project will be able to offer advice on letting, supervision of repairs, rent reviews, etc., and how these factors are likely to have an impact upon the design of the development. Today, there is an emphasis upon low costs in use; certainly, where this approach is adopted as a design principle, it will reduce the occupants' ongoing costs. This factor will then have the likely effect of enhancing the rental values of the property. Generally, higher-value locations attract a kind of occupier that will expect commensurate high standards, with consequently higher prices. For example, new office projects constructed in London are unlikely to find purchasers unless they are equipped with air-conditioning systems, computing and telecommunications networks.

16.2.4 When to build

It is desirable to build when construction costs are at their lowest. This implies a general scarcity of construction work and consequently a better availability of skilled craftspeople. In times of recession there is also a better chance of obtaining a higher-quality project at a less expensive price. Taking into account that projects may take at least two to three years to bring to fruition and, assuming that a recession will not last for ever, then this seems to be a good time to develop. Finance charges may also be lower during a recession. But all development relies on analysing forecasted trends, and it is essential that projects are available to meet these trends, otherwise the opportunity is lost. In some countries, the boom and slumps in construction activity are kept to a minimum by the public sector only building when prices are lower in times of recession.

One of the secrets of effective property development is anticipating the building needs of tomorrow and in the future. The developer must be able to anticipate when projects will be required for occupation, be able to assess long-term trends and then backdate them to the development process. During the 1980s there seemed to be an insatiable appetite for high-quality office accommodation in and around London, with needs mirrored elsewhere throughout the UK. There seemed no end to the possible demand for such types of property, with site values and rents rising accordingly. The square mile of the City, with the Bank of England at the centre, could not accommodate new offices. Many were too old-fashioned, too small and most importantly did not have the kind of space required to install the maze of cables for computer terminals. These were now the tools of the financial services industry. This was 1982. However, by the end of the 1980s, the country and world markets were heading into one of the worst recessions that had been seen that century, and that was to last for more years than many analysts predicted. Later forecasters predicted that any upturn in the economy would only result in moderate levels of economic activity for the foreseeable future. Consequently, new development proposals were curtailed, and in their wake several well-known development companies ceased to trade, most notably the developers of Canary Wharf in London's Docklands. This experience emphasises the importance of taking the longer-term view and accepting that slump follows boom as night follows day.

16.3 Customer satisfaction

Table 19.1 represents a survey of customer satisfaction published in *Constructing the Team* (Latham 1994). The data is based upon attitudes of clients who commission construction projects and have an interest in the ownership and use of property. It seeks to compare buildings with motor cars against a number of selected criteria. Some will not recognise the results of such a comparison, but the points made cannot easily be ignored. For example, in 1982 the Building Research Establishment commissioned a survey into housing defects. This found that, on average, there were forty-two faults per dwelling after the dwellings were handed over to the client. This statistic compares poorly against the number of faults found in a new motor car.

Note that where clients are satisfied with their building projects, they are more likely to want to build again or to recommend the process to others. This information should help to inform the industry on the qualities expected and how they are perceived to occur in practice.

16.4 Site identification

The development of land and buildings is promoted by:

- property developers who have identified the need for entrepreneurial projects
- commercial and industrial corporations, developing for their own use
- public and statutory authorities developing for their own functions
- house builders and housing associations
- charitable organisations
- government-sponsored development agencies
- other private clients.

The most important factor that influences development is the developer's perception of market conditions. This knowledge is important in identifying the type of development, the region in the country in which the development is likely to take place and finding a suitable site that satisfies its own criteria. The developer's ability and knowledge are important in identifying areas of potential growth for development, considering that it will take some time before such a project can become operational. Whilst there are always likely to be risks involved, the developer will use a range of market research techniques in order to identify levels of supply and demand, trends in expectations and measures to stay ahead of possible market competition. There are many considerations of sites to be undertaken for different development purposes (see Ashworth 2008).

16.5 The development process

There are several ways of describing and presenting the property development process. In its simplest analysis, it consists of design, construction, use and disposal. The process should be looked upon as a cycle of activities where land, once brought into development use, goes through these different stages. In prime city centre locations, where sites are more valuable, disposal and demolition are usually only a precursor to further property development. Returning land to its greenfield status is not a common occurrence in a society that continues to expand its range of activities, with a consequent need to use more of this scarce and limited resource.

The development cycle for a construction project can be best classified into five different stages and their associate phases (Table 16.1). These are not discrete activities. The respective functions of the architect, quantity surveyor, engineer and contractor are shown in Table 16.2.

Table 16.1 Comparison of RIBA work stages, OGC gateway process and RICS formal cost estimating and cost planning stages

RIBA work stages 2007		*RIBA work stages 2013*	*NRM1 cost plans*	*OGC gateway applicable to building projects*
Preparation	A Appraisal	1 Preparation	Order of cost estimate	1 Business justification
	B Design brief			2 Delivery strategy
Design	C Concept	2 Concept design	Formal cost plan 1	3A Design brief and concept approval
	D Design development	3 Developed design	Formal cost plan 2	
	E Technical design	4 Technical design	Formal cost plan 3 Pre-tender estimate	3B Detailed design approval
Pre-construction	F Production information		Bill of quantities	
	G Tender documentation			
	H Tender action		Post tender estimate	
		5 Specialist design		3C Investment decision
Construction	J Mobilisation	6 Construction (offsite and on-site)		
	K Construction to practical completion			4 Readiness for service
Use	L Post practical completion	7 Use and aftercare		5 Operations review and benefits realisation

Sources: RIBA 2008; RICS 2009; Office of Government and Commerce 2008.

Table 16.2 The development process

Architect	*Quantity surveyor*	*Engineer*	*Contractor*
Pre-contract client brief			
Identifying the client's needs and wants in terms of space, function and aesthetics	Guide cost information Financial factors such as taxation and funding Order of cost estimate		Only involved if negotiation, design and build or two-stage tendering
Investigation			
Feasibility and viability Planning possibilities Alternative sorts or types of construction Outline design approval	Approximate estimate Formal cost plan 1	Site survey and investigation Preliminary structural calculations	
Sketch design			
Outline planning permission Major planning problems solved	Formal cost plan 2 Life cycle cost plan		
Design			
Development of sketch plans Constructional details determined Constructional methods decided	Cost implications and cost checking Formal cost plan 3	Constructional methods decided Preliminary schemes for services Design schemes for structure	
Working drawings			
All construction drawings and details completed	Quotations from specialist firms Final cost checks Bills of quantities		All construction drawings and details completed
Tender stage			
All members of the design team check their calculations			
	Receipt of tenders and tender checking Post tender estimate		Contractors prepare tenders Materials and subcontractor quotations Determination of market factors Contractors' method statements
Post-contract construction			
Inspections of quality Subcontractor nomination Issue of instructions Issue of certificates	Valuations, forecasts and payments	Inspection of quality	Construction programme agreed Planning and progress control Subcontractor control
Maintenance			
Contractually up to the end of the defects liability period in-use repair and maintenance, changes in use, conservation, demolition, alteration and reconstruction, asset valuation, etc.			

These tables are based on the RIBA work stages. The titles of most of the work stages have been revised, largely to take into account required output expectations. The different functions merge into each other as the project moves through its life cycle. Emphasis should always be placed on securing those developments which best satisfy the criteria that will have been identified by the developer or client at the beginning of the development. A brief will have been written that identifies the type and scale of the project, its standard of construction, funding availability, its costs and the date when the project should be available for handing over for occupation. During each of these activities, those involved with the development will have different tasks to perform, in respect of designing, costing, forecasting, planning, organising, motivating, controlling and coordinating. These are some of the roles of the professions involved in managing property and construction, whether it be new build, refurbishment or maintenance. To these activities should be added research, innovation and improving quality and standards.

16.5.1 Appraisal

This first stage in the process involves identifying the employer's requirements and the possible constraints on development. The development process is initiated principally by either a project looking for a suitable site, or an available construction site looking for a project. The available site may sometimes be an existing building awaiting demolition or redevelopment. The development project may cover the whole spectrum of building types constructed for the public or private sector, including housing, commercial, industrial, recreational, social or activities as remote as forestry and agriculture. The initiative may come from a developer, the site owner or a client seeking a site for a proposed development. The planning authority, too, may make recommendations or designate revised land use patterns in an area. Many projects arise from long-term programmes where clients consider the scheme as a part of the overall objectives of their own organisation. Studies will be undertaken to enable a client to decide on whether to proceed and, if so, which procurement route should be selected.

16.5.2 Strategic briefing

The strategic briefing stage, which is done by or on behalf of the client, identifies the key requirements and constraints involved. It identifies the procedures, organisational structure and the type and range of consultants to be used. It is important during this early part of the process to consider a range of issues that are going to determine whether the project has any chance of coming to fruition. There is little point in expending large sums of money or time on a project that will never be constructed. The client or the developer will prepare an outline brief of the proposals and issues that need to be settled. These will include an analysis of the market potential and the costs of the development in very broad terms. For example, it may be necessary to consider any public funding that might be available in respect of grants or loans for a project in the private sector. During this stage it will also be necessary to determine whether the proposed site, if one has been located, is suitable for the project envisaged in terms of its location, size and ground conditions. It will be important to establish as soon as possible whether the project will receive planning consent from the local authority. It would also be prudent to establish matters of ownership, rights of way and other factors that might affect the whole process of development. When these are established, it will be necessary to purchase the site, generally with the condition that outline planning permission will be granted.

The developer will need to arrange for land transfer and the finance for the development. Two sorts of finance are generally sought. The first, known as short-term finance, will cover the costs of the development until the project is disposed of to a client. The second, long-term finance, sometimes referred to as funding for development, is to cover the costs of owning the development as an investment property.

16.6 Feasibility and viability

During the feasibility and viability phases, the client's or the developer's objectives for the project become established. Clients involved in single projects often come to their chosen consultants with a broad outline of their aspirations, a sum of money, which is often insufficient, and a timescale for occupation, which is often impossible. The better-informed clients, those involved in frequent capital development, usually have more realistic expectations of what can and cannot be achieved. The type of project will often determine who the client or promoter appoints as designer. On building projects this has traditionally been the architect. However, whilst traditions die hard, the building surveyor is increasingly being appointed to oversee smaller works and schemes of refurbishment on behalf of the client. As the different combinations of procurement are employed, such as design and build or management contracting, clients are now often appointing the construction firm direct, choosing an alternative consultant as a main partner in the venture or appointing a project manager in overall charge of the scheme.

The feasibility phase seeks to determine whether the project is capable of execution in terms of its physical complexities, planning requirements and economics. The available site, for example, may be too prohibitive in terms of its size or shape, or the ground conditions may make the proposed structure too costly. Planning authorities may refuse permission for the specific type of project or impose restrictions that limit its overall viability, perhaps in terms of the return on capital invested. Schemes may be feasible, but they might not be viable.

16.6.1 Outline proposals

The various options of choosing a separate designer from the contractor, or preferring design and build, are well documented in Chapter 11. Each has its own advantages and disadvantages. In any event, it is necessary to prepare schematic outline proposals for approval prior to a detailed design. These proposals will need to be accepted by the client or developer in terms of the requirements outlined in the brief. They will also need to be accepted by the relevant planning authorities for their official permission, and in terms of funding through the preparation of an initial cost budget. An early estimate of the proposed costs and a developer's budget will be required.

16.6.2 Detailed proposals

As the scheme evolves and receives its various approvals, a number of different specialist consultants will be employed. Some may be public relations consultants, particularly where a sensitive scheme, such as a new road, building in the green belt or project out of character with the locality, is being proposed. During the sketch design, the main decisions regarding the project's layout and form and the quality of materials and standards of construction will be agreed. A cost plan of the proposed project will be prepared by the quantity surveyor in order to guide the designers during the later stages of this process. An architect or planning consultant will be responsible for providing the well-thought-out scheme in order to secure planning permission. It has been estimated that more than 60 per cent of all planning applications are now dealt with by professionals other than architects.

16.6.3 Final proposals

When the scheme has been agreed and approved by the client, further investigations will then be undertaken in order to prepare the detailed design. Different solutions to spatial and other design problems will be considered, and some of these will require revisions to other aspects of the project which have already been agreed. Each solution will need to be costed to ensure that the cost plan remains on target and, where it significantly affects the client's proposals or developer's budget, it will require agreement before proceeding further with the design. It will be necessary during this stage to consult firms who supply or install any specialist equipment required.

In the UK, any change in use requires the approval of the appropriate planning authority. This is normally the local authority. Where planning permission is rejected, there is recourse to the Secretary of State for the Environment, Food and Rural Affairs, who may initiate a planning inquiry. The acquisition of planning permission can be a highly complex and technical activity, needing a detailed knowledge of legislation and government policies as well as local knowledge relating to the site of the proposed development. Obtaining planning permission may also involve the developer in additional planning agreements with the local authority, where additional conditions, described as planning gains, are sometimes required before planning permission is granted. These agreements inevitably increase the development costs for the proposed project. In some circumstances, it will be necessary to obtain further approvals, such as listed building consent, that is, the right to demolish or alter a protected building. There are about 500,000 structures in Britain protected because of historical or architectural importance. Listed buildings are grouped into three categories, graded 1, 2* and 2. Grade 2 offers minimal protection, but it is virtually impossible to alter or make any changes at all to a Grade 1 property. Information on this can be obtained from the Department for Culture, Media and Sport.

Clients and developers now need certainty about the costs of the proposal in order to input realistic and reliable information into their budgets. When it is known that planning permission will be forthcoming, the plans should then quickly achieve a level of detail in order for the quantity surveyor to provide a detailed estimate of the likely costs of construction. The cost plan will already have been prepared and this will be frequently updated to take into account modifications arising from planning and changes in design.

16.6.4 Production information

The production information is considered in two parts. The first part is concerned with providing adequate information that is sufficient to obtain tenders. The second part includes the balance of information that will be required under the building contract to complete the information for construction purposes.

16.6.5 Tender documentation

The documentation required for tendering purposes will be prepared at the end of this process. This will depend to some extent upon the procurement method that has been selected. When the project is approaching the tender stage, the firms which may be interested in constructing the project should be invited to tender. The long periods of time that elapse reflect the design and planning complexity required for solutions to bespoke designs in construction projects.

16.6.6 Tender action

Upon receipt of the documentation, the contractors enter their estimating phase, since the awarding of the works of construction is most frequently done through some form of price competition. The contractors' bids will be evaluated against price and other considerations (see Chapter 7) and a recommendation made to the client.

16.6.7 Mobilisation

This is the letting of the building contract to the successful firm and appointment of the contractor. The production information is issued to the contractor, and arrangements are made in respect of handing over the site to the contractor.

16.6.8 Construction to practical completion

This is the stage when the contractor commences the work on-site. It has often been referred to as the post-contract period, since it commences once the contract for the construction of the project has been signed and work has started on-site. Where the project is on a design and build arrangement or a system of fast-track procurement, this stage may start before the design is finalised, and then run concurrently. Contractors are critical of the traditional arrangements since they are frequently required to price works which, although assumed to be fully designed, are in reality not so. Throughout this stage, formal instruction orders are given to the contractor for changes in the design, and valuations are prepared and agreed for interim payment certificates. Contractual disputes all too frequently arise, all too often due to misunderstandings or incorrect information being made available to the contractor. The contractor is also sometimes overambitious and enters into legal agreements that become impossible to fulfil. These create grounds for damages on the part of the client. Project completion times can last from a few months up to ten years or more. Upon completion, the formal signing over of the project into the responsibility of the client is made.

16.6.9 After practical completion

One of the main tasks is now to ensure that the project can be completed to the specified quality, the calculated costs and the appropriate timescale. Commitments may have been made to future purchasers or occupiers, who will themselves have prepared their own plans for taking over the property. Anticipated problems need appropriate action to ensure that the project stays on target in respect of time and budget. Changing circumstances may mean that some variations to the scheme need to be instructed in order to maximise the potential for the finished product. Some factors remain outside the control of any of the parties involved, but the essence is how effectively and quickly these are resolved. The satisfaction of clients and developers centres around completion on time, at the agreed price and to a quality and standard that has been specified in their original brief. Satisfied clients are likely to recommend the company to others, and thus offer a great marketing potential.

16.7 In use

This is the longest phase of the project's life cycle, but one that the developers will keep at the forefront of their minds. The immediate aims of development are now hopefully satisfied and the project can be used for the purpose of its design and construction. However, no development is complete until occupiers or purchasers have been found who are willing and able to pay the rents or purchase the property. Forecasting future demand for development projects is difficult owing to the long time lag between inception and completion. The collapse in the need for property, due to sudden changes in the economy, can create financial disaster for a developer. Shrewd developers will attempt to make allowances for everything, even the unknown.

Routine maintenance will be necessary during this stage. The correct design, selection of materials, proper methods of construction and the correct use of components will help to reduce maintenance problems and their associated costs. A sound understanding, based upon feedback from project appraisals in practice, will help to reduce the possible future defects. Defects are often costly and inconvenient, and minor problems sometimes require a large amount of remedial work to rectify, sometimes out of all proportion to the actual problem that has arisen. Many projects have only a limited life expectancy before some form of refurbishment or modernisation becomes necessary. The introduction of new technologies also makes previously worthwhile components obsolete. City centre retail outlets have a relatively short life expectancy before some form of extensive refitting becomes required. Fifteen years seems to be an optimum age. Whilst the shell of buildings may have a relatively long life of up to sixty years, and some are able to last for centuries, their respective components wear out and need frequent replacement. Obsolescence may also be a factor to consider in respect of component replacement.

16.8 Demolition

The final stage in a project's life is its eventual disposal, demolition and a possible new beginning of the life cycle on the same site. Demolition becomes necessary through decay and obsolescence and when no further use can be made of the project (see Ashworth 2010). Some buildings are destroyed by fire, vandalism and explosion, or may become dangerous structures that require demolition as the only sensible course of events, years before the end of their expected lives. Other projects may need to be demolished because they are located in the middle of a redevelopment area. There are relatively few projects that last for ever and become historic monuments. Whilst some of this is attributable to decay, the style of living and the changing needs of space are constantly evolving to meet new challenges. Some projects of notoriety become listed buildings. The Secretary of State for the Environment has powers under the planning acts to compile lists of buildings of special historic interest. It then becomes difficult to demolish, alter or extend these buildings in any way that would affect their character. Where non-listed buildings are thought to have special historic or architectural interest, a planning authority may also serve a building preservation notice upon the owner. Whilst the planning regulations are onerous, their aim is to allow development to take place in an orderly fashion. In the long term, this must be the best policy for society. Projects that might have taken several years to plan and develop, and then cherished for decades, are finally removed from the urban landscape by demolition. In some cases, this is swift, where a lifetime's project can be reduced to a heap of rubble in a matter of minutes.

16.9 Environmental impact assessment

Environmental impact assessment was established in the USA as long ago as 1970. This is now a worldwide concept and a powerful environmental safeguard in the project planning process. The original EC (now EU) Directive 85/337 was adopted in 1985 and since then the individual member states have implemented the Directive through their own regulations. In the UK, the resulting environmental impact statements increased more than tenfold between the early 1980s and the early 1990s. As a result of the Directive, more than 500 statements are now prepared annually in the UK. The required contents of a statement are given in Annex III of the Directive:

- Description of the project: physical characteristics, production processes carried out, estimates of residues and emissions.
- Appropriate details of alternative sites and their possible effects.
- Description of aspects of the environment that are likely to be affected, such as population, fauna, flora, soil, water, air, climatic factors, material assets, architectural and archaeological heritage, landscape, and their interrelationship.
- Description of the likely effects on the environment of:
    - existence of the project
    - use of natural resources
    - emission of pollutants.
- Description of measures envisaged to prevent or reduce any adverse effects on the environment.
- A non-technical summary of the above.
- An indication of any difficulties encountered by the developer in compiling the above information.

In the UK, the Directive is implemented in the Town and Country Planning (Assessment of Environmental Effects) Regulations 1988. Further guidance is included in *Environmental Assessment: A Guide to Procedures* (Department of the Environment 2008).

16.10 Neighbourhood environmental data

Information on potential environmental issues relating to a particular development site or an existing building should include a consideration of the following:

- *Flood*: flood data is provided by the Environment Agency, which uses the best information currently available, based on historical flood records and geographical models. The data indicates where flooding from rivers, streams, watercourses or the sea are possible. However, the data does not show flood defences that offer vital protection in many areas. Nor does the data cover flooding from other sources, such as burst water mains, road drains, run-off from hillsides, sewer overflows, etc. Flood forecasting is not a precise science and the data can only give a general indication of risk areas.
- *Subsidence*: there is a risk of foundation damage to properties in some areas from natural subsidence hazards. Damage may be a combination of the type of building and forms of mining. The effects of more localised foundation damage from trees or other vegetation or from man-made hazards such as excavations or leaking drains may also have a detrimental effect on properties in an area. Data can be obtained from the *British Geological Survey.*
- *Radon*: radon is prevalent only in certain parts of the country. Contamination is determined by the action level and data can be obtained from the National Radiological Protection Board (NRPB).
- *Coal mining*: by existing within a coal mining area does not mean that all properties have been or may be affected by coal mining. In certain areas, it may indicate the presence of

workable seams of coal. A coal mining search can be obtained from the local authority. Further information can also be obtained from the Coal Authority.

- *Landfill*: information on the existence of current landfill sites is extracted from public registers maintained by the Environment Agency. Information on past landfill sites is compiled from a variety of sources by Sitescope Limited and is subject to an ongoing quality assurance exercise. The locations of landfill sites are estimated from grid references shown on the licence; the Environment Agency does not have the boundaries of landfill sites available. This data does not include illegal dumping or tipping.
- *Waste*: information on the existence of waste processing sites is provided by the Environment Agency. The data does not include the processing by unlicenced scrap yards or sites.
- *Historical land use*: an analysis of Ordnance Survey maps published from about 1880 up to the current day will indicate the previous uses, and especially a history of past industrial uses. Historical land use data is captured by Sitescope Limited.
- *Air quality*: air quality readings are based on a one-kilometre area in which a particular property is located, usually a postcode classification. Data is supplied by AEA Technology.
- *Pollution*: the identification of possible risks can be made from a number of public registers held by the Environment Agency, the Valuation Office, the Health and Safety Executive and the Department of the Environment, Food and Rural Affairs.

A concern indicator represents the composite degree of concern for one or more sources of risk being present. Where the concern level is low, this does not necessarily mean that pollution is present; rather, it indicates that it would be prudent to make further enquiries. The sources and their presence are:

- Pollution inventory: location of industrial sites.
- Integrated pollution control consents.
- Sites licenced for radioactive substances.
- Bulk fuel and petrol storage sites.
- Control of industrial major accident hazard sites (CIMAH).
- Sites for planning permission for hazardous substances.
- Installations handling hazardous substances.

17 Parties involved in the construction industry

The main purpose of construction activity is to provide a completed project for the building owner. This project may include the substance of a building contract, a project constructed speculatively for a developer, or civil engineering infrastructure works such as roads, bridges or pipelines. The employers, clients or promoters of the construction industry are many and varied. They include the public sector bodies, such as central and local government, private companies involved in building for domestic, commercial, industrial and retailing purposes and now a range of the quasi-public companies that were formerly part of the nationalised industries, those of coal, electricity, gas, water supply, sewage treatment, etc. Many of these latter companies are virtual monopolies with limited competition from other organisations.

Construction activity is typically divided on a 60–40 per cent basis in favour of new projects compared with repair and maintenance activities. This figure remained almost at a constant level throughout the late 1990s. However, the division between public and private sector workloads has since changed markedly, largely due to the privatisation programme of different governments during this time. Typically, the public sector used to account for about 50 per cent of the total output; by the start of the 1990s, this had fallen to less than 25 per cent. This figure still remained by 2015. Civil engineering works are worth about 20 per cent of the output of the industry. Work done by British companies overseas currently accounts for about 10 per cent of the total output of the construction industry. Some of the largest clients of the construction industry, e.g. British Airports Authority and Tesco stores, each spend in excess of £1 million per day on construction work.

Organisations are given a considerable amount of autonomy by the government of the day, and it is interesting to note the wide diversification in the methods used for the procurement and execution of major and minor capital works projects. Little uniformity exists either in the design procedures employed or the contract conditions used.

Clients in the public sector may be influenced by both social and political trends and needs, and the desire to build may be limited by these factors. They will nevertheless be restricted in their aspirations by the amount of capital they are permitted to borrow for these purposes. The private sector, which encompasses private housing ownership and the large multinational corporations, directs capital spending to the ventures considered to be money-making. However, in both sectors there has recently been a particular emphasis upon securing added value, and this has tended to be viewed on a building's life cycle rather than initial construction costs alone.

17.1 Employers

The employer is one of the parties to the contract, the other being the contractor. Each client will have different priorities, but essentially there will be a combination of:

- *performance* in terms of quality, function and durability
- *time* available for completion by the date agreed in the contract documents
- *cost* as determined in the budged estimate and the contract sum.

If employers are to be satisfied with the product, i.e. their construction project, then these three conditions must be critically examined. Box 17.1 includes some of the more important references to the employer in JCT contracts.

17.1.1 Developers

Property development involves a range of activities and, by their nature, they will involve developers with different objectives. All of them are concerned with future projections and expectations. Property developers work on margins between the cost and the sale price, sometimes based on cost and sometimes on sale price. In the absence of this margin, there would be no financial incentive for development. There are many different arrangements that can be provided for development; however, developers can essentially be broadly classified in two categories. Each may undertake work speculatively based upon market intelligence:

- *The investor developer* aims to retain ownership of the project. Short-term bridging finance may be required for construction and then a long-term loan, often from one of the institutional sources.
- *The merchant developer* completes projects using short-term funding and then sells them to an owner or occupier. The tax advantages of merchant development are examined later in this chapter.

The aims and objectives of property development are wide and diverse. The different types of developer have their own particular needs and desires regarding the project.

Box 17.1 Important factors to the employer in JCT

- Appointment of the architect
- Powers regarding insurances
- Duty to give possession of the site to the contractor
- Powers in respect of damages for non-completion
- Powers to determine the employment of the contractor
- Powers to engage directly employed contractors
- Duties regarding certificates
- Procedure in respect of adjudication, arbitration and litigation

Source: Joint Contracts Tribunal suite of contracts.

Some property developers may choose to specialise in terms of location, whilst other companies may offer only certain types of property, such as offices, industrial premises or housing. In some cases, the developer may provide a package deal arrangement to supply and construct factory-made (industrialised) units, that have the advantage of being available very quickly. Others will choose to concentrate on a particular process, such as conservation or refurbishment, in order to develop a niche market for their services. The advantages of specialisation enable the company to gain an above average level of knowledge and expertise.

17.1.2 Occupiers

Occupiers require buildings that suit their particular needs. The prime objective is to provide a building that best serves these particular needs, with benefits achieved from occupation and less concern for its market valuation. An industrialist, for example, will require premises that allow the production process to be carried out in the most effective, efficient and economic manner. The profits from such a business far outweigh any changes in the market value of the property.

A developer who also intends to become the occupier is able to specify a building that meets personal requirements as closely as possible. In some cases, the development may be so inflexible in its design that it cannot be easily utilised by others and its relative market value may therefore be small. Buildings such as schools and hospitals are essentially designed and constructed to suit their functions, with only limited concern for site values and possible resale opportunities. Where such projects are adapted for other uses, extensive conversion work is often required. It may be necessary to build such projects on sites that might otherwise have only limited development potential. Occupiers are generally more concerned about spatial arrangements and function than the possible long-term investment.

17.1.3 Investors

The investor's view of property is similar to that of the property company – financial gain. However, investors tend to take a longer-term view, expecting both the capital and income to increase over the invested life, which may be several years. The acquisition and disposal of property investments can be costly, and it is therefore necessary to have allowed an investment at least some maturity before converting it into other assets. When investors become involved in the development itself, they expect higher returns. The risks involved in project development are greater than those of a building for a client.

Investors are generally cautious, disliking unconventional investments that, because of their nature, may be unpredictable and difficult to dispose of at some future date. Property that has an unusual design, has been constructed with new methods or materials, has unconventional lease terms or involves substantial management capability will not be favoured as an investment potential.

Investment companies generally manage a portfolio of investments to spread risk across a number of geographical areas as well as different industry markets and commodities.

17.1.4 Builders and contractors

A building company may seek to enlarge its range of activities by carrying out development work of a speculative nature. This is often done for taxation and legal reasons through a separate company charged solely with this task. In this case, the company will become involved with the additional risks associated with land purchase, finance acquisition and sales or lettings. In this respect, the building firm is largely acting in the same way as a development company, with the added bonus of being able to profit from the building construction operations. When the firm acts only as a building company, the profits accrued largely result from the activities relating to the construction work being performed. Additional benefits of combining both development and construction are that the resources the contractor employs in terms of the workforce and expertise might be able to be retained through, for example, undertaking more development work when contracting work is not available or at too competitive prices. The building developer is therefore the reverse of the property company who employs a building contractor for a proposed development.

17.1.5 Public sector

Over the last century, the public sector has had a growing involvement with development projects, usually as a client. Their projects include housing, hospitals, education, roads, public utilities, etc. However, restrictions on public sector borrowing and spending and the privatising of many of the nationalised industries has accounted for some decline in recent decades.

Public sector policies are influenced by political ideals and the government holding power and control, both nationally and locally; however, there is the need for accountability in terms of raising and spending finance and communities' social requirements. Many public sector development projects would not be undertaken by the private sector, since they are unable to show financial profitability, but their provision can often demonstrate benefits that accrue from the expended costs; benefits that are of tangible worth to the country and the community. It is important that a public authority can demonstrate that it has acted lawfully and that all of its dealings are free from even a hint of suspicion of corruption. Public accountability often includes some element of public participation.

Most public works building projects, undertaken through central or local government or other government-controlled agencies, are directly related to the particular interests of the authority or government department. Traditionally, much of this work would be commissioned by, for example, a county or borough council. The design might have been undertaken by its own staff or private consultants and, for projects other than the smallest, a contractor would be appointed. In more recent years, some local authorities have undertaken development projects, such as the provision of industrial units or business parks, in order to attract commerce and industry to their locality. These organisations are, however, different from the private developers in several ways. They can only undertake work within their powers, otherwise they will be acting *ultra vires*.

17.2 Landowners

Landowners include the traditional groups such as the Crown Estates, the Church Commissioners and the landed aristocracy. Although certainly motivated by economics, such as return on capital invested, these landowners are also concerned with political and social issues to do with the land. Another category of landowners are the industrial corporations, who use land because it is incidental to their production processes. This group also includes farming and agriculture, retailers, etc. The former nationalised industries also fit comfortably into this group. Their principal motives for landownership are with the use of land, rather than for any financial investment that the land might otherwise provide. A further category is the financial institutions, where land is seen solely as a means of investment and this remains their prime, or only, reason for owning land. These are the most informed group regarding land and property values.

Landowners can have a major influence over the type of development that might be undertaken. Only the state has a bigger influence, through its planning procedures. It can encourage and discourage development and has the power to prohibit development that does not fit in with the plans produced by the local planning authorities.

17.2.1 Crown Estates

The Crown Estates is one of the most important landed estates in the UK and includes substantial urban, rural and marine interests. It is part of the hereditary possessions of the sovereign *in right of the Crown* that is managed by the Crown Estates Commissioners. Its origins go back to the reign of Edward the Confessor, and it has more than 300,000 acres of agricultural land, making it the largest agricultural landlord in the UK. Until the time of George III, who came to the throne in 1760, the reigning sovereign received its rent and profits. The net surplus is now paid to the Exchequer.

In the mid 1990s, the estate achieved a revenue surplus of almost £80 million. Its property values are now worth in excess of £3,000 million. It has a wide-ranging and quality property portfolio, with ownership of more than 1,000 listed buildings, 750 of which are located in London. Almost 50 per cent are Grade 1 listed, compared with the national average of 2 per cent. Its London properties are located in Central London and the West End, with more than 8 million square feet of office space, 2.5 million square feet of retail space and more than 1 million square feet of miscellaneous property, including hotels, clubs and residential accommodation, making it one of the capital's largest landowners.

The estate does not have borrowing powers, which creates both a constraint and a discipline on its activities. It resisted the temptation to invest in Docklands and in 1990, the start of the decline in property prices, it introduced a moratorium on development.

17.3 Professional advisers

Because the development process is so complex, and because most employers do not have the range of skills and expertise required, it is necessary for them to employ a range of professional advisers to advise on funding, design, costs, construction, letting, etc. These advisers will vary depending upon the type, nature and size of the project being envisaged, and might include some or all of the following.

17.3.1 Architects

The architect has traditionally been the leader of the design team. In the building process, where design and construction are separate entities, it is the architect who receives the commission from the client. Because projects today require high levels of specialised knowledge to complete the design, the architect may require the assistance of consultants from other professional disciplines.

The architect's function is to provide the client with an acceptable and satisfactory building upon completion. This will involve the proper arrangement of space within the building, shape, form, type of construction and materials used, environmental controls and aesthetic considerations – all within the concept of total life cycle design.

The architect's duties and powers are described under JCT 2005 and JCT 2016. A contractor who believes the architect is attempting to exercise powers beyond those assigned under the contract can insist that the architect specifies in writing the conditions that allow such powers.

The architect will generally operate under the rules of agency on the part of the employer. This means that instructions given to the contractor will be accepted and paid for by the employer.

In some forms of contract, the architect is termed the supervising officer. This is the name used in the GC/Works/1 form of contract, and is one of the alternative titles suggested in the JCT group of forms. It is used in the GC/Works/1 form since the designer may be an engineer rather than an architect, and the terminology has been extended to the JCT form under a similar assumption.

The scope of the work undertaken by the architect may be broadly divided into pre-contract and post-contract duties. Although it is more common for an architect to provide a fully comprehensive design and supervision service, a design-only service may be required by the employer. It is less common to expect the architect to supervise construction work only, although it could arise in situations where prefabricated buildings are used. However, even in these circumstances the architect is more likely to be asked for advice on a particular system building during the design stage.

In the normal pre-contract stage, the architect's basic duty is to prepare a design for the works. This may involve three facets: architectural design, constructional detailing and administration of the scheme. This latter aspect will entail integrating the work of the various job architects and other consultants, and ensuring that the information is available for a start on-site when required. During the work, the architect must exhibit reasonable skill and care in the design of the works. This duty may be established in accordance with normal trade practice. The architect will also generally be held responsible for any work delegated to another. In the JCT form of contract, for example, although the quantity surveyor is responsible for most of the financial arrangements, the architect is ultimately responsible regarding the certification of monies to be paid. If part of the design is undertaken by a nominated subcontractor, some protection may be afforded by a warranty from that firm.

During the post-contract stage, the work undertaken by the architect is largely supervision and administration. Some drawings and details may still need to be prepared, particularly where such information is reasonably requested by the contractor. The purpose of supervision is to ensure that the works are carried out in accordance with the contract. The amount of supervision necessary will vary from project to project. A complex refurbishment project will require more frequent visits than the construction of a large warehouse shed. On very large contracts, the architect may even be resident on-site. The duties of administration are used to describe the various functions that must be carried out during the progress of the works, such as issuing instructions to the contractor. The post-contract stage involves those duties described in JCT contracts. Boxes 17.2, 17.3 and 17.4 refer to responsibilities in JCT contracts.

17.3.2 Surveyors

There are several types of surveyor, which include general practice surveyors, building surveyors and quantity surveyors. Many laypersons confine the term 'surveyor' to a land surveyor. Land surveyors are involved in mapping and surveying the land in terms of its location, line and level.

General practice surveyors are employed in four main areas of work: agency, valuations, management and investment. Their knowledge and understanding of the local property market, land and property values are the particular attributes of this profession. Valuation is one of the main skill bases vital to investment work. These may be required for a variety of purposes, such as sale, lease, insurance, investment or loans, and for a range of clients, such as developers, purchasers and property owners. General practice surveyors may be involved at the outset of a

Box 17.2 Architect's instructions under JCT

- Discrepancies and divergencies between documents
- Justification of instructions
- Instructions to be in writing
- Confirmation of verbal instructions
- Divergence between statutory regulations and project documents
- Opening up of work for inspection
- Removal from site of work, materials or goods which are not in accordance with the contract
- Exclusion from the works of any person
- Instructions given to person in charge
- Variations requirements
- Expenditure of prime cost and provisional sums
- Sanction in writing of variations created by the contractor
- Defects in the contractor's work
- Postponement of any work
- Execution of protective work after an outbreak of hostilities
- War damage
- Antiquities
- Nominated subcontractors
- Nominated suppliers

Source: Joint Contracts Tribunal 2016.

Box 17.3 Architect's responsibilities in respect of certificates under JCT

- Practical completion of works
- Completion of making good defects
- Estimate of the approximate total value of partial possession
- Completion of making good defects after partial possession
- Failure to complete the works by the completion date
- Determination
- Interim certificates
- Final certificate

Source: Joint Contracts 2016.

new development project and are sometimes the client's first point of contact on a proposed development. They also advise the financial institutions on investment in order to yield the best result for their shareholders or members.

Traditionally, the building surveyor's role was in assisting other colleagues and clients with the maintenance and repair of buildings and preparing survey reports for prospective purchasers and users of real estate. Building surveying today is a rapidly expanding profession. Some of

Box 17.4 Architect's other responsibilities

- Provision of documents, schedules and drawings
- Stating levels and setting out the works
- Access to site and workshops
- Limitation of assignment and subletting
- Granting an extension of time
- Reimbursement of loss and expense to the contractor
- Arbitration

Source: Joint Contracts Tribunal 2016.

this is owing to the growing popularity of building conversion and renovation and the poor and deteriorating nature of our buildings stock. It is also to some extent because of the nature of our society, with its 'make do and mend' approach and the desire for the conservation of older properties. As a profession, it is somewhat unusual in being largely restricted to the UK and some of its ex-colonies at the present time. Building surveyors are, however, rarely concerned with projects of a large size, and are not specifically referred to in the forms of contract.

17.3.3 Quantity surveyors

The quantity surveyor has developed from the function of a measurer to a building accountant and a cost adviser. The emphasis of the quantity surveyor's work has moved from one solely associated with accounting functions, to one involved in all matters of forecasting finance and costing, and cost and value management.

The function of the quantity surveyor in connection with construction projects is therefore threefold:

- as a cost adviser, attempting to forecast and evaluate the design in economic terms on an initial cost basis and a life cycle cost basis
- preparing much of the tendering documentation used by contractors
- in an accounting role during the construction period; where the quantity surveyor will report on interim payments and financial progress and the preparation and control of the final expenditure for the project.

Quantity surveyors are employed on behalf of both the building owner and the building contractor that tends to specialise in post-contract functions or commercial management. Boxes 17.5 and 17.6 list the duties of the quantity surveyor in connection with a building contract. The contractor's surveyor will be involved in the agreement of subcontractors' work, other duties of a commercial nature and possible bonus payments and ancillary functions.

The quantity surveyor is seen as an essential member of the construction team. The quantity surveyor's work frequently extends beyond that described above. Loss adjusting, arbitration and auditing are other areas where quantity surveyors are employed. The roles of the quantity surveyor within the JCT form of contract are listed in Box 17.7. Whilst the Institution of Civil Engineers (ICE) form allocates these duties to the engineer, in practice it is frequently the quantity surveyor who carries them out.

Box 17.5 Pre-contract role of the quantity surveyor

- Initial cost advice
- Approximate estimating
- Cost planning, value engineering, life cycle costing
- Bills of quantities and tender documentation
- Specification writing (where the bills are not required)
- Procurement
- Tender evaluation

Box 17.6 Post-contract role of the quantity surveyor

- Valuations for interim certificates
- Final accounts
- Remeasurement of the whole or part of the works
- Measuring and valuing variations
- Daywork accounts
- Adjustment to prime cost sums
- Increased cost assessment
- Evaluation of contractual claims
- Cost analysis

17.3.4 Engineers

A wide range of engineers are employed in the construction industry, from civil and structural engineers to building services engineers. Civil engineers are responsible for the design and supervision of civil and public works engineering, and are employed in a similar way to architects employed on a building contract. In addition, the engineer's counterpart working for the contractor is often a civil engineer. Their work can be very diverse, and may include projects associated with transportation, energy requirements, sewage schemes or land reclamation projects. Structural engineers are usually employed by the architect on behalf of the client. They act as consultants to design the frame and the other structural members in buildings. The building services engineers are responsible for designing the environmental conditions that are required in today's modern buildings. There has been an upsurge in their membership in recent years as greater attention is paid towards this aspect of building design.

Engineers are not mentioned by name in JCT, although the supervising officer mentioned in some of the forms may equally be an engineer, a surveyor or an architect, depending upon who is largely responsible for the design and supervision of the works.

17.3.5 Clerks of works

The clerk of works is employed under the direction of the architect as an inspector of the works under construction. The clerk of works may give instructions to the contractor, but these are of no effect unless they are subsequently authorised by the architect.

Box 17.7 Quantity surveyor's responsibilities under JCT

- Confidential nature of contractor's prices
- Valuing variations
- Calculation of loss and expense
- Preparation of interim valuations
- Preparation of the final account
- Accounts of nominated subcontractors and suppliers
- Fluctuations

Source: Joint Contracts Tribunal 2016.

The contractor must give the clerk of works every reasonable facility to carry out all duties. The clerk of works is the counterpart of the person in charge that is employed on behalf of the contractor. Duties include ensuring that the contract is fully complied with in terms of the specification and further instructions from the architect. The clerk of works will attempt to make sure the materials used and the standards of work are in accordance with the contract requirements. This will involve: inspecting the materials prior to their incorporation within the works; obtaining samples where necessary for the approval of the architect; testing materials such as concrete, bricks and timber to the specified codes of practice; and generally ensuring that the construction work complies with accepted good practice.

17.3.6 Other professions

A wide range of other professions are associated with the construction industry. One's viewpoint will determine whether these represent an unnecessary fragmentation of the industry or a desirable specialisation. The clear demarcation of activities has now become blurred, particularly as we look forward and consider the characteristics of the EU and the USA. Certainly, the amount of knowledge now available and the skills required are too great for a single person to control, and some specialisation even within a conglomerate of the professions is essential. Here are some of the other professions involved:

- *Building control officer*: normally employed on behalf of the local authority to ensure that the building plans and proposals comply with the building regulations and by-laws made under the public health and building laws.
- *Estimator*: responsible for calculating in advance of building the cost of the project to a particular contractor based upon the total costs of all the labour, materials and plant that will be needed.
- *Interior designer*: developing the internal shell of buildings to provide good aesthetic and working conditions to create an acceptable ambience for the owner and user.
- *Landscape architect*: helps to create the all-important context and space in which the building is set. Increasingly in modern buildings, this often includes internal spaces.
- *Planner*: involved with the legislative aspects of the building's location in interpreting the structure and district plans of local authorities. Ensures that the building fits into the environment.

17.4 Contractors

17.4.1 Main contractors

The majority of construction work in the UK is undertaken by a main contractor. The term 'general contractor' is now outdated since relatively few of these firms undertake the work themselves. These firms, which will be public limited companies (PLCs), will vary in size, having from just a few to many hundreds of employees. Many of the larger companies are household names and have developed only since the beginning of the twentieth century. Although there is no clear dividing line between building and civil engineering works, many firms tend to specialise in only one of these sectors. Even in the larger companies, separate divisions or companies exist, often trading and structured in entirely different ways depending upon the sector in which they are employed. Even the operatives' unions and the rules under which they are engaged are different.

The smallest building firms may specialise in one trade, and as such may act as either domestic subcontractors or jobbing builders carrying out mainly repairs and small alterations. The medium-sized firms may be a combination of trades operating as general contractors within one town or region. These firms may specialise in certain types of building projects or be speculative house builders. The largest firms may be almost autonomous units, although it is uncommon even in these companies to find them undertaking a complete range of work. On the very large projects, it is usual to find specialist firms for piling, steelwork and high-class joinery.

It has been suggested that one-quarter to one-third of the work of the construction industry is minor in nature, being largely of repairs and maintenance. This work is often carried out by the smallest companies. A recent survey also indicated that one-third of the building firms in Britain do not employ any operatives, but the work is carried out by the partners of the firm. The larger companies may be represented by fewer than 100 firms throughout the country. In more recent years, there has been a trend away from the multi-million-pound project, resulting in a slight reduction in the number of these firms. Overseas projects of this size have helped to keep such firms viable within the UK. The reduction in the size of projects has also meant the breaking down of some of the larger firms into smaller-sized units working on a more localised basis.

The contractors under JCT and the other forms of contract agree to carry out the works in accordance with the contract documents and the instructions from the architect. They agree to do this usually within a stipulated period of time and for an agreed amount of money. The main contractor must also comply with all statutory laws and regulations during the execution of the work, and ensure that all who are employed on the site abide by these conditions. The contractor will still be responsible contractually for any defects that may occur for the period of time stipulated in the conditions of contract, which is normally six months. However, the responsibility of the contractor for the project does not end here. In common law, the rights of the employer will last for six years and twelve years, respectively, depending upon whether the contract was under hand or seal.

The contractor is mentioned extensively in the conditions of contract largely because, along with the employer, they are one of the parties to the contract. Some of the more important provisions are listed in Box 17.8.

17.4.2 Person in charge

The person in charge is responsible for the effective control of the contractor's work and workpeople on-site, and also for organisation and supervision on the contractor's behalf and for

Box 17.8 Contractor's responsibilities under JCT

Quality

- Contractor's obligations
- Compliance with architect's instructions
- Duties in setting out
- Compliance with the standards described
- Responsibility for faulty work standards
- Duty to keep on-site a person in charge
- Requirement to give the architect access to the site and workshops
- Limitations on assignments and subletting
- Right to object to nominated subcontractors
- Duty to employer's directly employed contractors

Time

- Procedure for partial possession
- Necessity to proceed diligently with the works
- Liabilities in the event of non-completion
- Duty to inform the architect of any delays
- Rights in cases of determination of the contract

Cost

- Duty to ask for any loss or expense
- Responsibility for payment to nominated subcontractors
- Procedure for certificates and payments

Others

- Liability for injuries to persons and property
- Duties regarding insurances

Source: Joint Contracts Tribunal 2016.

receiving instructions from the architect. Depending upon the size and nature of the works and the type of firm, this may be a general overseer, site agent or project manager. The person in charge may have received initial training as a trade craftsperson or be a chartered builder or engineer. The responsibilities will vary with the size of the project and company policy. On the larger projects, considerable assistance will be received from other site staff. The person in charge, whatever the title, is the site manager on behalf of the main contractor and is very often a member of the Chartered Institute of Building.

17.4.3 Suppliers

Building materials delivered to a site may be described under one of three headings: materials, components and goods.

Materials

Materials are the raw materials to be used for building purposes and include cement, bricks, timber and plaster. The items included within this description will, in total, probably represent the largest expense on the traditional building site. As more and more of the construction processes are carried out off site, so the value of this section will diminish.

Components

Components represent those items delivered to site in almost 'kit' form. They may include joinery items such as door sets or joinery fittings to be assembled on-site. The industrialised building process is based to a large extent on the assumption that many items can come to site in component form to be assembled very quickly for a very small amount of money.

Goods

Goods include those items that are generally of a standard nature and can be purchased directly from a catalogue, for example, sanitary ware, ironmongery, electrical fittings.

The contractor's source of supply for these items may vary, but must in all circumstances comply with those specified in the contract documents regarding quality and performance. The contractor will probably make extensive use of builders' merchants, because they stock a wide variety of items that can be purchased at short notice. Some of the items will need to be obtained directly from the manufacturer, and specialist local suppliers of timber or ready-mixed concrete will be used. Contractors are able to secure trade discounts for the items that they purchase, and such discounts will be increased either to attract trade or because of large orders. Some clients, undertaking extensive building work, are able to arrange with suppliers a bulk purchase agreement. This helps to reduce their own costs of construction because they are able to secure very reasonable rates for the items. The contractor must then obtain the appropriate items described from such suppliers.

JCT refers to two types of suppliers, i.e. named suppliers and nominated suppliers. A named supplier is a firm specified in the contract documents from whom the contractor should obtain certain materials, components or goods. It is usual to suggest a list of alternative suppliers or sometimes to add the words 'other equal and approved'. The contractor would then need to show that the items proposed to be purchased for the work complied with those specified. In some circumstances a single supplier or manufacturer may be named where the intention of the architect is not to diverge to any other alternative. This is often the case in respect of sanitary ware, where a particular design is selected, or for suspended ceilings and kitchen fittings.

In some situations, the architect may choose to nominate a particular supplier to provide some of the various materials, components or goods. In these circumstances, the architect chooses to include these items as a prime cost sum in the bills.

17.4.4 Subcontractors

It is very unusual today for a contractor to undertake all the contract work with their own workforce. Even on minor building projects, the main contractor is likely to require the assistance of some trade or specialist firms. Works undertaken by firms other than the main contractor are often described as subcontractors, although in some situations it is not uncommon to find specialist firms working on the site beyond the normal jurisdiction and confines of the main contractor, and hence not a subcontractor within the generally understood description.

The employer may, for example, choose to employ such firms directly, and in this context these firms are not to be considered as subcontractors of the main contractor. Provision is made in the conditions of contract for such firms in order that they may have access to the contractor's site. For example, a firm constructing a sculpture may come into the confines of this clause. Secondly, the employer may choose to nominate particular firms to undertake the specialist work. In these cases, the employer may adopt this approach in order to gain a greater measure of control over those who carry out the work. These subcontractors enjoy a special relationship with the employer. Although after nomination they are often supposedly treated like one of the main contractor's own subcontractors, they do have some special rights, for example, in respect of their payment.

The architect may also choose to name subcontractors in the bills of quantities or specification who will be acceptable for the execution of some of the measured work. This procedure avoids the lengthy process of nomination, but still provides a substantial measure of control on the part of the architect. Provisions are found within the conditions of contract, and a requirement is to name at least three firms who will be acceptable to the architect. This provides for some measure of competition and allows the contractor a choice of firms. The contractor may also add to this list with the approval of the architect, and this should not be refused unreasonably.

All the remaining work is still unlikely to be carried out by the main contractor, and provision is also made for the use of the contractor's own subcontractors. These subcontractors are referred to in the conditions as the main contractor's domestic subcontractors. Named subcontractors are employed to undertake that work when contractors either make no provision within their companies or their own employed workers are busy elsewhere. The contractor must seek the approval of the architect in this respect, but it is unusual for this approval not to be given.

A further group of subcontractors are those described as statutory undertakers, e.g. gas, water and electricity. These are separately described for the following reasons: the employer and contractor often have no choice in employing them because of their statutory rights; they sometimes require payment in advance; and they refuse to give any cash discount for prompt payment.

The intermediate form of contract (IFC) identifies another type of subcontractor, known as the named subcontractor, who is akin to the nominated subcontractor. Chapter 6 covers the IFC and gives more information on this.

17.5 Regulation

17.5.1 Planning and control of development

Government has wide powers of control over the development of construction works. It seeks to resolve the conflicting demands of industry, commerce, housing, transport, agriculture and recreation by means of a comprehensive statutory system of land use planning and development control. The government's aim is for the maximum use of urban land, sometimes called *brownfield land*, for new developments. It also aims to protect the countryside, sometimes called *greenfield sites*, by assisting urban regeneration in towns and cities.

The system of land use planning in Britain involves a centralised structure under the Secretary of State for the Environment. The strategic planning is primarily the responsibility of the county councils, whilst the district councils are responsible for local plans and development control, the main housing function and environmental health. The development plan system involves the structure plans and local plans. The structure plans are prepared by the county planning authorities and require ministerial approval. They set out the broad policies for land use and ways of improving the physical environment. Local plans provide detailed guidance, usually covering a ten-year period. These are prepared by district planning authorities, but must conform with the overall structure plan.

Before a building can be constructed, application for planning permission must first have been obtained. This is made to the local authority in the form required by the Town and Country Planning (Making of Applications) Regulations. In the first instance, outline approval is sought to avoid the expense of a detailed design which could fail to secure approval. Full planning permission must still be obtained. If a scheme fails to obtain approval, then a planning appeal can be made to the Secretary of State.

The Building Act was introduced onto the statute book in 1984. The 1985 Building Regulations are framed within the Act and allow two administrative systems to be applied: one through the local authority building control department and the other via certification. Such controls are necessary to:

- secure improved standards of design and construction
- ensure the safety and health of the occupants
- provide for the proper location of buildings and industry
- make the best use of the land that is available
- provide for the safety, health and welfare of those engaged in the construction process and those affected by it.

Whilst successive Acts of Parliament have introduced more legislation, and this now also needs to comply with the wider EU legislation, there has been a desire to speed up the process. This has tended to be achieved through increasing flexibility, but at a cost of increasing the uncertainty of the outcome.

17.5.2 Local authorities

Local government is broadly divided into county councils and district councils, each with their own function. A major reorganisation of local authorities was introduced by the Local Government Act 1972, which came into operation in 1974. Further reorganisation occurred in 1996. The allocation of the various functions appropriate to construction are as follows:

- *Building regulations*: district councils.
- *Highways*: county councils:
    - District councils do maintenance by agreement.
- *Refuse collection*: district councils.
- *Refuse disposal*: county councils.
- *Structure plans*: county councils.
- *Local plans*: district councils.
- *Water and sewage*: regional water authorities.

County councils have to find tipping sites for refuse. Local councils receive structure plans from their county council, then they complete the local details. Local government is restricted in the way it can exercise its powers:

- *Area*: it is generally confined to a defined geographical area. It works with adjoining authorities on matters of common interest through joint committees and joint boards.
- *Central government*: the control of a local authority is exercised through the Secretary of State for the Environment. In some governments, a local government minister has been appointed to carry out this control. Central government control may be exercised by supervision where a statutory inspection can take place, and because a local government relies to a large extent on the finances from central government. All the local authority's accounts must be properly audited. In some cases, the minister concerned has powers to carry out a local authority's functions in cases where it defaults.
- *Judicial control*: because local authorities are created by statute, they derive their powers and functions in this way. They must therefore act within the laws afforded to them. Building owners and contractors are likely to encounter the local authority concerned initially for the approval of planning permission, where an appropriate fee is payable, and then during construction, where they must conform to the building regulations and other statutory documents.

17.5.3 The Crown

The Crown is used in this context to denote the governmental powers exercised through the Civil Service. This means central government offices rather than local government. Central government has many powers conferred upon it by Parliament. For example, it exerts a certain measure of control over local government. This may be achieved by:

- *Supervision*: the Secretary of State has power through statutes to supervise the activities of local authorities. For example, some town planning decisions may need to be referred to the Secretary of State. Some local services, such as fire, police and education, are subject to statutory inspection.
- *Finance*: much of a local authority's finances are obtained from central government. The minister can therefore exercise considerable pressure on proposed expenditure, and also through the auditing of the local authority's accounts.
- *Supersession*: in some cases, the minister concerned has power to supersede in certain functions in case of default by a local authority.

Because government is a major spender in the construction industry, it is considered to be a very important client. It can also severely restrict and regulate how the work will be carried out. Although the privatised Property Services Agency (PSA) still undertakes a large amount of central government's building work, other departments also have extensive construction programmes. Much of the central government's work is undertaken on the GC/Works/1 form of contract.

17.6 Direct labour

The public sector employer remains a large client of the construction industry. It undertakes work ranging from multi-million-pound engineering projects to the minor repair and maintenance of local authority dwellings. There have been discussions regarding a national building corporation operating on a regional basis, but this has yet to be developed. However, there remain the direct labour construction departments employed within a reducing number of local authorities. The expansion or contraction of all of these organisations is a politically sensitive issue, with differing opinions expressed on their necessity and efficiency. In some local authorities, the direct labour departments may be responsible for no more than building and highway maintenance. In a few examples, they may be large enough to undertake major projects. Whilst they were established to work only within the confines of their geographical area, they may now tender for work in other local authorities' areas. Where capital projects are envisaged, the direct labour departments are usually required to tender for this work against private contractors. Maintenance work may be undertaken by the direct labour department as a matter of course, although even within this sphere of work some element of competition is now required.

17.7 Professional bodies

Designing, costing, forecasting, planning, organising, motivating, controlling and coordinating are some of the roles of the professions involved in managing construction, whether it be new build, refurbishment or maintenance. These activities also include research, development, innovation and improving standards and performance. Appendix A provides a list of the main professional bodies in the built environment.

17.7.1 The Royal Institute of British Architects (RIBA)

RIBA is the main professional body for architects in England and Wales. In Scotland, there is a similar body, the Royal Incorporation of Architects in Scotland (RIAS). Under the Architects (Registration) Act 1938, it continues to be illegal for anyone to carry out a business describing themselves as an architect unless they are registered with the Architects' Registration Council (ARCUK) established under an Act of 1931. Registration involves appropriate training and education as evidenced by the possession of qualifications set out in the Act or approved by the council. However, this does not prohibit anyone from carrying out architectural work such as the design of buildings.

17.7.2 The Royal Institution of Chartered Surveyors (RICS)

The RICS was formed in 1868 and incorporated by Royal Charter in 1881. It originated from the Surveyors' Institute and has grown as a result of a number of mergers with other institutes, most notably the Land Agents (1970) and the Institute of Quantity Surveyors (1982). The RICS is now administered in faculties which represent the varying interests of different chartered surveyors. The general practice surveyors (42 per cent) and quantity surveyors (37 per cent) divisions account for more than three-quarters of the membership.

17.7.3 The Institution of Civil Engineers (ICE)

The term 'civil engineer' appeared for the first time in the minutes of the Society of Civil Engineers, founded in 1771. It marked the recognition of a new profession in Britain as distinct from the much older profession of military engineer. The members of the Society of Civil Engineers were developing the technology of the industrial revolution. The Royal Charter of the institution contains the often quoted definition of civil engineering as being 'the art of directing the great sources of power in nature for use and convenience of mankind'.

17.7.4 The Institution of Structural Engineers (IStructE)

The IStructE began its life as the Concrete Institute in 1908, was renamed in 1922 and incorporated by Royal Charter in 1934. Its aims include promoting the science and art of structural engineering in all its forms and furthering the education, training and competencies of its members. The science of structural engineering is the technical justification in terms of strength, safety, durability and serviceability of buildings and other structures.

17.7.5 The Chartered Institution of Building Services Engineers (CIBSE)

The grant of a Royal Charter in 1976 enabled the Institution of Heating and Ventilating Engineers, which was founded in 1897, to amalgamate with the Illuminating Engineering Society of 1909 to create an institution embracing the whole sphere of building services engineering. The objectives of the CIBSE are set out in their charter as the promotion of the art, science and practice of building services engineering for the benefit of all, and the advancement of education and research in this area of work.

17.7.6 The Chartered Institute of Building (CIOB)

The CIOB was formed in 1834, registered as a charity in 1970 and granted a Royal Charter in 1980. It originally started as the Builders' Society, a small and exclusive club with a wide influence that was responsible for helping to produce the early forms of contract. The CIOB objectives are the promotion, for public benefit, of the science and practice of building, the advancement of education and science including research, and the establishment and maintenance of appropriate standards of competence and conduct for those engaged in building. The institute encourages the professional manager and technologist to work together with their technician counterparts in order to achieve a ladder of opportunity as a main objective of the training and examination structure of the institute.

17.7.7 The Landscape Institute (LI)

This is the chartered institute in the UK for landscape architects. Landscape architects work on all types of external space, large or small, urban or rural and with hard or soft materials. It received its charter in 1997.

17.7.8 Chartered Institute of Architectural Technologists (CIAT)

The CIAT was formed in 1965 as the Society of Architectural and Associated Technicians (SAAT) representing architectural technicians, and received its charter in in 2005 as the CIAT. It is primarily a UK-based organisation instrumental in developing an architectural technologists' profession and accredits suitable degree programmes.

17.7.9 Chartered Institution of Civil Engineering Surveyors (ICES)

ICES was established in 1969 as an institution dedicated to surveyors practising in the civil engineering construction sector. The institute received its Royal Charter in 2009 has reciprocal membership agreements in place with the ICE.

17.7.10 Membership

A proportion of members in the construction professions hold membership of more than one institution. Several members of the profession are also represented in other professional bodies that are more broadly based than construction alone. For example, a large proportion of the Chartered Institute of Arbitrators (CIArb) are members of the construction professions, whereas in the Chartered Institute of Management (CIM) membership includes only a small proportion of those connected with the construction industry. Table 17.1 lists the chartered professional bodies in the construction industry and their relative sizes; although all of these are increasing the size of their membership. There are also a number of non-chartered professional bodies with memberships ranging from as few as 1,000 to nearly 10,000. Mergers and acquisitions are taking place and new organisations are being formed. Appendix A provides a comprehensive list of these in the built environment.

17.7.11 Europe and the USA

There are wide cultural considerations to be taken into account in any comparison between the construction industry professions in the UK and those in other parts of the world, notably mainland Europe and the USA. Historically, the industry and practice developed differently. In much of the rest of the world, architects and engineers dominate the construction industry. The ratio between building and civil engineering works in other countries is similar, at about 80:20. The various professional disciplines in the UK are not mirrored elsewhere, other than in Commonwealth and former Commonwealth countries. The role of the professional bodies also varies. In the UK, a professional qualification is one by which to practice. In Europe, a professional body is more of an exclusive club, of which relatively few of those engaged in practice are members. In the USA is the emerging discipline of construction management alongside those of architect and engineer. In the UK, the architect is the only registered profession, and this is also the case in mainland Europe, but the practice there is more controlled; for example, an architect must be employed to sign the plans in mainland Europe, otherwise a building cannot be constructed. Many of the professional bodies in the UK have royal charters and enjoy charitable status. Some organisations question the need for so many bodies, some with overlapping interests and others that have a silo mentality.

Table 17.1 Chartered professional bodies in the construction industry

	Total membership	*Percentage*
Royal Institution of Chartered Surveyors (RICS)	120,000	31
Institution of Civil Engineers (ICE)	88,000	23
Chartered Institute of Building (CIOB)	48,000	12
Royal Institute of British Architects (RIBA)	44,000	11
Institution of Structural Engineers (IStructE)	27,000	7
Royal Town Planning Institute (RTPI)	23,000	6
Chartered Institution of Building Services Engineers (CIBSE)	21,000	5
Chartered Institute of Architectural Technologists (CIAT)	9,000	2
Landscape Institute (LI)	6,000	2
Chartered Institution of Civil Engineering Surveyors (ICES)	4,000	1

17.8 Construction associations

17.8.1 Construction Industry Council (CIC)

The CIC was formed in 1988 with five founder members. Since then, it has grown to be the largest pan-industry body concerned with all aspects of the built environment. Its members represent more than 500,000 professionals working for, and in association with, the construction industry and more than 25,000 construction firms. It is the representative forum for the industry's professional bodies, research organisations and specialist trade associations. The mission of CIC is to:

- Serve society by promoting quality and sustainability in the built environment.
- Give leadership to the construction industry, encouraging unity of purpose, collaboration, continuous improvement and career development.
- Add value and emphasis to the work of its members.

The executive board acts as the main policy and strategy vehicle of CIC through its five electoral colleges. There are currently six standing committees covering lifelong learning, finance, industry improvement, innovation and research, national regions and sustainable development.

17.8.2 Construction employers' associations

The construction industry has typically been seen as two sectors, building and civil engineering. This distinction was also seen at employer level, with the separate Building Employers' Confederation (BEC) and the Federation of Civil Engineering Contractors (FCEC). These bodies merged towards the end of the 1990s to form the Construction Confederation (CC), but in 2009 this body was dissolved. The UK Contractors Group (UKCG) is now the primary association for large contractors working in the UK. It currently has thirty members, who deliver about one-third of total construction output. The UKCG has two main objectives: to promote the interests of the construction industry and to take leadership to raise standards within the industry. The UKCG works closely with the Confederation of British Industry (CBI) Construction Council on its first objective and with its supply chain partners to meet the second. Its current key priority is promoting the case for infrastructure investment and working with government to improve the efficiency of public sector construction procurement.

In addition, there are a number of subsidiary groupings of firms, such as the house builders. Numerous activities undertaken by the different organisations are replicated and add to the costs of administration. Many of the contractors believe that the construction industry's poor educational framework, lobbying strength, image and relationships with clients and the professions will improve only through having one united voice.

17.8.3 Construction Industry Training Board (CITB)

The CITB was established by an Act of Parliament in 1964 to improve the quality of training, improve the facilities available for training and help to provide enough trained people for the construction industry. It is partially funded through government, but most of its income is derived from a levy system on contractors based upon the number of their employees. In addition to being appointed as the primary managing agent for the construction industry's craft apprentices, the CITB also provides skills trainings at its national training centres. Whilst the emphasis of its activities is on practical craft skills training, it also offers courses in supervisory management. The CITB is one of the few remaining publicly funded industrial training boards.

In 2003, a partnership was formed between CITB GB, CITB Northern Ireland and the CIC, to be known as CITB-Construction Skills. ConstructionSkills has an important role to play in providing the industrial and commercial dimension that will help universities and colleges when designing their programmes of study. The group is formally recognised by the Sector Skills Development Agency (SSDA) and has been granted Sector Skills Council (SSC) status. ConstructionSkills has three important challenges:

- *Improving business performance* through a multi-agency approach to business support, particularly for the small businesses prevalent within the industry.
- *Qualifying the workforce* by working with, and through, the industry's major clients, supply chain networks and its widely accepted certification scheme.
- *Improving image and recruitment* into the sector through the development of high-quality programmes and collaborative action to attract and retain women and ethnic minorities.

ConstructionSkills has close links with some of the other twenty-three SSCs, most notably AssetSkills (property, housing, cleaning and facilities management) and SummitSkills (building services engineering).

17.9 Industry skill needs

The UK construction industry occupies a critical position in the UK economy. As well as representing about 7 per cent of the gross domestic product and being a very substantial employer, it provides an infrastructure that is necessary for all other sectors to succeed. It has delivered a sharp increase in productivity over the last few years, growing faster than any other major sector of industry. Inevitably, this has put pressure on the industry's capacity, particularly in terms of its labour market. It is a national industry carrying out work throughout the UK and also has a sizeable market overseas, so is therefore a good export earner. Concentrations of activity occur in urban areas. London and the Southeast, in particular, account for almost one-third of the industry's current workload.

Within the construction industry are a range of sectors in the development and maintenance of the built environment. In addition to work performed by contractors and subcontractors, it also covers the professional and design work of architecture, engineering and surveying. The broad sectors of its activity are shown in Table 17.2. The construction industry is also important to the property industry, since some see this as a different industry focused on property values, finance, development and the ongoing management of the built assets.

The distribution of activities shown in Table 17.2 includes both new build and the refurbishment of existing buildings and structures. The work is carried out by contractors and consultants who are employed by both the public and private sectors of the economy.

17.9.1 Construction industry

The construction industry is essentially a national industry, although separate markets co-exist within this definition. There is the localised market, within very small geographical boundaries, together with regional and national markets. The industry has a labour force of about 2.9 million, although these numbers are swelled by those in the professions and also increased by a number of others, most notably those involved in the quarrying and manufacture of

Table 17.2 Construction industry activity (value of construction output 2015)

	Value £	*Percentage*
Housing	31,175	33
Infrastructure	20,248	22
Industrial	4,896	5
Schools and universities	10,314	11
Health	2,853	3
Offices	10,153	11
Commercial	11,613	12
Other	2,216	2
Total	93,468	

Source: Office for National Statistics 2016/

building materials, goods and components. In addition, a large number of secondary employers also rely on the work produced in the construction industry; large projects being constructed over a number of years generate their own growth of secondary employers.

The UK has about 280,000 construction companies, several of whom are the large construction firms and many that are household names. The number of firms fluctuates from year to year and their trends can be found in *Construction Statistics Annual 2016* (Office for National Statistics 2016). More than 90 per cent of the construction industry firms include employers with fewer than thirteen people. These micro, small to medium-sized employers (SMEs) include professional firms, subcontractors, building material and component suppliers as well as small contractors.

Table 17.2 provides an overview of the construction industry in terms of its value addition to the economy. The overall profile of the UK construction industry can be understood from the infographic developed by the Coalition government (Department of Business, Innovation and Skills 2013b). It indicates how the 2.9 million workers of the construction industry are employed and how the global current construction output would grow by $4 trillion from 2012 to 2025 (Figure 17.1). It also states the increasing significance of the green construction sector, which is bound to further increase with the UK's strong commitment to the Paris agreement on Climate Change.

The larger firms have well-established training programmes offering their employees appropriate career pathways and development. The smaller firms' activities are often related to general updating and to meet specific needs such as changes in statutory requirements, e.g. health and safety obligations or maintaining an awareness of current best practices. The requirements for people working in built environment technical, management and professional occupations are indicated in Figure 17.2. CITB Construction Skills Network has forecasted a growth in each of these occupational areas from 2017 up to 2021. The growth is required to:

- increase the shortfall of new entrants into the industry
- upskill manual occupations
- replace those who leave the industry through retirement
- replace those who change occupations
- reflect changes in the supply change.

Construction

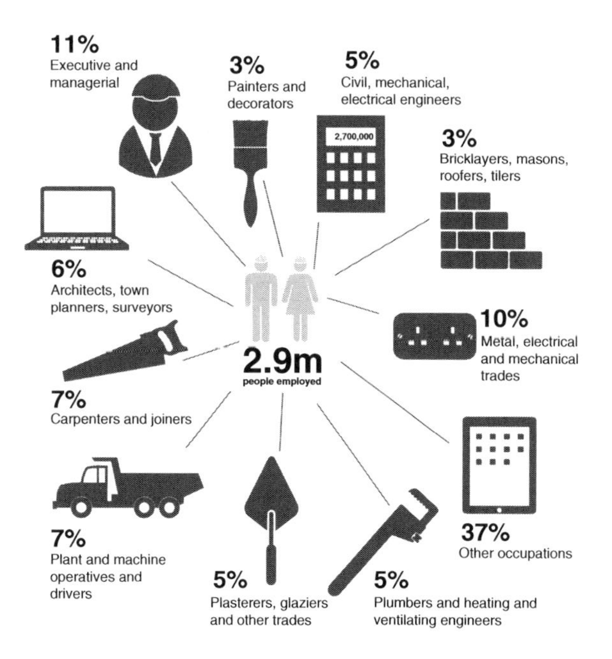

There are **2.9 million** jobs filled in the Construction Industry, circa 10% of all jobs (in over 280,000 businesses)

Construction contributes nearly **£90bn** to the UK economy, 6.7% of the total

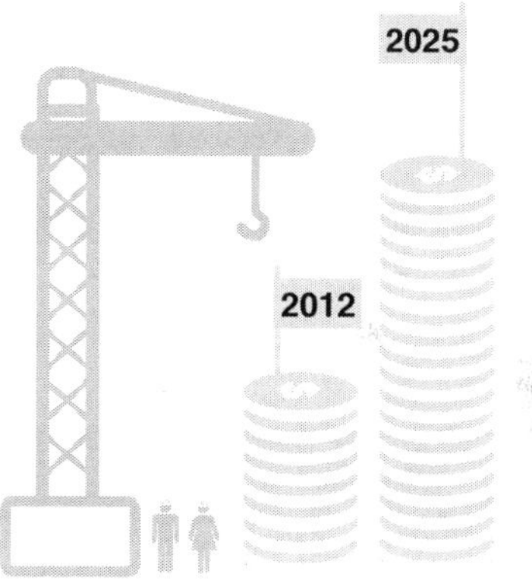

Global construction output is forecast to increase from around $8.5 trillion today to **$12 trillion in 2025***

*Source: Global Construction 2025

The **UK has the sixth largest green construction sector in the world**. Around 60,000 jobs are expected to be supported by the insulation sector alone by 2015

#indstrategy

Figure 17.1 The construction industry profile.

Source: Department of Business, Innovation and Skills 2013b.

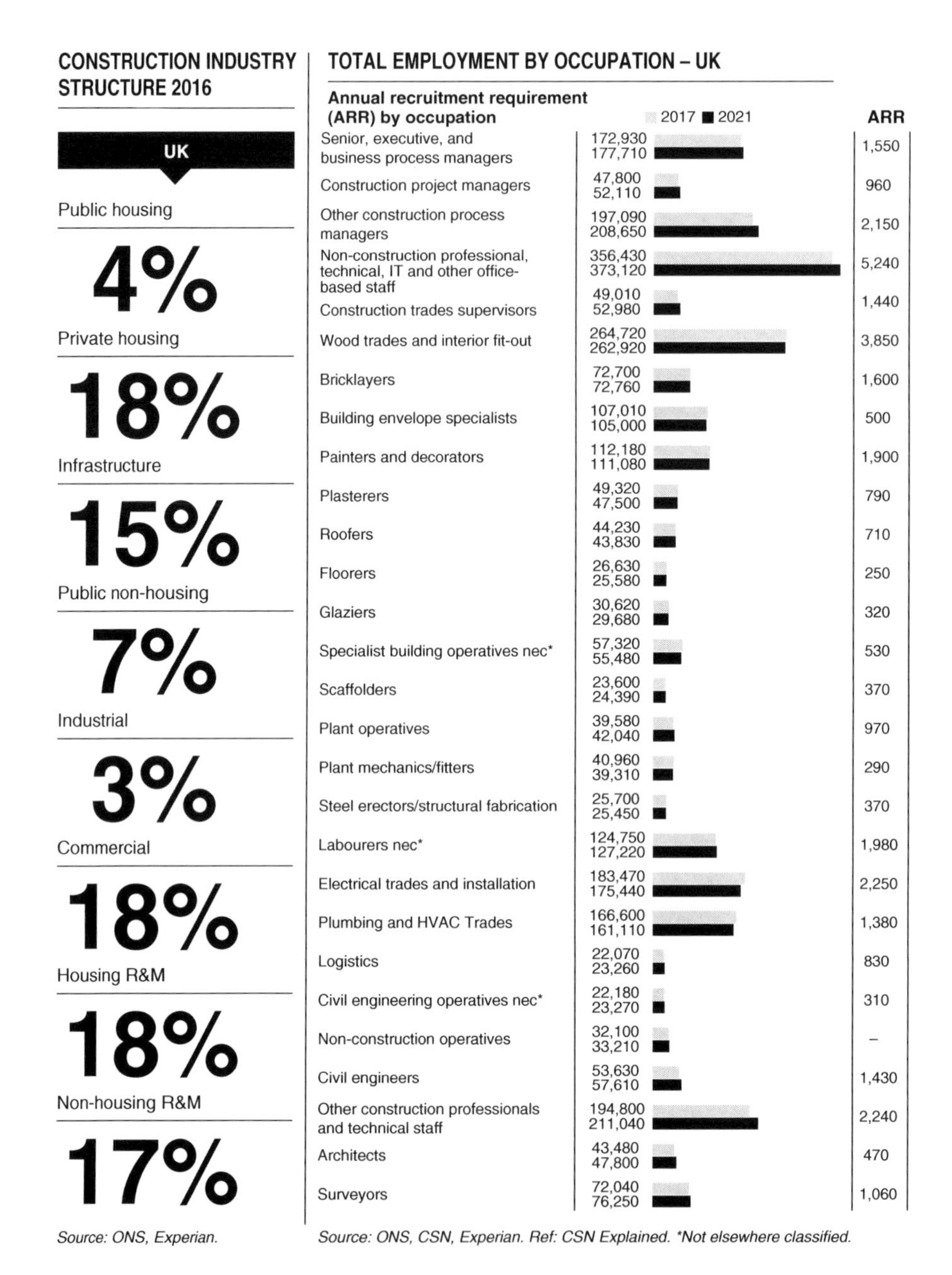

Figure 17.2 Construction skills network forecasts 2017–2021.

Source: CITB 2005.

The total construction workforce requirement is also shown in Figure 17.2, for comparative purposes, and a breakdown of the different occupations involved have been indicated from the Industry Insights: *Construction Skills Network Forecast Report 2005* (CITB 2005). In terms of construction related occupations, manual workers, building services trades and wood trades, dominate the industry, representing about three-quarters of the workforce.

There is some evidence of regional variation in skill levels across the UK. For example, London and the Southeast in particular have an overall higher level of skills because of the concentration of head offices of construction firms and professional practices.

There are many factors that act as the main drivers for sector change. These are discussed in the following subsections.

Economic

The level of demand for construction industry products has the greatest influence on the ongoing performance of the industry:

- Historically it has suffered substantial booms and slumps in workloads.
- The economic cycle is believed to use the industry as an economic regulator.
- Recession during the mid 1990s and again in 2009–11 laid off about one-third of the workforce.
- Within the subsets of the industry, fluctuating levels of economic activity have been even more pronounced.
- In 2009, the global recession had a severe impact on the UK construction market.
- In 2017, the ongoing prospects for construction industry were generally positive.
- In recent years, the UK construction industry has improved its performance in international markets. Its construction contracting exports rose to £1.65 billion with trade surplus of about £590 million.
- In 2011, exports of architecture and surveying services had a trade surplus of about £530 million.

Customer needs

An effort is being made to help the industry better manage itself and so improve capacity and performance:

- An important part of this is in attempting to gain a better understanding of market need and to develop longer-term relationships with customers and suppliers.
- The larger companies have fared better at this through supply chain management, and this has led to their improved financial position.
- The government is seeking improvements in its public services and infrastructure.
- The private finance initiative (PFI) and public private partnerships (PPP) have an increasing role in fulfilling these objectives.
- Government bodies are moving from being owners and occupiers to becoming clients purchasing long-term services.
- The construction industry is seeking to become less adversarial.
- There is an important focus towards whole life costing and facilities management.
- The capacity to take on these roles is difficult for a fragmented industry.
- Better coordination and improvements in standards and performance are required throughout the supply chain.
- Clients increasingly expect to see evidence of better standards, improved skills and qualifications and a refusal to accept a no-barriers approach.
- The different sectors will change at different rates; the smaller works projects and repairs and maintenance sector being slower to adapt to new practices.

Government regulation

The trend within governments across Europe is to increase their intervention in the way that business operates. The construction industry is under legislative pressure from all levels of government, including European, central and local government, in respect of employment legislation, the improvement of public services, economic redevelopment and planning. The policies that are likely to have the greatest impact on the construction industry include:

- the focus on improving public services
- the introduction of procurement frameworks that focus on best value
- employment legislation, especially in respect of health and safety
- procurement directives relating to open tendering, sustainability and environmental impact
- consultation with employees and works councils
- an increased need to interpret legislation at all levels.

Technological trends

New technologies and innovations are generally adopted if there is a sympathetic set of business, legislative or cultural conditions:

- There has been previous under-investment in research and development.
- There is now an increased need to improve competitiveness and productivity through technology and innovation.
- The current labour and skills shortage has been the biggest catalyst for technological change.
- Significant developments have occurred in:
    - prefabrication and off-site manufacture of structures and building components
    - standardisation of production
    - development and use of new building materials
    - continued development in heavy and hand-held plant
    - better integration of information and communications technology (see Chapter 22).

Demographic changes amongst consumers and the workforce

Population characteristics, such as size, growth, density, distribution, age, gender and ethnicity, drive both supply and demand:

- Demographic changes shape the expectations of customers and the ability of the industry to meet their demands.
- The needs of the population in respect of infrastructure and buildings are only achievable if there is sufficient capacity in terms of labour and skills.
- An increasing life expectancy and more culturally diverse population within a growing rate of household formation present the industry with some of its greatest challenges.
- The changing nature of the population not only affects what might be required but also what the industry is able to provide in terms of the built environment.
- The ageing workforce can be partially attributed to the decline in recruitment and training during the early 1990s.

- The government's ambitious target of a 50 per cent participation rate in higher education further limits the pool of available labour from which the construction industry is able to recruit.
- Coupled with this is the long-standing trend towards early retirement.
- Construction has been seen as a white male dominated industry. At all levels, it has failed to recruit sufficient women and ethnic minorities.

Sustainable development issues

The government's sustainability strategy is expected to have a major impact on the construction industry and vice versa:

- The principles of protecting the environment by adopting sustainable construction policies emerged from the 1992 Earth Summit in Rio and the subsequent 1997 Kyoto Summit.
- This resulted in a commitment to reduce greenhouse gas emissions.
- It encouraged participating countries to develop national sustainable development strategies at local and national government levels.
- In the UK, Agenda 21 has resulted in published reports, most notably on sustainability and building a better quality life.
- Recent changes in legislation, affecting building regulations, planning and an aggregate tax, collectively encourage the use of:
    - brownfield sites
    - energy efficiency
    - waste management
    - recycling
    - whole life costs.
- There is an increasing requirement for construction solutions to be sustainable and a driving force in technological change and innovation.

Changing practices

The construction industry is seeking a rapid change in its organisation, culture and practices through agencies such as Constructing Excellence (formerly the Rethinking Construction and the Construction Best Practice Programme; see Chapter 18). Hence, the changing needs and practices of the construction industry will continue to evolve and be different from those that are required and valued today. The construction industry is changing:

- Clients are demanding change to a whole range of practices, including the way in which the work is procured.
- The economy, especially the EU, is creating new opportunities for foreign competition both at home and abroad.
- Task forces, such as Constructing Excellence, are encouraging firms to think more smartly.
- The industry is developing new technologies and innovation with a greater emphasis on off-site manufacture.
- Greater attention is being given to employment conditions and methods of working as further evidence of change in culture and image.

- The focus is on improving public services through, for example, best value methods.
- There is greater synergy between the public and private sectors, such as PFI and PPP.
- There is an increased emphasis being placed on sustainable practices.
- There is an imbalance between the supply and demand of skilled people in the construction industry.
- The Better Together project is demonstrating improvements in interdisciplinary education and practices.
- Advancement of technologies, such as building information modelling (BIM), are changing the way buildings are procured.

18 Site communications

The construction of buildings and engineering structures utilises a range of resources, such as people, machines, money, materials, mechanical plant, management and methods, and needs to combine them in the most effective, efficient and economic manner.

The construction industry is a unique industry. It is one of the few that separate design from production, although this is changing with the proliferation of design and build projects, and it undertakes the bulk of its work on the client's premises. The industry docs not build replicas; whilst projects may appear to be similar, they are different in certain respects. Even comparable housing units on an estate will differ in their location, levels, amount of work below ground level, plot size and, hence, external works required.

Site managers do not move materials, operate plant or carry out the physical production function. Site management organises, informs, coordinates, orders, instructs and motivates others to undertake these tasks. The effectiveness of their performance will depend upon the ability to listen, read, speak and write. More importantly, their effectiveness will be evident in being understood by others and understanding others' point of view.

Site communications therefore include gathering information, to ascertain the needs of those involved with the project, as a necessary tool of management to integrate the functions of departments and as a vital link between manager and subordinate in order to get the job completed.

18.1 Communication principles

Communication and the social interactions involved are keys to success in this work. Communication skills and strategies are necessary to share ideas and experiences, to find out about things and to explain to others what you want. Developing ways of communicating, including language and body signals, is essential in expressing feelings and insights. Learning to communicate effectively may mean the difference between barely coping with life as you know it and actively shaping your world as you would like it to be. Some of the components of good communications are:

- knowing what to say
- gaining the attention of the listener
- establishing and maintaining relationships
- knowing the listener's likes and interests
- choosing how to communicate from a range of options
- being skilful in communicating
- choosing when and where to communicate

- being clear, brief and coherent
- listening actively
- understanding and clarifying any messages you receive
- trying not to be easily distracted
- knowing how to close conversation or communication.

Effective communication starts with a purpose or objective in transferring information. This should be done in the most effective way by using an appropriate form to suit the receiver. Provision should be allowed for feedback, and the need for action on that feedback should be appreciated. The elements that should be present in all communications are:

- *Clarity*: i.e. being easily understood.
- *Presentation*: i.e. creating a favourable impression.
- *New information*: i.e. attracting interest.
- *Drive*: i.e. demanding the necessary action.
- *Tone*: i.e. creating a responsive attitude.
- *Feedback*: i.e. ensuring the information is transferred.

18.1.1 Forms of communication

- *Written*: letters, reports, bills of quantities, specifications, site instructions, British Standards. This method is used when the subject matter is complex, important or likely to have possible legal implications and where a permanent record is required for future reference.
- *Visual*: films, slides, posters, graphs, charts. It includes the project drawings and works programmes. This sort of communication often has a greater effect upon the receiver. Where messages are less complex, this form of communication is the most effective.
- *Oral*: speaking to the other person. This method is instant and often generates an immediate response. Attitudes and behaviour can also be observed. The face-to-face approach is appropriate when the subject matter may be difficult or disagreeable. It is also used in those circumstances for simple, less important and informal messages.

18.1.2 Barriers to communication

An organisation that is suffering from poor morale, or lacks confidence in its future, is likely to see an increase in communications problems. In these circumstances, there is sometimes a lack of will to communicate effectively. However, nearly all barriers to good communication can be classified as follows:

- *Physical barriers* might include disability on the part of the receiver, such as deafness or blindness. They might include noise on-site, poor telephone lines and layers of site management. These can sometimes introduce distortion and inaccuracy into the information that is being transmitted.
- *Psychological barriers* are the biggest cause of communication breakdown. They affect the attitudes, feelings and emotions of the receiver. Sometimes, individuals will only hear what they expect to hear. Preconceived ideas are used to interpret the information in ways not anticipated by the giver of the information. Feelings and emotions affect the ability to receive a true message. When worried, the receiver feels threatened. When angry, the information may simply be rejected.

- *Intellectual barriers* affect concepts and perceptions that may have been built up over a lifetime of work. Difficulties may arise because the information was not transmitted properly in the first place. It may use excessive technical jargon or attempt to use words that are out of context.

18.2 Site information

In addition to the everyday running and organising of a contract, it is also the site overseer's responsibility to maintain accurate records of the important happenings on-site. This information should be properly recorded, so that it can be quickly retrieved whenever necessary. The site records that are normally kept include:

- daily reports
- site diaries
- materials received sheets
- advice of variations
- daily labour allocation sheets
- drawing registers
- confirmation of verbal instructions
- weather reports
- daywork records
- subcontractor files
- site correspondence.

If this information is to be valuable for future reference then it must be:

- *Precise*: clear, straightforward, as simple as possible and well thought out.
- *Accurate*: true and correct in all its details.
- *Definite*: future users should be left in no doubt as to what the message means.
- *Relevant*: to the particular situation and the people who will use it.
- *Referenced*: linked wherever possible to other documents being used by the contractor.

18.3 Site meetings

Site meetings are held for various purposes. They may be held with the contractor's staff only, to establish how the works should be carried out or deal with an internal matter in connection with the project. They frequently involve other parties, such as subcontractors and suppliers. The general site meeting would usually involve the client, architect, quantity surveyor, other consultants, subcontractors and clerk of works as well as the main contractor. These meetings frequently take place on a monthly basis, depending upon the size of the project and the issues that need to be resolved. The site meeting is a vehicle to pass information from one party to another or for consultation on how to tackle a problem that has arisen. The normal site meeting can also be used for decision-making purposes, or perhaps to persuade the parties involved of a different method of solving a particular problem. In essence it is: What do we need to know? How are we getting on? What is wrong, and how can this be corrected?

Here are some suggestions on meetings preparation:

- Determine the meeting's objectives.
- Produce an attendance list.

- Decide on the time and place.
- Determine the meeting's style.
- Circulate an agenda and other information.
- Identify if special facilities are required (e.g. audio-visual).
- Rank and set times for each item.
- Assess possible areas of conflict.
- Pre-discuss items with individuals where necessary.
- Determine the sorts of minutes required.
- Adequately brief the participants.
- Other matters: refreshments, car parking, etc.

Frequently, the architect will chair the site meeting and prepare an agenda, often in consultation with the contractor. The agenda will follow the format of a business meeting, with apologies for absence, matters arising from the previous minutes, any other business and the date and place of the next meeting. In addition, items of a routine nature, such as the main contractor's report and special items that address current issues, will also be included.

The minutes of the meeting are the follow-up and, as such, are vital to the effectiveness of the meeting. The minutes ensure a record of the collective views of the meeting. They should also identify the areas where responsibility for further action lies, and may include an action plan in order to minimise future problems or difficulties that might be encountered. Here are some suggestions that should help meetings to flow more easily and become more productive:

- Prepare an agenda and establish rules of procedure.
- Start promptly and use brevity.
- Practise active listening.
- Keep replies to the point and avoid wasting time.
- Clarify issues by asking relevant questions.
- Summarise progress.
- Restate important points.
- Be prepared to change strategy if necessary.
- Be supportive: That sounds like good idea.
- Confront issues: Are we really prepared to do that?
- Question critically: What exactly do you mean?
- Provide accurate supporting information.
- Do not allow the meeting to be interrupted by telephone calls, etc.
- Avoid interrupting.
- Do not be afraid to make your feelings known.
- Refrain from distracting behaviour.
- Do not talk to others during presentations.
- Never lose your temper.
- Do not embarrass others.

18.4 Site diaries

The site diary, if well maintained with the correct sort of information, is a very useful document maintained by the contractor. It will contain information that can be used where disputes occur between the different parties concerned with the project. Site diaries have been used in litigation as valuable evidence to help substantiate a case put forward by the contractors. The diary should record information that generally does not warrant separate records being kept. This will vary with the size and type of project, the nature of the client and the information records kept by the contractor. Figure 18.1 is a typical form that can be used for this purpose.

18.4.1 Weather conditions

Weather conditions will have a major influence upon the progress of the works. They must therefore be recorded, especially where they cause a delay or suspension of the works.

18.4.2 Drawings received

The site manager must ensure there is an adequate system for the receipt and recording of architect or engineer drawings. Although there will be a drawing register, the site diary should also record when this information is received. There will be times when information is requested from the consultants. This should always be in writing, giving sufficient notice to avoid the possibility of delays. The late receipt of information is one of the reasons for an extension of time where this is considered to be a relevant event. In addition, if the delay in

Details	Date
1. Weather conditions	Temperature
2. Drawings received	Number
3. Instructions: oral/written	Given by
4. Variation orders received	Number
5. Dayworks: reasons/descriptions	Sheet numbers
6. Delays: reasons	Labour on-site required
7. Lost time: reasons	
8. Urgent requirements	Visitors to site
9. Unusual occurrences	

Figure 18.1 Site diary.

Source: Adapted from Davies 1982.

receipt of such information causes the works to be suspended for longer than the period named in the appendix, usually three months, then the contractor may determine the contract. The contractor in each case must have requested further instructions and drawings and the date for the period of delay will commence from that date.

18.4.3 Architect's instructions

During a visit to site by the architect, and more frequently through the clerk of works, instructions will be given both orally and in writing. These need to be recorded in the site diary under the heading of site instructions. Remember that, in order to be valid, all instructions must eventually be in writing. The JCT form lays down procedures to be followed in the case of dealing with oral instructions (see Chapter 24). The site manager therefore needs to record precisely when the instruction was given.

18.4.4 Variation orders/change orders

Variation orders (VOs) are a special type of architect's instruction since they may alter or modify the design, quality or quantity of works shown on the drawings and in the bills. A variation may also alter the conditions under which the work is to be carried out. In many other respects, VOs are similar to those of other architect's instructions (see Chapter 24).

18.4.5 Dayworks

Daywork is 'work which cannot be properly valued by measurement'. The site manager should record in the site diary reasons why some of the work should be paid for on this basis. Prior to arriving at this decision, the contractor's own quantity surveyor should be consulted for an opinion. Where dayworks are envisaged, then the architect should be informed in order that the clerk of works can carry out the necessary checks on the labour and materials used on such work. The inclusion of large numbers of dayworks in the site manager's diary may lead to the conclusion that the nature of the project has changed from that originally tendered.

18.4.6 Delay

The contractor will have prepared a master programme for the project for use by both the contractor and the architect. The contractual implications of this are slightly different under the ICE Conditions of Contract, nevertheless the architect should have assessed its appropriateness. The programme will indicate how the contractor intends to carry out and complete the works. It also assists the architect on the dates when the contractor requires the various pieces of information. The programme provides the contractor with the best indication that the works are being constructed as planned when progress is measured against it. Where the progress shown is different, then the reasons for such variation will need to be established.

18.4.7 Lost time

A record of any delays will be made in the site diary at the time they actually occur. This will be the most accurate record of these events. Where the delays result from factors that the contractor should have controlled, these will need to be remedied at the contractor's own expense. This will then avoid liquidated damages being applied for late completion.

18.4.8 Urgent requirements

The site diary will be used to remind the site manager of any urgent requirements. These may include a note, in advance, of items that are expected for delivery or action on a certain date. This information can be included in the following manner:

- Sanitary fittings for delivery next Tuesday; storage provision required.
- Latest date for forwarding design details of external screens, since the manufacturer requires six weeks prior to delivery.
- Inform refuse chute subcontractor when they will be required on-site. They require three weeks' notice.

18.4.9 Unusual occurrences

Any unusual happening on the site will need to be recorded for possible future reference by the contractor. Disputes may occur between the parties concerned and it is important to establish as soon as possible the reasons why such disputes have arisen. Here are some typical examples:

- Local stoppage on-site for two hours due to a disagreement over bonus payments.
- Mr R. Taylor fractured his ankle. An accident report has been completed.
- Intruders entered the site last evening. No damage or theft occurred. Police have been informed.
- Raining-in below patent glazing. Architect and subcontractor have been informed.

18.4.10 Labour records

The site manager will need to keep a careful record of labour that was required and labour employed on-site.

18.4.11 Visitors to the site

Visitors should always be recorded, especially those who might have a direct effect upon the work. The diary record may show:

> The architect visited the site at 10.00 am. A tour was made with the site manager and clerk of works. Satisfaction with the quality and progress of the works was noted.

The contractor cannot refuse reasonable entry to the site for the architect or representatives. The building owner owns the site and therefore has rightful access. Building control officers, health and safety inspectors and similar officials have a legal right to entry to the site at any reasonable time. All visits must be made causing as little inconvenience to the contractor as possible.

18.4.12 Recording of information

If the site diary is to be of any value, then it is important that the events are entered each day in a logical, careful and legible manner. The information will thus provide an accurate assessment of the site's daily progress together with a record of the labour and plant that have been used. Any matters that affect the following items must be regularly and accurately recorded:

- Completion date.
- Costs.
- Quality and standards of work.
- Contractor's performance.

The site diary must be completed daily in order that the salient points are not forgotten for ever. The completion of the task will require some self-discipline on the part of the site manager, particularly when other aspects of work are requiring attention. The information should be accurate and represent a fair picture of the day's events. Exaggerated information or hearsay remarks should not be included. Where proper and adequate records are not maintained, the contractor may suffer the following consequences:

- The loss of reputation of a well-run organisation.
- Liquidated damages being imposed through late completion of the works.
- Being refused additional payments for losses incurred.
- Having the contract unfairly terminated due to client dissatisfaction.
- Being levied with further damages to redress the client's loss.

18.5 Planning and programming

A programme or schedule is developed by breaking down the work involved in a construction project into a series of operations that are then shown in an ordered stage-by-stage representation. Without a programme of work which specifies the time and resources allocations to undertake each stage of the project, the execution of the contract will be haphazard and disordered. There are several methods that may be used for this purpose, such as bar charts (sometimes referred to as the Gantt chart) or network analysis (critical path analysis). A detailed description of these methods is outside the scope of this book. The bar chart is perhaps the best known of all the planning techniques. In its simplest form, the sequential relationships between activities are not completely prescribed. However, they can be linked to show the relationship between an activity and the preceding and succeeding activities. Thus, dependency between the activities highlights the effect of delays. The resources required can also be calculated.

The linkages between preceding and succeeding activities, combined with a set of arrows to represent the bars of the bar chart, give rise to a simple network diagram. This forms the basis of a network analysis, which identifies the longest irreducible sequence of events. It also quickly defines those parts of the programme which could benefit from the use of increased resources and thereby benefit the project. As networks are rarely the best method for communication, the output of the analysis is often presented as a bar chart. Network analysis has been used for many large and complex projects. It is claimed that network analysis can reduce project times by up to 40 per cent.

18.5.1 Resources

The time taken to complete an activity in the programme depends on the resources allocated to that activity. Approaches used in assessing the required resources can be based on completing the project in a given time or completing the project with specified limited resources. Once the level of resourcing has been finalised, the overall resource demands are smoothed, if necessary, by rescheduling activities to ensure an acceptable overall demand for the project.

There is a measure of uncertainty in estimating the time for each activity, particularly as delays in delivery of materials and adverse weather can delay work in progress. Probabilistic distributions have been used for the generation of the most likely times for activities. An approach that was used in early projects was the project evaluation and review technique (PERT).

The flexibility of the contractor's workforce has changed in recent years. Previously, a contractor, using its own workforce, would operate a number of sites in an area. There were thus ample opportunities for transferring workpeople between the different sites. However, the widespread use of labour-only subcontracting has to some extent curtailed this opportunity. This has reduced the adaptability of day-to-day site management, resulting in a greater measure of pre-site planning. Also, the construction industry, when compared with manufacturing, remains a labour-intensive industry. The widespread use of plant and machinery on-site is perhaps less developed than it ought to be.

18.5.2 Monitoring and control

Once a programme of work and the resource requirements for each activity have been determined, it is possible to monitor the construction work as it progresses. In practice, updating will take place and control will be exercised. This will involve the rescheduling of activities and the revision of resources. The planning model is often used to explore the overall development of the project before work on-site is undertaken. This assists in investigating the influence of different construction techniques and the timing of the individual activities to better optimise the use of resources.

Ensuring that the works are constructed to the specified level of quality is essential. This extends from the initial setting out of the project to the inspection, storage, handling and incorporation of the specified materials within the finished building. All site operations must be governed by appropriate safety measures, which start with safe design and erection procedures for permanent and temporary works. The construction industry remains one of the largest single contributors to fatal accidents (see risk identification in Chapter 12).

18.5.3 Planning into practice

The construction industry is usually involved in one-off projects; these are invariably managed with a new team. As the location of each project varies widely, the workforce is largely new and the conditions under which the project is undertaken can differ depending on the site conditions, climate, etc. Unlike the manufacturing industry, the approach to a construction project is rarely uniform. The size of the main contractor's site organisation, which comprises technical and non-technical staff, often depends on the size of the works. However, as the use of subcontractors has become more widespread, the main contractor's team has reduced in size.

The work can be carried out efficiently if the site is laid out in such a way that the temporary buildings – offices, stores and workshops – are conveniently located with respect to the permanent works. This results in an orderly arrangement that facilitates the economy of construction and administration. The construction of a high-rise building on a confined city centre site, for example, requires the efficient storage of materials, off-peak deliveries, high-speed vertical travel and site facilities placed to minimise operative travel.

Table 18.1 Delays in construction of industrial buildings

Cause of hold-up and delay	*Percentage (%)*
Subcontracting	49
Tenant/client variations	45
Ground problems – water, rock, etc.	37
Bad weather	27
Materials delivery	25
Sewer/drains obstruction or rerouting	20
Information late	20
Poor site management and supervision	18
Steel strike	16
Statutory undertakers	14
Labour shortage	10
Design complexity	6

Table 18.1 indicates some of the reasons for delays that have occurred in the construction of industrial buildings together with the frequency of their occurrence. The influence of these delays on the total site time varies. For example, ground problems or bad weather occurred on sites which achieved both fast and slow site times, suggesting there was less overall effect. However, poor site management and supervision, late information and deliveries, design complexities, difficulties with statutory undertakers and labour shortages were usually associated with construction overruns of two months or more. These factors were judged to be the most damaging to the more effective organisation of the construction process.

19 Constructing excellence in the built environment

Constructing Excellence is a movement based in the UK. It has wide support from its various stakeholders in the construction industry. Its aims are to achieve a step change in construction productivity by tackling the market failures in the sector and selling the business case for continuous improvement. Constructing Excellence has developed a clear strategy to deliver the process, product and cultural changes needed to drive major productivity improvements in the sector. There are focused programmes in:

- innovation
- best practice knowledge
- productivity and engagement.

Constructing Excellence was formed in 2004 through the amalgamation of Rethinking Construction and Construction Best Practice. These two initiatives followed the publication of *Constructing the Team* (Latham 1994) and the later review of the Construction Task Force that produced the report *Rethinking Construction* (Egan 1998). These reports followed many other construction industry reviews that started with *The Placing and Management of Building Contracts* (Simon 1944). Several such reports are referred to in Chapter 12.

In August 2016, Constructing Excellence (CE) became a subsidiary of the Building Research Establishment (BRE), but retained rights to maintain its brand, functions and portfolio of activities. The CE foundation was established under the BRE trust enabling any trading profits generated by CE to be channelled into research and education related to its member organisations. Currently, CE has nine regional centres and thirty best practice clubs, making it one of the leading movements for change in the built environment.

19.1 *Constructing the Team* (the Latham Report)

This report began by stating that previous reports on the construction industry (see Chapter 12) have either been implemented incompletely or the problems that were identified have continued to persist. Looking back over the last ten years, since the Latham Report was published, whilst significant changes have taken place, these have not occurred throughout the industry. In a number of cases, individuals and organisations have resisted several changes and recommendations. Elsewhere, implementation has been minimal and progress has either been questioned or ignored. Changes in practice have been more in evidence amongst major contractors and on large projects, and much less so amongst small and medium-sized enterprises and the smaller construction projects. At the extreme of the small, domestic subcontractors, implementation has been negligible.

The Latham Report has sought to put the client in the position of driving change. Clients, including government, continue to have an important role in promoting excellence, both through design and construction. The larger client organisations have sought to bring about beneficial change, particularly since they undertake a considerable amount of construction work on a regular and routine basis. A number of clients now spend in excess of £1 million per day on their new and much-needed construction projects. These clients are able to instil change in practices and procedures expected from their professional advisers and the contracting firms and organisations whom they employ. For example, when the Latham Report was first published, the British Airports Authority (now BAA Airports Limited), a major client, contested its view of a 30 per cent cost reduction, suggesting that this was too small a reduction.

The report recognised that the state of the wider economy remains crucial to the construction industry. The booms and the slumps of the construction industry have restricted investment and this has had a negative effect on productivity (see Harvey and Ashworth 1997). The booms and slumps in the economy have been much less pronounced in the last ten years and this has assisted the construction industry in moving forward the agenda outlined in the Latham Report. Over this time, the economy has enjoyed low interest rates and increased employment opportunities; however, coupled with this has been a lack of personnel at almost every level in the industry. School leavers and graduates have failed to recognise the benefits to be obtained through employment in the construction industry. However, building craft courses are now as full of trainees as they have ever been, largely due to the publicity surrounding plumbers' wages. There has also been an upturn to technician and graduate programmes, although it is recognised that the supply will take a number of years to fulfil.

The Latham Report made the following recommendations:

- Preparing the project and contract strategies and brief requires patience and practical advice.
- A checklist of design responsibilities should be prepared.
- The use of coordinated project information should be a contractual requirement.
- Design responsibilities in building services engineering should be clearly defined.
- Endlessly refining existing conditions of contract will not solve adversarial problems.
- A set of basic principles is required on which modern contracts can be based.
- The role and duties of project managers needs clearer definition.
- Tender list arrangements should be rationalised.
- Tenders should be evaluated by clients on quality as well as price.
- A joint code of practice for the selection of subcontractors should be drawn up.
- The industry should implement recommendations that it previously formulated to improve its public image.
- Existing research initiatives should be coordinated and involve clients.
- A productivity target of 30 per cent real cost reduction by the year 2000 should be launched.
- A construction contracts bill should be introduced to give statutory backing to the standard forms.
- Adjudication should be the normal method of dispute resolution.
- Mandatory trust funds for payment should be established for construction work governed by formal conditions of contract. 'BUILD' insurance should become compulsory for new commercial, industrial and retail projects.

The report drew a stark comparison with the modern motor car (see Table 19.1). Incidentally, in 1900, a motor car and family house cost about the same to purchase. Now, some 100 years later,

Table 19.1 Construction industry performance compared to the car industry

Wants	*Motor car*	*Modern buildings*		
		Domestic	*Commercial*	*Industrial*
Value for money	****	*****	***	****
Pleasing to look at	****	****	***	***
Mainly free from defects	*****	***	*	**
Timely delivery	****	****	****	****
Fit for purpose	*****	****	**	***
Guarantee	*****	****	*	*
Reasonable running costs	****	****	**	***
Durability	****	***	**	**
Customer delight	*****	***	**	**

Source: Latham 1994.

a family house costs about ten times more than a car. Had the motor car industry not made the considerable advances in manufacture, quality, cost and productivity, relatively few of us would be able to own a car.

19.2 *Rethinking Construction* (the Egan Report)

This report of the Construction Task Force, on the scope for improving the quality and efficiency of UK construction, was for Deputy Prime Minister John Prescott.

It recognised that a successful construction industry is essential for all. Everyone benefits from high-quality housing, hospitals or transport infrastructure that are constructed efficiently. At its best, the UK construction industry displays excellence. However, there is no doubt that substantial improvements in quality and efficiency are desirable and possible. Indeed, they are vital if the industry is to satisfy all of its customers and reap the benefits of becoming a world leader.

The report examined leading-edge practices in the construction industry and also in other industries, most notably manufacturing, which has also undergone considerable change, and suggested that continuous and sustained improvement was achievable if efforts could be focused on delivering what customers need and desire. Existing wasteful practices and poor quality, often arising from existing structures and working practices, needed to be reviewed. The Egan Report identified five key drivers for change, as shown in Table 19.2.

The Egan Report made the following observations and recommendations:

- The construction industry has delivered some of the most difficult and innovative projects and matches than any other construction industry in the world.
- There is a deep concern that the industry is underachieving.
- It has low productivity and invests too little in capital, research and development.
- Too many of the industry's clients are dissatisfied with its overall performance.
- There have been radical changes and improvements in other industries.
- Any future improvements can be spread throughout the construction industry.
- Ambitious targets and the effective measurement of performance are essential to deliver improvement.
- A series of targets for annual improvement are suggested.
- The targets are based on experience and evidence obtained from projects within the UK and overseas.

- In order to achieve these targets, the industry will need to make radical changes to the processes through which it delivers projects.
- These processes should be explicit and transparent to the industry and its clients.
- The industry should create an integrated project process around the four key elements of:
    - product development
    - project implementation
    - partnering the supply chain
    - production of components.
- If the industry is to achieve its full potential, substantial changes in its culture and structure are also required to support improvement.
- The industry must provide decent and safe working conditions and improve management and supervisory skills at all levels.
- The industry must design projects for ease of construction, making use of standard components and processes.
- Major clients of the industry must give leadership by implementing projects that will demonstrate the approach identified.

19.2.1 Performance indicators

It is easy to suggest that improvements in processes and practices are being achieved based on subjective judgment and anecdotal evidence alone. There is often also resistance to want to measure or attempt to quantify such changes. It is all too easy to distort the data, unless clear and precise guidelines are employed. In some cases in the past, improvements have occurred and their effect has then been attributed to a particular cause. Upon further investigation, the cause and effect are not linked. For example, the 30 per cent reduction in cost identified by the Latham Report (1994) may appear to be achieved largely because of the suppressed costs, of both labour and materials, from the recession in the middle of the 1990s. When looking back to the start of the twenty-first century, it may be difficult to identify whether improved methods of working actually achieved this goal, or whether, without it, natural effects of improved technologies were the real reasons. The debates that raged throughout the 1980s about the poor time performance of UK building when compared with countries abroad were only partly remedied through productivity agreements. Other improvements in time performance were often restricted because of the different regulations and organisation adopted in the UK.

In this context, it is important that the construction industry sets itself clear and measurable objectives. These might be achieved through the use of performance indicators or quantified targets. Measures of improvement will be required in terms of cost, time and quality, relevant to the aims and objectives of the individual client. The targets must be real and composite. They must not

Table 19.2 Five key drivers for changes

Committed leadership: the need for vision and change at all levels.
Focus on the client: the view is that the customer is always right.
Integration of processes and teams: fragmentation of operations affects success.
Quality driven agenda: about getting it right first time.
Commitment to people: providing appropriate health and safety, wages and site conditions.

Source: Egan 1998.

Table 19.3 Performance indicators

Indicator	*Improvement per year*	*Definition*
Capital cost	Reduce by 10%	All costs excluding land and finance
Construction time	Reduce by 10%	Time from client approval to practical completion
Predictability	Increase by 20%	Number of projects completed within time and budget
Defects	Reduce by 20%	Reduction in the number of defects at hand-over
Accidents	Reduce by 20%	Reduction in the number of reportable accidents
Productivity	Increase by 10%	Increase in value added per head
Turnover	Increase by 10%	Turnover of construction firms
Profits	Increase by 10%	Profits of construction firms

Source: Egan 1998.

be achieved through cutting corners in other respects, such as safety and wages. In order to make such gains last, and thereby add value, continuous improvement must be implemented.

The Egan Report identified a number of measures designed for sustained improvement. These are shown in Table 19.3.

19.2.2 Other industry comparisons

It is always relevant, when examining a subject like procurement, to see how it is done elsewhere. This comparison may be made against similar or competing firms, perhaps in the form of a benchmarking study. Alternatively, comparisons can be made with firms or organisations overseas, in countries that mirror UK practices and in countries where different traditions are employed. It is also important to consider other industry comparisons, as illustrated in *Constructing the Team* (Latham 1994), which compared the performance of the construction industry with that of the motor car industry (see Table 19.1), and comparisons made with the aerospace industry (Flanagan 1999). The outcome of such studies acts as a guide to good practices found elsewhere, but which might have been overlooked. Current comparisons do not place the construction industry in a good light, but do act as motivators to help change its culture. The motor car manufacture and aerospace sectors include the following attributes that are generally absent from the construction industry:

- The recognition of a manufacturing culture.
- The integration of design and with production.
- The importance of the supply chain network.
- A focus on innovation and that this will only be secured through adequate research and development.
- An acceptance of standardisation in design, components and assembly across the product range.

19.3 Rethinking Construction achievements

This report (Rethinking Construction 2002) outlined how this movement within the construction industry was progressing and highlighted some of its achievements. The principles remained the same:

- client leadership
- integrated teams throughout the delivery chain
- respect for people.

The Rethinking Construction movement can be described as being about values and visions. It has six key targets:

- reduced capital costs
- reduced construction time
- fewer defects
- fewer accidents
- increased productivity
- increased turnover and profits.

At the heart of Rethinking Construction is the demonstration projects programme. This provides the opportunity for leading-edge organisations to promote projects that demonstrate innovation and change and which can be measured and evaluated. These are either site-based projects or organisational change projects. There are now more than 400 of these projects, which, when taken together, outperform the average UK industry against a key set of indicators. They provide evidence that continuous business improvements are being achieved. There are four key strategic objectives:

- *Proving and selling the business case for change*: this is largely achieved through the evaluation of the demonstration projects and the collection of KPIs (key performance indicators).
- *Engage clients in driving change*: encouraging clients to promote Rethinking Construction through involvement in demonstrations and commitment to the clients' charter.
- *Involve all aspects of industry*: ensuring that every sector of the construction industry is involved at some level.
- *Create a self-sustaining framework for change*: ensuring that the industry takes responsibility for developing and maintaining continuous improvement, nationally and regionally.

Because of the varied nature of the construction industry and its products, there are a number of separate streams of activity that have been identified. These are as follows.

19.3.1 The Movement for Innovation (M4i)

This focuses generally on the construction industry. It has developed regional networks focusing on off-site prefabrication, knowledge management and lean construction (see Chapter 13).

19.3.2 The Housing Forum

This concentrates on both the public and the private housing sectors. It has worked with many organisations who are seeking improvements in quality, efficiency, sustainability (see Chapter 21) and value for money.

19.3.3 The Local Government Task Force

This promotes an agenda amongst local authority clients, who still represent one of the largest clients in the UK. It has focused on whole life costs and best value obligations.

19.3.4 The Respect for People Steering Committee

This has trialled a number of toolkits to help improve recruitment, retention and health and safety. These are fundamental to achieving a world-class industry. In 2000, the then construction minister commissioned a report, *A Commitment to People, our Biggest Asset* (Rethinking Construction 2000). The respect for people initiative is based around seven themes:

- diversity in the workplace
- on-site working environment
- health
- safety
- working conditions off-site
- career developments and lifelong learning
- behaviour.

19.3.5 The Construction Best Practice Programme

This has been the main dissemination arm for Rethinking Construction. It is an integral part of the initiative. Best practice is the adoption and development of ideas, systems or methods in a way that measurably improves a business or enterprise so that it continually offers or secures best value for money.

Two reports are worthy of reference. *Rethinking Construction, Innovation and Research: A Review of Government R&D Policies and Practices* (Fairclough 2002) focused on the innovative capacity of an industry in influencing its long-term competitiveness and effectiveness. Research and development (R&D) is the important driver of innovation. No valid argument suggests that the construction industry is any different from any other industry, but it is not given the same priority, as measured in R&D expenditure, as a proportion of turnover. The construction industry organises its resources around projects and, although it is evident that considerable innovation occurs and is funded within projects, there is a problem with institutional learning to capture such innovation for future projects. *Accelerating Change* (Strategic Forum for Construction 2002) is the title of a report by the Strategic Forum for Construction chaired by Sir John Egan. This report further encourages change to take place following the principles of Rethinking Construction. As its title suggests, it implies that progress has been too slow and that the industry needs to change more quickly to meet the targets of performance required.

In 2002, key performance measures were published that indicated the following:

Clients were happier	11 per cent increase in client satisfaction.
Quality is increasing	30 per cent more projects were reporting fewer defects.
Safer place to work	Demonstration projects are 25 per cent safer than the industry average.
Keeping promises	15 per cent more demonstration projects are finishing ahead of programme and on budget.
Workforce productivity	Average added value is £10,000 more than the typical industry figure.
Quicker projects	Demonstration projects are completing schemes quicker than a year ago.

19.4 Construction Best Practice

The main drive has been to improve the business management of construction through the delivery of services to the sector and the dissemination of best practice information.

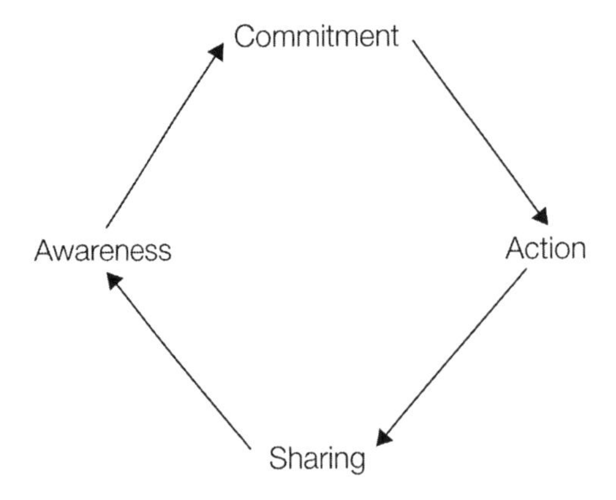

Figure 19.1 CBPP cycle.

Source: Constructing Excellence 2004.

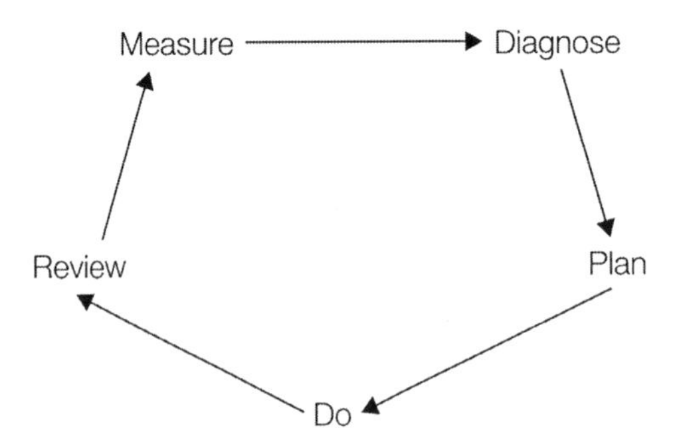

Figure 19.2 CBPP action plan.

Source: Constructing Excellence 2004.

The Construction Best Practice Programme (CBPP) has played a specific role in continuous business improvement, providing opportunities for individuals, business teams, entire companies and supply teams to engage in best practice. CBPP is about raising awareness, gaining commitment and facilitating shared knowledge.

Its 1,500 publications include case studies, profiles, guides and director's briefing workshops around the theme of learning by doing. The key points of the Construction Best Practice Cycle (shown in Figure 19.1) are the following:

- *Awareness*: encouraging people within the construction industry to rethink their approach to business and understand the benefits of best practice.
- *Commitment*: helping companies to understand their potential and opportunities for improvement, and to get buy-in from the right people.
- *Action*: assisting committed companies to choose the right steps to take, and to identify and use appropriate tools and techniques.

- *Sharing*: providing opportunities for people to pool ideas and experiences to enable best practice to be adopted widely and quickly within the industry.

The CBPP provides a range of resources and services including website and helpdesk, best practice profiles, case studies, fact sheets, construction best practice (CBP) partners, CBP clubs, learning by doing, diagnostics, key performance indicators, company visits, partner's workshops and advisers.

In any company or organisation, whether large or small and regardless of sector or industry, there is always scope for improvement. Knowledge about how to make improvement is often best acquired through learning from others who have already faced the same issues. However, CBPP is not prescriptive since what may work within one company may not always work in another. What matters is that companies are consistently looking at how they can do their work better, more effectively, more efficiently and more economically. Figure 19.2 shows The CBPP action plan:

- *Measure*: a good place to start is to measure current performance. This will help to identify areas for improvement and provides a baseline by which to measure progress against.
- *Diagnose*: try to understand why things are done in a particular way. This may require focus on a small number of areas in more detail.
- *Plan*: decide what needs to be done. Decide what improvement is required, how it is going to be achieved and how progress will be measured.
- *Do*: make the necessary changes.
- *Review*: keep the progress under review to ensure that progress is being made.

The CBPP has run a number of workshops, some of which have been under the umbrella of the Construction Productivity Network. These have helped companies network with the belief that networking companies innovate. These networks have focused on a common range of issues, such as benchmarking, partnering, supply chain management and managing innovation. They have also addressed some less common themes, such as e-commerce, lean construction, design management and knowledge management.

19.5 Constructing Excellence

Constructing Excellence is a public body that aims to deliver business improvement services to the UK construction industry. It has responsibility for promoting the principles of Rethinking Construction and to continue with the work of the Construction Best Practice Programme.

The broad aims of Constructing Excellence build on the work that has been done previously, as described above. It is seeking a step change in the construction industry, rather than a gradual change, through continuous improvement. Its vision is for the UK construction industry to realise maximum value to all clients, end users and stakeholders. It also wants the industry to exceed expectations through the consistent delivery of world-class products and services. Stakeholder groups include government, private sector clients, the regional development agencies (RDA), the research community and the media. It provides a definition of what it understands excellence in construction to mean:

- Creating individual, community and national prosperity (wealth) through the provision of products and services.
- Creating opportunities for living, learning, recreation and development that will advance the interests of the community at large.

- Exceeding all community expectations for products and services offered, creating added value.
- Achieving expected margins and ensuring value is delivered.
- Earning community respect for aesthetic, safety and environmental standards.
- Having integrated teams, delivering world-class constructed products, buildings, facilities and infrastructure incorporating quality components, systems and products.
- Respecting its people and the wider community.
- Exporting a range of products and services to other industries.

Constructing Excellence has strategic objectives that support the Strategic Forum for Construction targets:

- Improving performance through increased productivity and competitiveness.
- Improving industry image by taking action to create a step change in culture, the development of people and enhanced engagement with the community and customers.
- Engagement and taking action with individuals, businesses, organisations and industry associations.

Constructing Excellence is organised into four complementary and integrated programmes of activity that consolidate the functions of Rethinking Construction and CBPP. The four programmes are:

- *Innovation*: identifying and promoting tomorrow's best practice.
- *Productivity*: improving the competitiveness of the UK construction industry.
- *Best Practice Knowledge*: creating continuous improvement through the exchange of best practice.
- *Engagement*: working with people, businesses and organisations to change the culture of the construction industry.

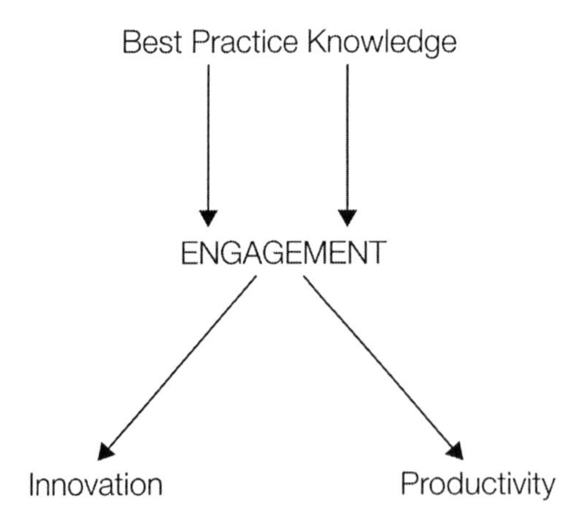

Figure 19.3 Engagement.

Source: Constructing Excellence 2004.

Engagement is the critical delivery mechanism and is fundamental to Constructing Excellence attaining its objectives, as shown in Figure 19.3.

As with Rethinking Construction, Constructing Excellence continues to support a number of specific initiatives to engage with key sector groups. These include:

- The Housing Forum
- Local Government Task Force
- Infrastructure Task Force
- Central government clients
- The movement for innovation.

Constructing Excellence is forging ahead to embrace the fourth industrial revolution (see Chapter 22) and proactively preparing the construction industry for the change. Generation 4 Change (G4C; www.g4c.org.uk/) is part of Constructing Excellence education and providing access to innovative industry thinking. It offers opportunity for construction industry professionals to network, peer learn and be involved in progressive and collaborative activities to make the construction industry future ready.

20 Health, safety and welfare

The construction industry is a dangerous environment and it has a poor health and safety record. Serious injury and death happen far too frequently as a result of construction activities and especially on construction sites. Injuries affect not only the construction workers employed on-site, but those visiting the site and members of the general public. Improving the management process is essential in helping to prevent accidents and ill health in the industry.

Different governments have initiated measures aimed at reducing accidents on construction sites. The most common legislation is the Health and Safety at Work Act 1974, which applies to all industries. However, it is recognised that some industries are more dangerous than others. The construction industry is just one of those industries.

20.1 The construction industry

The management of health, safety and welfare is without doubt a major and important activity of construction managers. The construction industry and the activities that it carries out are a dangerous occupation for the persons involved. In too many cases individuals are injured, sometimes seriously and sometimes fatally. The industry, although it takes no delight in the published statistics to support this statement, has nevertheless failed to adequately deal with the problems that are inherent in its activities. Poor health and safety also costs the construction industry a huge slice of its profits each year, through lost productivity, claims for compensation and high insurance premiums. These factors alone should encourage the industry to perform better in this area of concern. The Egan Report, *Rethinking Construction* (see Chapter 19), which was published in 1998, stated that 'the health and safety record of construction is the second worst of any industry' and suggested that accidents can account for between 3 and 6 per cent of the total project costs (Egan 1998).

The report *Accelerating Change* attempted to cost accidents (Strategic Forum for Construction 2002). It suggested that on its demonstration projects (see Chapter 19) accidents were 50 per cent lower than the industry average. Estimates put accident costs across the industry as 8.5 per cent of turnover. On these demonstration projects, which were worth £6 billion at the time of the report, the reduced costs from accidents were shown to be worth £255 million. If one-third of the industry followed the principles and recommendations of Rethinking Construction, this would achieve a cost reduction of £638 million.

A successful approach towards health and safety will recognise and value the important contribution made by individuals in support of a safe working environment. Whilst such contributions will focus on health and safety issues, these will also generate a wider culture of safety, health and welfare within a firm or organisation. This will also provide additional benefits in respect of improving the image and attractiveness of a working life in the construction industry.

20.2 Health and safety in the construction industry

In the last twenty-five years in the UK, according to the Health and Safety Executive (HSE), almost 3,000 people were killed on construction sites or as a result of construction activities. These included some members of the public. Many more have been injured or have developed prolonged illnesses. In the year 2001/02, seventy-nine workers died and thousands were injured as a result of construction site operations in the UK, but the trend over the years has been a considerable improvement in these rates. In 2016/17 this figure came down to thirty fatalities, though it is still a considerable number. The main causes were:

- falling through fragile roofs and roof lights
- falling from ladders, scaffolds and other work places
- being struck by excavators, lift trucks or dumpers
- overturning vehicles
- being crushed by collapsing structures.

Figure 20.1 provides a comparison of the number of fatal accidents across the major industries in the UK. This shows clearly that construction sites are more dangerous places in which to work than the manufacturing industry. 2016/17 was a good year for the construction sector, with lowest recorded levels of fatalities (thirty). However, according to the HSE (2017), over the last five years the number has fluctuated, with forty-three fatalities in 2015/16 compared with thirty-five in 2014/15. The annual average for the past five years is thirty-nine.

In 2009/10 the construction industry accounted for 4 per cent of all of the employees in the UK. This figure had declined considerably since its peak a few years earlier, indicating just how severely the recession had affected employment in this sector. By comparison, statistics from the HSE reported that in 2015/16 the construction industry accounted for 66,000 self-reported non-fatal workplace injuries. Slips, trips and falls accounted for 23 per cent, lifting and handling 22 per cent, falls from height 20 per cent and struck by object 11 per cent.

The construction industry has the largest number of fatal injuries of any of the main industry groupings. In 2009/10 there were forty-two fatal injuries that gave a rate of 2.2 per 100,000 workers. The CITB reports that the worker fatal injury rate in 2015/16 was 1.94 per 100,000 workers, which is more than four times the average rate across all industries of 0.46 per 100,000 workers. It shows that construction activities are inherently more risky

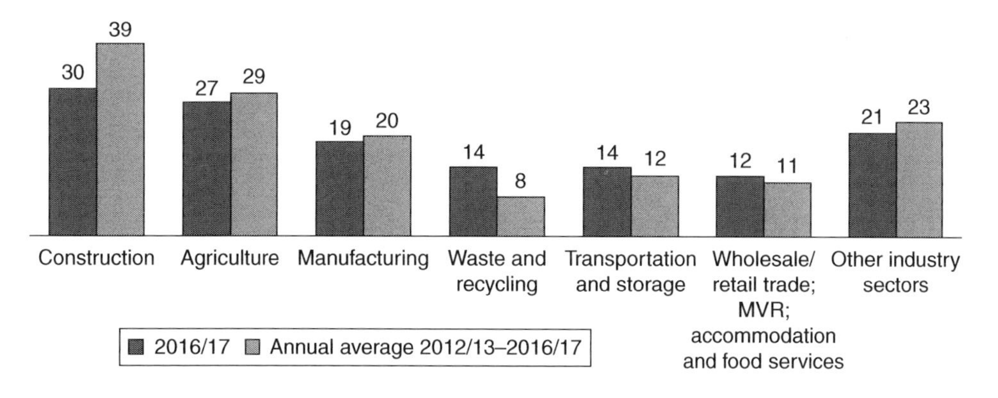

Figure 20.1 Fatal injuries to workers by main industry.

Source: www.hse.gov.uk/statistics/pdf/fatalinjuries.pdf (accessed 24 August 2017).

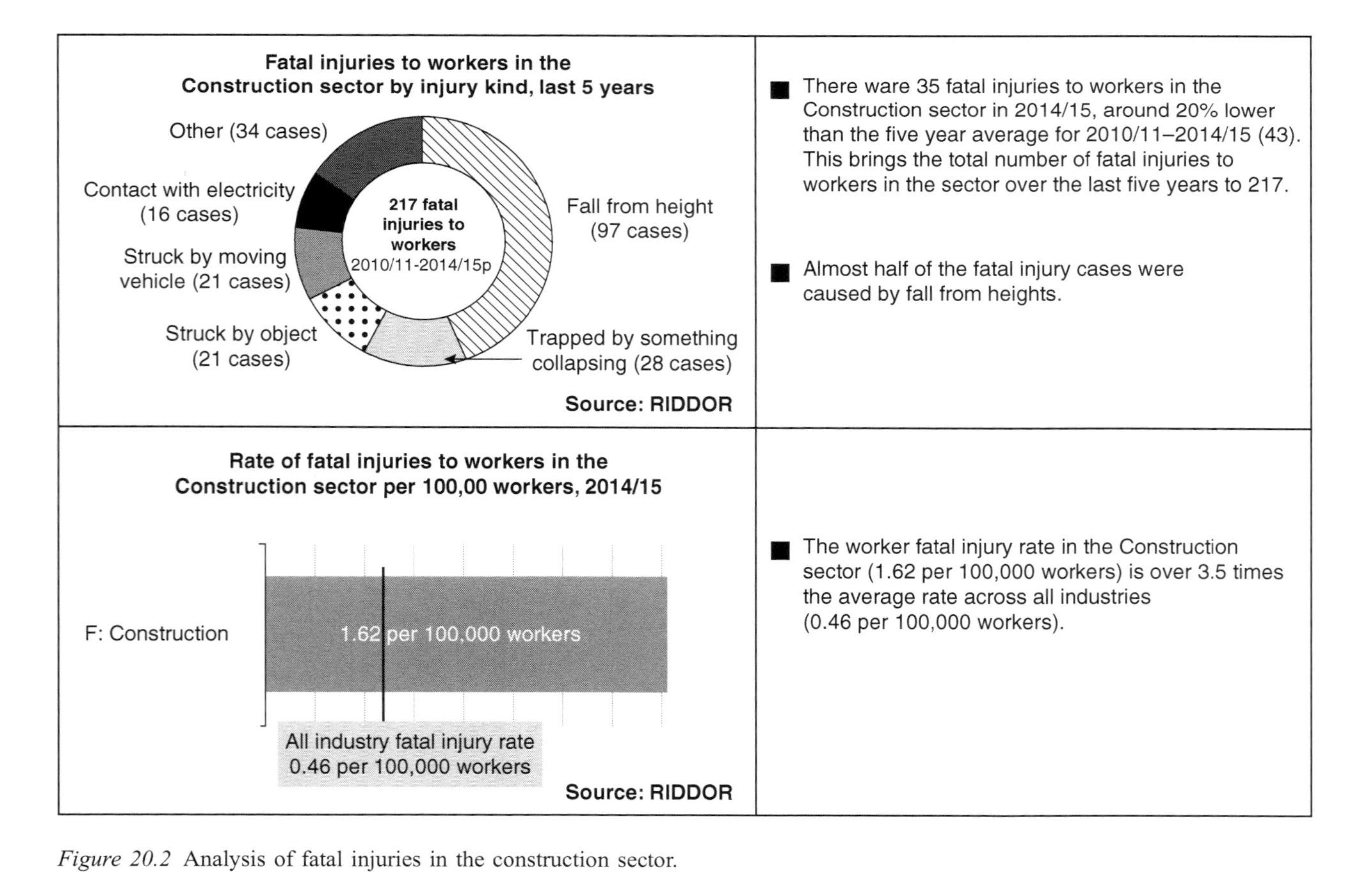

Figure 20.2 Analysis of fatal injuries in the construction sector.

Source: www.hse.gov.uk/statistics/pdf/fatalinjuries.pdf (accessed 24 August 2017).

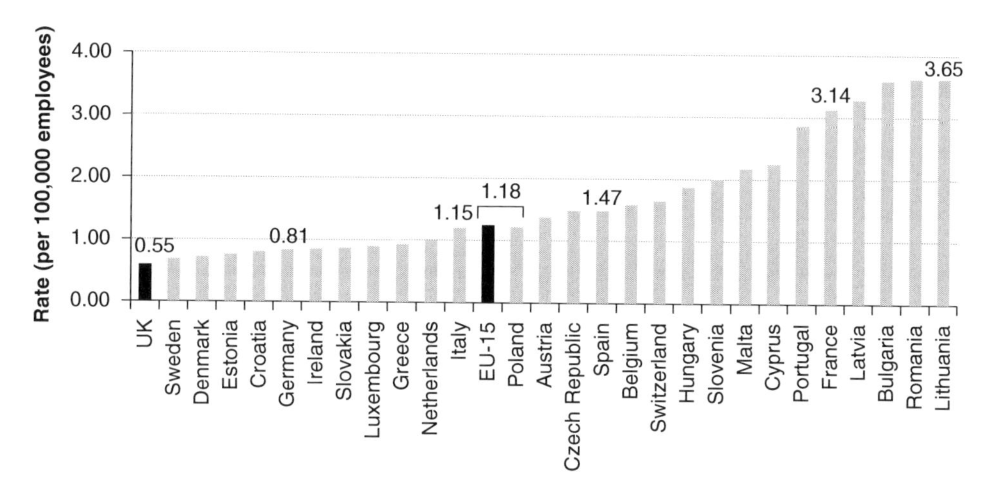

Figure 20.3 Standardised incidence rates (per 100,000 employees) of fatal injuries at work for 2014.

Source: www.hse.gov.uk/statistics/pdf/fatalinjuries.pdf (accessed 24 August 2017).

compared to most other industries. The good news is that all of the incidence rates in the construction industry have fallen over the years to 2016/17. However, there were 217 fatalities between 2010/11 and 2014/15, which is much lower than the 700 fatalities that occurred between 2000/01 and 2009/10 (Figure 20.2). The most common cause of fatalities in the construction industry is falling from a height.

The UK performance in health and safety stands as world leading, as indicated in Figure 20.3. The standardised rates published by Eurostat are based on fatalities occurring across all main industry sectors (excluding the transport sector).

The HSE reports around 65,000 annual non-fatal injuries from the construction sector (see Figure 20.4). Specialised construction activities relating to specialised trades reported the most number of casualties, followed by building construction and civil engineering.

It should be noted that the data in Figure 20.4 is based on survey estimates as only a fraction of injuries get formally reported. Accordingly, there were 5,245 reported non-fatal employee injuries in 2015/16. The same figure was 5,400 in 2014/15 and has been at similar levels for the past four years. Reported non-fatal injuries are categorised as either specified (a predefined list of certain injury types, including fractures, amputations and serious burns) or injuries resulting in more than seven days off work.

20.2.1 Health and Safety Executive

The HSE was formed on 1 January 1975 with a remit to undertake the requirements of the Health and Safety Commission and to enforce health and safety legislation in all workplaces, except those regulated by local authorities. The Health and Safety Commission (HSC) conducted a review of health and safety regulations in 1994. It found that people were confused about differences between:

- Guidance.
- Approved codes of practice.
- Regulations.

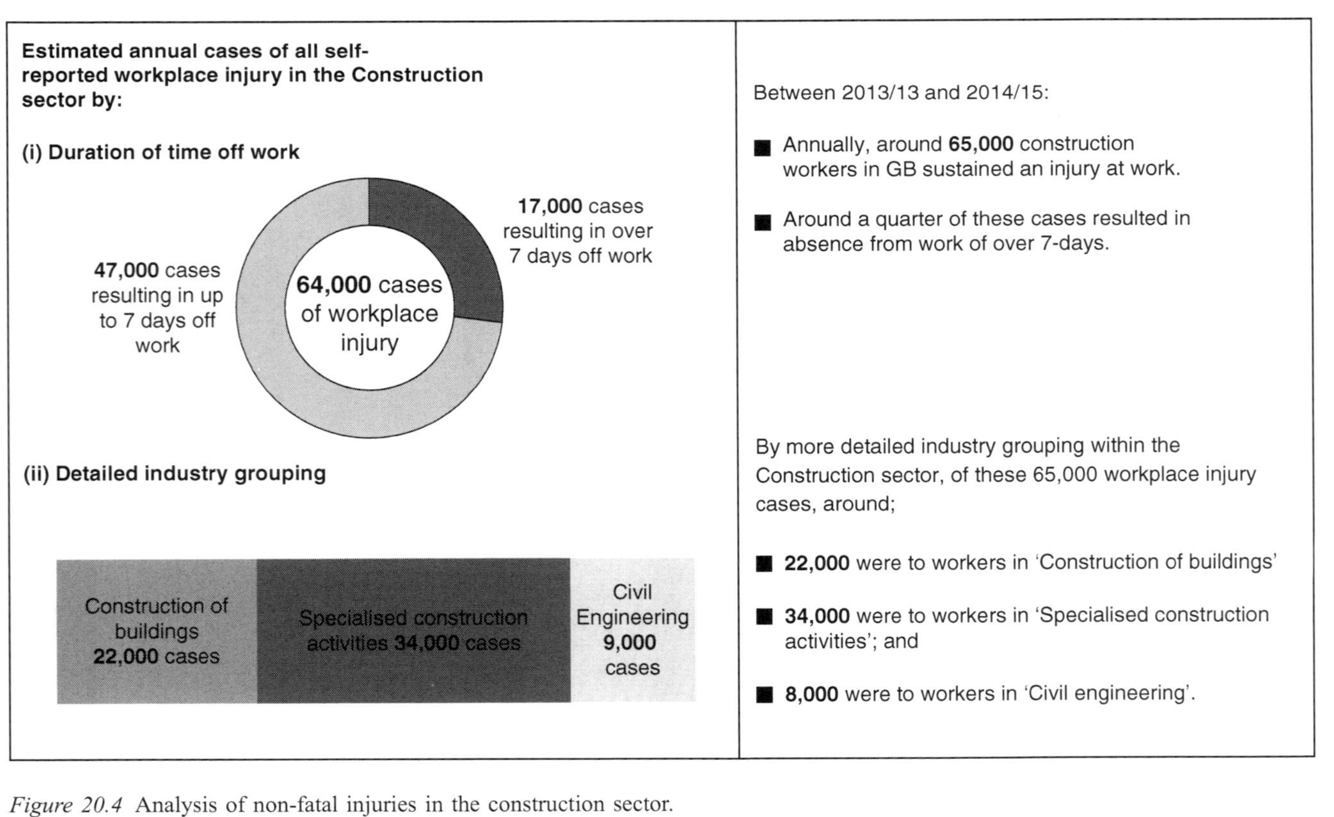

Figure 20.4 Analysis of non-fatal injuries in the construction sector.

Source: www.hse.gov.uk/statistics/pdf/fatalinjuries.pdf (accessed 24 August 2017).

The basic information of British health and safety law is contained in the Health and Safety at Work Act 1974. This Act sets out the general duties that employers have towards employees and members of the public, and that employees have to themselves and to each other. The duties are qualified by the Act by the principle of 'so far as is reasonably practicable'. The employer needs to make judgments in respect of time, cost and other measures in relationship to the risk that is involved. The law requires employers to provide good management and common sense and to incorporate sensible measures to avoid problems taking place.

The Management of Health and Safety at Work Regulations 1999 generally makes more explicit what employers are required to do to manage health and safety under the Health and Safety at Work Act 1974. As with the Act, the Regulations apply to every work activity. The main requirement on employers is to carry out a risk assessment and to record the significant findings from this.

A risk assessment will be fairly straightforward in a workplace such as an office. It is obviously more complicated where serious hazards exist, such as a nuclear power station or an oil rig. Employers must also:

- Make arrangements for implementing the identified health and safety measures.
- Appoint competent people.
- Set up emergency procedures.
- Provide clear information and training for employees.
- Work together with other employers sharing the same workplace.

In recent years, much of Britain's health and safety law has originated in Europe. Proposals from the European Commission may be agreed by the member states, who are then responsible for making them part of their own country's laws. Modern health and safety law is based on the principle of risk assessment.

Guidance

The HSE publishes guidance on a range of subjects concerning health and safety problems relating to a particular industry, such as construction. The main purposes of the guidance are to:

- interpret, by helping people to understand the requirements of the law and European commission directives
- help individuals comply with the law
- offer technical advice.

Guidance is not compulsory and employers are free to take other action. Following the guidance will normally be sufficient to comply with the law.

Codes of practice

These offer practical examples of good practice. They give advice on how to comply with the law by, for example, providing a guide to what is reasonably practicable. Approved codes of practice have a special legal status. Where an employer is prosecuted for a breach of health and safety law, and it is proved that they have not followed the relevant provisions of the approved code of practice, a court can find them guilty.

Regulations

These are laws approved by an Act of Parliament. These are normally made under the provisions of the Health and Safety at Work Act following proposals from the HSC. Guidance and approved codes of practice give advice. Regulations identify risks and set out a specific action that must be taken. Often, these requirements are absolute, i.e. to do something without qualification, rather than whether it is reasonably practicable.

20.2.2 Health, safety and welfare legislation

In addition to the Health and Safety at Work Act, there are many other pieces of health and safety legislation. These include the following:

- Management of Health and Safety at Work Regulations 1999 requires, for example, employers to carry out risk assessments.
- Workplace (Health, Safety and Welfare) Regulations 1992 covers a wide range of basic health, safety and welfare issues such as ventilation, heating, lighting, workstations and seating and welfare facilities.
- Personal Protective Equipment at Work Regulations 1992 requires employers to provide appropriate protective clothing and equipment for their employees.
- Provision and Use of Work Equipment Regulations 1998 requires that equipment provided for use at work, including machinery, is safe.
- Manual Handling Operations Regulations 1992 cover the moving of objects by hand or bodily force.
- Health and Safety (First Aid) Regulations 1981 cover the requirements for first aid in the workplace.
- Reporting of Injuries, Diseases and Dangerous Occurrences Regulations 1995 (RIDDOR) requires employers to notify certain occupational injuries, diseases and dangerous events.
- Control of Substances Hazardous to Health Regulations 2002 (COSHH) requires employers to assess the risks from hazardous substances and take appropriate precautions.
- The Construction (Design and Management) Regulations 2015 (CDM 2015) came into force on 6 April 2015, replacing CDM 2007. It provides guidance on the legal requirements for construction, design and management.

20.2.3 Risk assessment

A risk assessment is nothing more than a careful examination of what in the work could cause harm to people. This helps employers to consider whether enough precautions have been taken or whether more should be done to prevent harm.

- *Hazard*: means anything that can cause harm (e.g. chemicals, electricity, working from ladders, etc.).
- *Risk*: is the chance, high or low, that someone will be harmed by the hazard.

The following five steps will help to assess the potential risks in the workplace:

- *Look for the hazards*
    - Look around the workplace and reconsider what hazards may exist.
    - Concentrate on significant hazards that could result in serious harm.

  - Ask employees for their opinions.
  - Use manufacturer's data sheets or instructions to help identify hazards.
  - Examine accident and ill-health records.

- *Decide who might be harmed and how*
  - Consider trainees, young and new workers.
  - Remember visitors and cleaners, who might not be on-site all the time.
  - Do not forget the public, who could be hurt through site activities.

- *Evaluate the risks and decide whether existing precautions are adequate or whether more should be done*
  - Consider how likely it is that any hazard could cause harm.
  - Evaluate further whether more needs to be done to reduce risks.
  - Decide whether the remaining risks represent a significant hazard.
  - Ask whether you have done all that is required to satisfy the law.
  - Ensure that accepted industry standards are in place.
  - Reduce all risks to a minimum.
  - Draw up an action list and give priority to any remaining risks.
  - Control the remaining risks so that harm is unlikely.

- *Record your findings*
  - Prepare written information for record purposes.
  - Write down the significant hazards and conclusions.
  - Tell employees about your findings.

- *Review your assessment and revise it if necessary*
  - Repeat the above when new machines, substances or procedures are introduced.
  - Add any significant changes to the assessment.
  - Do not amend the assessment for trivial changes.
  - Review practices from time to time.

The HSE has produced a simple guide to assist employers and self-employed people to assess the risks in the workplace. This is titled *Risk Assessment: A Brief Guide to Controlling Risks in the Workplace* (HSE 2014). Table 20.1 identifies different categories of harm.

20.2.4 Health and safety toolkit

The HSE has provided an easy to use toolkit in respect of health and safety on construction sites. This is in the form of a checklist to help contractors manage and avoid problems occurring.

Table 20.1 Evaluation criteria for severity of harm

Assigned value	*Evaluation criteria of hazard*
1	Minor injury – no first aid attention
2	Illness – chronic injury
3	Accident – needing first aid attention
4	Reportable injury – under RIDDOR
5	Major injury – under RIDDOR
6	Death

Source: Health and Safety Executive.

Table 20.2 Site health and safety checklist

- Access on-site
- Welfare
- Scaffolds
- Ladders
- Roofwork
- Excavations
- Manual handling
- Traffic, vehicles and plant
- Tools and machinery
- Hoists
- Emergencies
- Fire
- Hazardous substances
- Noise
- Hand-arm vibration
- Electricity and other services
- Protecting the public

Source: HSE.

It seeks to ensure the safety of the contractor's own workpeople, subcontractors, clients and their advisers and the public. The toolkit also acts as a signpost to more detailed advice.

The toolkit begins by examining the contractor's business by focusing on the planning, organising and controlling the work that is involved. It asks whether a firm knows what is actually happening on its sites and what procedures are being followed to avoid dangerous practices occurring. The HSE is able to provide advice to contactors on health and safety issues. Where major incidents occur resulting in death or serious injury, then the accident must be reported to the HSE Incident Contact Centre.

The toolkit recognises the importance of employing properly trained and competent people on construction sites and making sure that they are given clear instructions and are properly supervised. They must have access to washing and toilet facilities and have the right tools, equipment, plant and protective clothing. There must also be opportunities to discuss health and safety issues either from a management or a worker's perspective. This same approach must also be adopted in respect of subcontractors working on the project. The site health and safety checklist is shown in Table 20.2. A copy of the full toolkit can be obtained from the HSE's website at www.hse.gov.uk.

20.3 The Health and Safety at Work Act 1974

This Act was introduced in response to the constantly expanding laws on health and safety in the UK. The Act consolidated much of the previous legislation and good practices. It provided for the development of personal responsibility for health and safety. The Act places duties upon a number of parties including employers, the self-employed and employees.

Employers have a duty to ensure, as far as is reasonably possible, the health, safety and welfare of their employees. This includes the safe use of mechanical equipment, safe methods of working and providing instruction, supervision and training. They also have duties and responsibilities towards others working on a construction site and for the safety of the public and other third parties.

Employees have a responsibility to take reasonable care for their own health and safety and for others who may be affected by their acts or omissions. They are also responsible for any duty or requirement imposed upon them by their employers or relevant statutory provisions.

The enforcement of the Act is carried out by HSE inspectors and by local authority inspectors. The nature of the main activity of the business determines the enforcing authority. When considering an action, an inspector uses discretion, but considers the following:

- The risk involved.
- The gravity of the alleged offence.
- The history of the business in respect of previous events and their compliance.
- The inspector's confidence in the management of the firm.
- The likely effectiveness of a particular action.

The decision to bring about a prosecution rests with the enforcing authority. Should a breach be established of sufficient consequence, then legal action will follow. This action may result in a fine or imprisonment or both depending upon the nature of the offence.

20.4 The Construction (Design and Management) Regulations 2007 and 2015

The CDM Regulations are of significant importance in the health and safety management of construction projects and their sites and became operative in 1995. They combine the CDM Regulations 2004 and the Construction (Health Safety and Welfare) Regulations 1996 into one regulatory package. This is an attempt to lessen the complex and somewhat bureaucratic approach taken by many duty holders. The aim of the CDM Regulations is to reduce the risk of harm to workers who build, use, maintain and demolish structures.

The Regulations emphasise the management of health and safety throughout all stages of construction projects. This has introduced a step change in the approach towards health and safety management in the UK construction industry. Rather than being reactive to problems as they arose, this Act has resulted in a more proactive approach towards planning, design and production. The Regulations place specific duties and responsibilities upon the various parties involved. More importantly, all parties must now be coordinated and managed throughout the project's stages, ensuring that no gaps in practice occur.

A European study of the construction industry revealed that the primary cause of 37 per cent of accidents were failures attributed to construction site management. Of the remaining accidents, 28 per cent were attributed to poor planning and 35 per cent to unsafe designs. An important conclusion resulting from this analysis is that almost two-thirds of accidents were due to decisions prior to the work commencing on-site.

The CDM Regulations apply to all construction projects, with few exceptions. The exceptions include the very small projects that are often of a domestic nature.

The two fundamental components of health and safety management introduced in the CDM Regulations are the development of a health and safety plan and the compilation of a health and safety file. The key features of health and safety management within CDM Regulations are:

- *Risk assessment*: parties must identify and assess project health and safety risk to comply with their duties.
- *Competence and adequate resources*: all of those involved must be pre-qualified by assessment to ensure that they are competent and have the necessary resources to fulfil their duties for health and safety.

- *Cooperation and coordination*: all parties involved must work together to identify and minimise health and safety risks.
- *Provision of information*: all parties have a duty to share information that is pertinent to health and safety. This will contribute towards the project's health and safety plan and health and safety file.

The CDM Regulations place responsibilities on clients, planning supervisors, designers and contractors to plan, coordinate and manage health and safety throughout all stages of a construction project. Anyone who appoints a designer or contractor has to ensure that they are competent for the work and will allocate adequate resources for health and safety. The Regulations apply to construction projects and everyone associated with their design and construction. The Regulations are about the management of health and safety. Two documents must be created:

- *The health and safety plan*: this is prepared in two stages, prior to and following the appointment of the contractor, and:
  - provides the health and safety focus for the construction phase of the project
  - shows that adequate resources have been allocated to the project.
- *The health and safety file*: this holds information about health and safety matters that will assist those carrying out construction, maintenance, repair or demolition work at any time before or after completion. It:
  - is a record for the client and user, identifying those responsible for the structure and the risks that have to be managed during maintenance, repair or renovation
  - describes the services that are installed in the building, the materials used and the building construction
  - is given to the client when the project is complete
  - is provided by the client to anyone carrying out work on the structure in the future.

The health and safety file is a document that is updated during the design process as more information becomes available. The health and safety plan must be sufficiently developed to form part of the tender documentation. It must do three things:

- Clarify the health and safety issues specific to the project.
- Identify where the principal risks are likely to occur and alert tenderers to any possible unexpected hazards.
- Clarify the parameters against which to judge the selection of competent and properly resourced contractors.

There are five key parties, firms or individuals involved; each has specific duties to perform.

Clients must ensure:

- They are able to allocate sufficient resources, including time, to enable the project to be carried out safely.
- Only competent and adequately resourced people are employed.
- Reasonable enquiries about the land or premises being developed are made and are passed to the planning supervisor.

- Construction work does not start until an adequate health and safety plan is in place.
- A health and safety file is available for inspection.

Designers produce drawings and written documents, such as specifications. Where risks cannot be avoided, adequate information must be provided. *Designers* must ensure:

- Structures are designed and specified to minimise any possible risks to health and safety during construction and subsequently during their maintenance.
- Adequate information is provided on possible risks.
- Cooperation is made with planning supervisors.

Planning supervisors have the overall responsibility to:

- Coordinate the health and safety aspects of the design and planning phase.
- Prepare the early stages of the health and safety plan.
- Advise clients of the competence and adequate resourcing of the principal contractor.
- Ensure that a health and safety file for each structure is delivered to the client on completion.

Principal contractors carry out the following duties and responsibilities:

- Take account of health and safety issues when preparing tenders or estimates.
- Exclude unauthorised persons from the site.
- Cooperate with the planning supervisor.
- Develop the health and safety plan.
- Coordinate activities of all contractors to ensure that they comply with the health and safety plan.
- Provide information and training to employees and self-employed workers on health and safety.

Subcontractors and the self-employed should:

- Cooperate with the principal contractor on health and safety matters.
- Explain how they will control the health and safety risks in their work.

Subcontractors also have duties towards the provision of other information to the principal contractor and to employees. The self-employed have similar duties to contractors. Employees on construction sites should be better informed and have the opportunity to become more involved in health and safety matters.

Whilst these new procedures add extra costs to the industry, and hence its clients, through implementation and monitoring, they will help to reduce costs by developing better construction practices over the longer period. They should help to save lives and reduce accidents, and reduce the disruption sometimes caused to work on-site. They will reinforce the need for coordinating and managing health and safety from construction inception to completion and during the use of the completed project.

Breach of the Regulations is a criminal offence and as such can result in both fines and prosecution.

The CDM Regulations 2007 were replaced by CDM 2015, to provide regulations governing the way construction projects are designed, planned and delivered and to improve the overall health, safety and welfare of those working in construction. It places legal duties on all involved in a construction project, which are enforceable by criminal law. CDM 2015 forces even occasional construction clients (DIY projects) to conform to these regulations as these have significant health and safety implications.

CDM 2015 requires a client to do the following:

- Appoint the right people at the right time.
- Ensure there are arrangements in place for managing and organising the project.
- Allow adequate time for the project.
- Provide information to the designer and contractor.
- Communicate with the designer and building contractor.
- Ensure adequate welfare facilities on-site.
- Ensure a construction phase plan is in place.
- Keep the health and safety file.
- Protect members of the public, including employees.
- Ensure workplaces are designed correctly.

In cases of domestic clients, the only responsibility a domestic client has under CDM 2015 is to appoint a principal designer and a principal contractor, especially where there is more than one contractor/builder. It then transfers all aforementioned duties of the client to the principal contractor.

20.5 The Construction (Health, Safety and Welfare) Regulations 1996

The Regulations apply to all forms of construction work. Other regulations should also be considered, as listed in section 20.2.2, especially RIDDOR and COSHH. The Construction (Health, Safety and Welfare) Regulations 1996 consist of thirty-five regulations and a number of appended schedules. The Regulations focus on safe methods of working that will protect the health of individuals working on construction sites. They consider the use of appropriate measures when carrying out demolitions or using explosives and recognise the danger associated with open excavations. They recognise the inherent dangers from falling objects and the fact that falls in general account for a large number of deaths in the construction industry (see section 20.2). Appropriate steps should be taken to prevent fires occurring, but sites must also have strategies in place to deal with emergencies of different types, and emergency exits should be suitably signposted. Individuals working on-site should have clearly identified places of safety pointed out to them in the case of such events happening.

The Regulations also cover the provision of welfare facilities, such as adequate sanitary conveniences, washing facilities and access to drinking water. Facilities should be provided for canteen and restroom areas, where these are reasonably practicable. Consideration should be given to ventilation, lighting, temperature and weather protection.

The contractor is also required to keep the site tidy and in good order, since a tidy site is much less likely to result in accidents occurring. Adequate training in matters of safety should be given to each person and they should be supervised appropriately to prevent accidents from occurring.

These Regulations replaced the existing Construction (Health and Welfare) Regulations that were introduced in 1966.

20.6 Management of Health and Safety at Work Regulations 1999

These Regulations were introduced to implement the provisions of a European directive to encourage improvements in the safety and health of workers at work. The Management of Health and Safety at Work Regulations 1999 consist of thirty regulations, and the HSE has provided a code of practice to support them. The various regulations focus on prevention measures and risk assessment, employers' and employees' duties and amendments to existing regulations.

20.6.1 Rethinking Construction recommendations on health and safety

The following recommendations were proposed in *Accelerating Change* (Strategic Forum for Construction 2002):

- Use the Construction Site Certification Scheme (CSCS) to ensure that those people who work at height are competent to do so, i.e. developing a specific test to evaluate their preparedness.
- Maximise the opportunities to develop solutions that involve less site processing and more pre-assembly and prefabrications.
- Develop transportation and materials distribution processes that reduce risk to personnel on-site.
- Develop an occupational health scheme for industry. The HSE is piloting a scheme. The pilot, and work towards the wider scheme, should be progressed as quickly as possible.
- Ensure that the workforce is consulted on health and safety matters. The Major Contractors' Group (MCG) is implementing a multi-step approach to workforce communication and the HSE is currently undertaking a worker safety adviser pilot. The opportunity to learn from and build on these and similar initiatives should be grasped.

20.7 Health, safety and welfare in contract documents

Bills of quantities for building works are based on the Standard Method of Measurement (SMM7). Within SMM7 there are a general set of rules applying to all the other sections. The general section has nothing particular to say about safety, health or welfare. The main section of the bills of quantities that make reference to these items are included in Section A: Preliminaries/General Conditions. These reiterate and reinforce the conditions of contract (JCT80). The following clauses are important:

- Clause A34 employer's requirements for security, safety and protection.
- Clause A36 employer's requirements in respect of facilities, temporary works and services. This clause makes particular reference to the provision of temporary sanitary accommodation and temporary fences, hoardings, screens and roofs.
- Clause A41 contractor's general cost items: site accommodation.
- Clause A42 contractor's general cost items: services and facilities. This is a particularly important clause, making reference to safety, health and welfare, the disposal of rubbish and the overall protection of works. This clause also makes particular reference to general attendance provided for other firms working on the project.
- Clause A47 contractor's general cost items: temporary works.

SMM7 provides a checklist of what should be covered in the preparation of the contract documents regarding health, safety and welfare. When compiling bills of quantities, quantity surveyors will need to add the specific project details of what the contractor is expected to provide and the work that needs to be carried out in respect of safety, health and welfare; a general comment sometimes found written in bills of quantities, 'to include everything necessary', is insufficient on its own. They also ensure that any subcontractors, of whatever classification, must comply with these and other conditions imposed upon the main contractor.

20.7.1 Conditions of contract

The conditions of contract are considered in detail in Part 4 of this book. Specific clauses in JCT refer directly to health, safety and welfare issues. For example, the principal contractor is to maintain within the CDM Regulations (14) a health and safety file for the project. The contractor is to carry out the works in a proper workmanlike manner and in accordance with the health and safety plan. Every instruction from the architect potentially carries a health and safety implication, which should be addressed under the Regulations.

21 Sustainable development and construction

Sustainability is defined in dictionaries as: the ability to carry forward or support or maintain for a prolonged period of time approaching perpetuity. In business terms, this is something that has traditionally been aspired to in terms of economic factors, as businesses strive to balance inputs and outputs to maintain profitable enterprise. In 1987, the United Nations Commission on Environment and Development, known as the Bruntland Commission, drew attention to the fact that economic development often has a detrimental effect on our planet and on society in general. It defined sustainability as: 'Development that meets the needs of the present without compromising the ability of future generations to meet their own needs'. The UK has defined sustainable development as:

- progress that recognises the needs of everyone
- effective protection of the environment
- prudent use of natural resources
- maintenance of high and stable levels of economic growth and employment.

21.1 The concept of sustainability

Sustainability adopts a triple bottom line approach of factors to consider, as shown in Figure 21.1. Achieving a balance of all three pillars of sustainability is important as a way of achieving equity.

The three pillars of sustainability and how they are measured, quantified or monitored is indicated in Figure 21.2.

Each of these factors must be considered in a holistic way to be effective, and sustainable construction goes some way towards achieving them within current constraints. Following on from this, there are five capital models that aid us in our thinking. These are shown in Table 21.1. They and their subsets are in no particular order, since each project may consider them individually and prioritise them in different ways.

A sustainable literate person will understand the need to change to a sustainable way of doing things, even at a cost. They will have sufficient knowledge and skills to decide and act in a way that favours sustainable development. They will also be able to recognise and reward sustainability literate decisions and actions of other people.

21.2 The construction industry

The construction industry is a global industry operating in all countries around the world. It is estimated to be worth annually more than \$8.5 trillion (\$12 trillion by 2025) and currently accounts for about 10 per cent of gross domestic product (GDP). It is a crucial activity in the

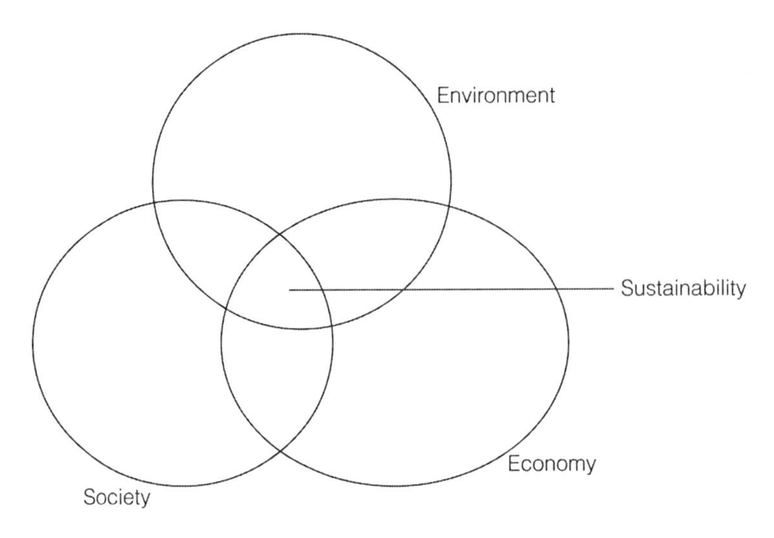

Figure 21.1 The triple bottom line approach.

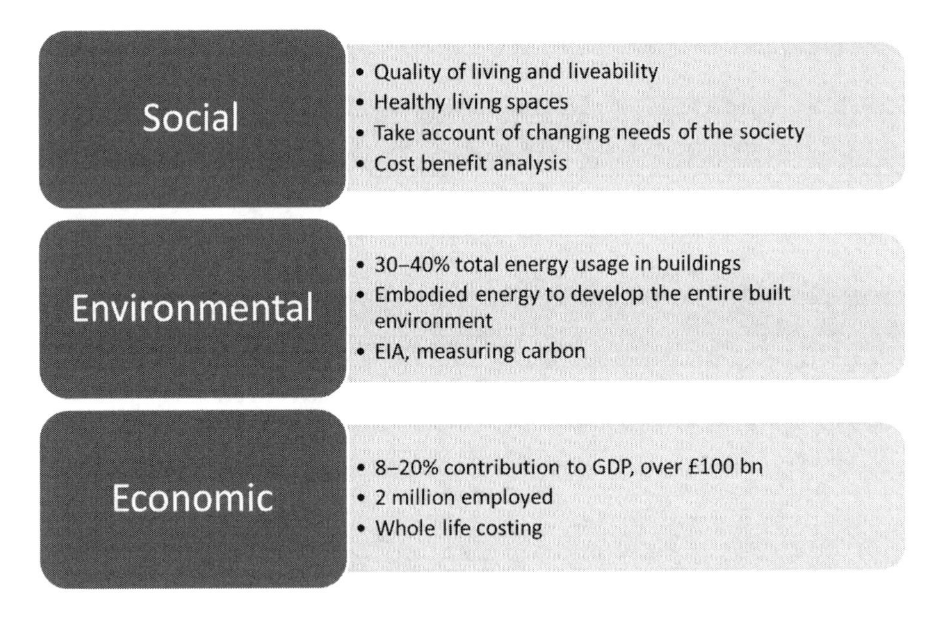

Figure 21.2 Factors to consider in sustainable construction.

regeneration and development of towns, cities and their communities. An effective construction industry must be a sustainable one. The scale of the industry means that it also has an immense potential to contribute towards the achievement of global environmental objectives. If it is to achieve this objective, it must:

- design with these longer-term objectives in mind rather than just considering initial solutions
- use construction materials in a sustainable way

- promote energy efficiency
- minimise pollution
- promote environmentally friendly solutions throughout the whole life of the project
- provide liveable building solutions for the society.

The construction industry faces many challenges. It is sometimes portrayed as an industry of underachievement, with low investment and profitability and significant client dissatisfaction. But the industry is changing through many different initiatives and organisations, such as Constructing Excellence and its predecessor body, Rethinking Construction (see Chapter 19). Table 21.2 provides a simplified analysis of the UK construction workload in 2015.

21.2.1 Analysing sustainable construction

Sustainable construction has several definitions. It includes construction practices that strive for integral building quality, including economic, social and environmental performance. Thus, the rational use of natural resources and the appropriate management of the building assets contribute towards reducing the damage to the natural and social environments, minimising the use of resources generally, reducing energy consumption in all of its forms and patterns and maintaining stable levels of economic growth. Sustainability involves the entire life cycle of

Table 21.1 Capital stocks and flow of benefits

Natural	Stock: land, sea, air, rivers, ecology Flow: energy, food, water, climate, waste
Human	Stock: health, knowledge, motivation, spiritual Flow: energy, work, creativity, love, happiness
Social	Stock: governance, community, family Flow: security, justice, social inclusion
Manufactured	Stock: infrastructure, buildings, equipment Flow: places to live, work and play
Financial	Stock: money, stocks, bonds Flow: means of valuing, exchanging, owning

Table 21.2 Overall estimated breakdown of industry work by types (UK)

	New housing		31,175		22%	
	Infrastructure		20,248		14%	
	Public	10,417				
	Private industrial	4,816			7%	
	Private commercial	26,812	42,045		3%	
Other new work					19%	
All new work				93,468		65%
Repair and maintenance				50,544		35%
	Housing	26,229			18%	
	Infrastructure	8,459			6%	
	Other work	15,856			11%	
All work				144,012	100%	100%

Source: Office for National Statistics 2017.

buildings by taking environmental and functional quality and future values into account. Historically, existing practices have often tended to focus only on the initial construction attributes, paying limited attention to the consequences of construction activity and the longer-term sustainability set against a number of identified criteria.

It should be easily recognisable that the construction and property sectors between them make a huge contribution towards the quality of everyone's lives. At their best, they can help to stimulate wider economic development, assist in building communities and enhance many aspects of the environment around us, especially the built environment. However, and in contrast, the construction industry alone is responsible for some of the most serious effects on environmental issues, which have a detrimental effect on society.

The vast bulk of human activity relies, in one way or another, on the provision of fabricated buildings and other structures. The supply of appropriate buildings and their associated infrastructure is crucial for the efficient operation of industry and commerce, the provision of dwelling houses, health, welfare and education and for social and recreation purposes. The task of providing for these important needs falls on the construction and property sectors for their inception, development, construction and long-term maintenance. The immediate goal must be to create and operate buildings that are able to function as desired, but in a sustainable way throughout their projected life cycle. This life cycle begins at inception and includes perhaps several changes of use, as fashion and needs evolve, major refurbishment and eventually, for all but a few buildings, demolition and site clearance.

What role does the construction industry play, therefore, in efforts to deliver more sustainable products? The industry employs around 2 million people directly and many others indirectly, and accounts for up to 10 per cent of GDP in times of major activity. The use of materials for construction purposes is around six tonnes each for every man, woman and child living in the UK. The construction and demolition of buildings and other structures produces around 70 million tonnes of waste materials each year. This is three times the amount of industrial waste products. Unused materials represent 20 per cent of this waste. More than 44,000 mega-litres of water are used in the process of construction every day, which in several regions and locations is close to exceeding the rainfall levels. Some estimates suggest that the UK is using three times our global share of available resources.

Sustainable development, at its heart, is a simple idea of ensuring, with a bit of thought and consideration of others' needs, a better quality of life for future generations. A widely used definition is 'development that meets the needs of the present without compromising the ability of future generations'.

21.2.2 Costs in use

Around £25 billion is spent annually in the UK on maintaining and repairing built environment assets. Also, every year about 1 to 2 per cent is added to the building stock, resulting in additional costs in use. However, through careful observation and analysis of whole life costs, the replacement of buildings should enable the less cost-efficient buildings to be disposed of through demolition. Around 75 per cent of the current building stock was constructed prior to 1980, and some of it has already undergone a series of refurbishments. Perceived wisdom indicates that the maintenance and people costs of using buildings are far greater than the initial capital costs of construction, but are sometimes given less consideration during the design phase. These comparative costs are shown in Table 21.3. The age of the built asset stock and the costs involved underline the importance of providing for their adequate management and maintenance.

Table 21.3 Average cumulative costs of an office over 20 years

People	68%
Rent and rates	25%
Maintenance and energy	4%
Capital construction costs	3%

Source: The Sustainable Buildings Task Group.

According to a guide published by the CBI, *Property for Business: An Essential Guide for Senior Executives* (CBI 2008), property issues should be higher on the boardroom agenda, and it explains that property is often the second highest cost after wages and salaries. A lack of strategic awareness of such expense costs UK businesses millions of pounds each year. The guide provides practical examples of how leading businesses have optimised the use, cost and operational value of property assets.

21.2.3 Government

The government's policies take account of ten principles that reflect key sustainable themes:

- Putting people at the centre.
- Taking a long-term perspective.
- Costs and benefits.
- Creating an open and supportive economic system.
- Combating poverty and social exclusion.
- Respecting environmental limits.
- The precautionary principle.
- Using scientific knowledge.
- Transparency, information participation and access to justice.
- Making the polluter pay.

Sustainable construction is an inherently complex subject, with a very large range of variables that interact and are frequently contradictory. Objective studies have often shown that many of the interactions are counter-intuitive and there are many misconceptions about sustainable construction issues perceived by the construction industry. For example, it is commonly held that recycling will substantially reduce the quantities of future materials extracted. Objective studies have shown that the quantities of waste arising from demolition are too small for the current demand for materials for recycling to dramatically reduce the quantities of materials extracted. Recycling is nevertheless very important as a strategy for preserving landfill capacity.

The UK strategy for more sustainable construction resulting in a better quality of life suggests key themes for action by the industry. These include:

- design for minimum waste
- lean construction
- minimise energy both in construction and use
- do not pollute
- preserve and enhance biodiversity

- conserve water resources
- respect people and local environments
- set targets and monitor and report in order to benchmark performance.

Most of the points simply make good sense, i.e. minimising wastage increases efficiency. Sustainability is of increasing importance to the effective, efficient, economic and responsible operation of business.

A number of construction sector sustainability strategies have been published. They aim to develop a common understanding of the issues and present effective targeted approaches for each sector to contribute towards achieving a more sustainable construction industry. Some of these sectors include civil engineering, brick housing, steel in construction, building services, cement and concrete, wood in property and construction products.

21.3 Sustainable development

Sustainability is the capacity to endure, maintain or keep in existence. Development, in a property context, can be defined as activities carried out on land, buildings or structures changing current use with employment of primary resources of land, labour, capital and entrepreneurship. The Bruntland definition of sustainable development (see beginning of this chapter) succinctly brings these two concepts to fruition. In other words, development activities need to be considered in a sustainable framework, making it sustainable in terms of environment, society and economy. The framework needs to satisfy current demands for development activities without unduly compromising resources and their availability for the future.

21.3.1 Sustainable development goals

In September 2015 the United Nations member states adopted a set of goals to end poverty, protect the planet and ensure prosperity for all as part of a new sustainable development agenda. There are seventeen goals in total, and each goal has specific targets to be achieved over the next fifteen years. The United Nations General Assembly resolution is formally known as 'Transforming Our World: The 2030 Agenda for Sustainable Development'.

The seventeen goals are:

- Goal 1: No Poverty
- Goal 2: Zero Hunger
- Goal 3: Good Health and Well-being
- Goal 4: Quality Education
- Goal 5: Gender Equality
- Goal 6: Clean Water and Sanitation
- Goal 7: Affordable and Clean Energy
- Goal 8: Decent Work and Economic Growth
- Goal 9: Industry, Innovation and Infrastructure
- Goal 10: Reduced Inequalities
- Goal 11: Sustainable Cities and Communities
- Goal 12: Responsible Consumption and Production
- Goal 13: Climate Action
- Goal 14: Life Below Water

- Goal 15: Life on Land
- Goal 16: Peace, Justice and Strong Institutions
- Goal 17: Partnerships for the Goals.

These seventeen goals are associated with 169 targets that describe what needs to be achieved in order to fulfil each of them.

21.3.2 The state of sustainable development

Environmental protection is currently administered through DEFRA (Department for Environment, Food and Rural Affairs) and is concerned with, for example, air quality, contaminated land, noise, nuisance, pollution, radioactivity, waste and recycling. Effective protection of the environment requires activity on wide-ranging fronts. There are four main objectives of sustainable development:

- Limit global environment threats, such as global warming.
- Improve the energy efficiency of buildings.
- Combat fuel poverty through social action.
- Provide economic growth through a more efficient use of resources, such as reuse, recycling and the recovery of waste products.

The concept of sustainable development includes four main strands:

- Social progress that recognises the needs of everyone.
- Effective protection of the environment.
- Prudent use of natural resources.
- Maintenance of high and stable levels of economic growth and employment.

The construction of buildings affects the environment in three main ways:

- The raw materials used for the manufacture of building materials is considerable. The quarrying of 250 to 300 million tonnes of material in the UK each year for aggregates, cement and bricks imposes a significant environmental cost. Currently, about 10 to 15 per cent of the aggregate used in construction is from recycled or alternative sources. Efficient use of this material can save money, reduce waste for disposal and reduce energy consumption and pollution from the supply process.
- Construction sites are often the cause of local nuisance, such as noise, dust, vibration and the pollution of watercourses and groundwater. Good environmental protection is required to reduce costs and nuisance to neighbours and the immediate environment. For example, some 70 million tonnes of construction waste, including clay and sub-soil, are generated annually. Materials recovery will happen when economics makes this possible. Constructing Excellence, and its predecessor Rethinking Construction, have carried out some interesting research and produced guidelines for how construction sites can get on better with their neighbourhoods.
- The construction industry uses about six tonnes of material per person per year in the UK. Approximately 20 per cent is for infrastructure and more than 50 per cent is used for repairs and maintenance. The use of buildings contributes significantly towards environmental problems, such as global warming.

21.3.3 Environmental aspects of building project development

The impact of the construction industry on the environment is substantial. During the extraction and manufacture of construction materials, their transportation, the process of construction and use of buildings, large quantities of energy are used. Major contributions are made to the overall production of carbon dioxide, which exacerbates the 'greenhouse' effect. The environmental impact of the construction of new buildings is a global issue, since it requires the use of raw materials from around the world. During the construction activities on-site, which often last for a number of years, communities and individuals can be severely affected by the process of construction, and in recent years there has been a much greater effort to reduce to a minimum these disruptions to normal life.

There is increasing concern in society with the effect of human activity on the environment, and greater pressure has been placed on clients and developers to state the possible effects of their project on the local area. Since 1968, a European Community directive has required an environmental impact assessment to be provided with all planning applications for major projects and for smaller schemes where the planning officials consider them to be important. The assessment requires a statement of the impact of the project on the surrounding area and details of how this can be limited; for example, soundproofing in the case of a noisy transportation system. A further requirement is that clients and developers should wherever possible undertake wide consultations involving the public and environmental groups. Despite the requirements of the directive, the quality of the assessments varies considerably, and the Department of the Environment is attempting to formulate appropriate standards.

Assessments also tend to be parochial and not to examine the wider issues involved beyond the confines of the particular project concerned. Alternative proposals should be considered that compare the environmental factors involved in the choice of different sites or locations and the different constructional methods that might be adopted.

21.3.4 Energy issues

In the 1950s and 1960s, building maintenance and running costs were largely ignored at the design stage of new projects. Today, the capital energy costs expended to produce the building materials and to transport them and fix them in place are often ignored in our so-called energy efficient designs, where the emphasis is placed upon the energy use of the building. In any given year, the energy requirements to produce one year's supply of building materials is a small (5 to 6 per cent) but significant proportion of total energy consumption, which is typically about 10 per cent of all industry energy requirements. The building materials industry is relatively energy intensive, second only to iron and steel. It has been estimated that the energy used in the processing and manufacture of building materials accounts for about 70 per cent of all the energy requirements for the construction of the building. Of the remaining 30 per cent, about half is energy used on-site and the other half is attributable to transportation and overheads. Although the energy assessment of building materials has still to be calculated and then weighted in proportion to their use in buildings, research undertaken in the USA has shown that eighty separate industries contribute most of the energy requirements of construction, and five key materials account for more than 50 per cent of the total embodied energy of new buildings. This is very significant, since considerable savings in the energy content of new buildings can be achieved by concentrating on reducing the energy content in a small number of key material producers.

Most buildings are designed to cope with the deficiencies of a light, loose structure, designed just to meet the building regulations thermal transmittance standards. Our building regulations are also some way behind those of Scandinavian and other European countries. About 56 per cent of the energy consumed, both nationally and internationally, is used in buildings and this should provide designers with opportunities and responsibilities to reduce global energy demand. Whilst there is a need to make substantial savings in the way that energy is used in buildings, there is also a need to pay much more attention to the energy used in the manufacture of materials and components and their fixing in place in the finished building. It has been estimated that this may be as high as five times the amount of energy that the buildings' occupants will use in the first year.

21.3.5 Some best practice

Investment in more sustainable construction can, at best, be described as patchy, both in the UK and worldwide. However, the number of projects seeking to demonstrate sustainability credentials is growing. Some investors and developers are providing incentives that will assist in moving this philosophy and practice further.

In 2000, the Japan Housing Loan Corporation provided premium loans for 180,000 energy efficient homes and also provides premium loans for using recycled materials.

In the UK, English Partnerships require a very good BREEAM (BRE Environmental Assessment Method) and eco-homes standard for all developments now. The Millennium Communities must additionally reach an excellent standard. Greenwich Millennium Village and Chatham Maritime are vibrant developments that have used such targets.

In supporting and encouraging sustainable developments, the South East England Development Agency has produced a sustainability checklist for developments. It enables local authorities and developers to implement sustainability in a considered and practical way, using a common framework. This increases certainty for the investor and opens up new areas for investment. Sustainable development is development that makes possible sustainable living in the following ways:

- Build development that sustains life and improves the quality of life for human beings.
- Work that removes the environmental or social damage from the past and improves the sustainability of the wider environment and ecosystems.
- Development of individuals and of societal quality of life.

Much work has been done by people in different parts of the world in respect of sustainable development initiatives. The following summarises some of the more important themes:

- Reuse and improve the performance of existing built assets.
- Locate any new development in appropriate localities.
- Relate land-use planning to the transport infrastructure.
- Design for minimum wastage and an improved use of resources.
- Design for life.
- Aim for lean construction.
- Reduce energy consumption.
- Utilise renewable energy resources.
- Do not pollute the wider environment.
- Preserve and enhance natural features and appropriate biodiversity.
- Conserve water resources.
- Respect people and their local environment.

21.4 Environmental impact assessment

Environmental impact assessment was established in the USA as long ago as 1970. Such assessments are now worldwide and a powerful environmental safeguard in the project planning process. The original EC (now EU) Directive 85/337 was adopted in 1985 and, since then, the individual member states have implemented the Directive through their own regulations. In the UK, the production of the resulting environmental impact statements have increased more than tenfold between the early 1980s and the early 1990s. As a result of the Directive, more than 500 statements are now prepared annually in the UK. The required contents of a statement are given in annex III of the Directive and are as follows:

- Description of the project: physical characteristics, production processes carried out, estimates of residues and emissions.
- If appropriate, details of alternative sites and their possible effects.
- Description of aspects of the environment that are likely to be affected, such as population, fauna, flora, soil, water, air, climatic factors, material assets, architectural and archaeological heritage, landscape and their interrelationship.
- Description of the likely effects on the environment of existence of the project, use of natural resources and emission of pollutants.
- Description of measures envisaged to prevent or reduce any adverse effects on the environment.
- A non-technical summary of the above.
- An indication of any difficulties encountered by the developer in compiling the above information.

In the UK, the Directive is implemented in regulations described as the Town and Country Planning (Assessment of Environmental Effects) Regulations 1988. Further guidance is included in *Environmental Assessment: A Guide to the Procedures* (Department of the Environment 2008).

21.5 Sustainable business

The business benefits of adopting a more sustainable approach to the construction of buildings are already being recognised by many far-sighted clients, designers and constructors. By preventing pollution, optimising energy and reducing waste, companies such as Xerox and BP have demonstrated significant opportunities for business. The business case for sustainable development is now more widely recognised by both organisations and clients who procure construction projects. Enlightened construction firms, contractors and subcontractors also recognise the benefits from more environmentally friendly construction methods.

For a business to be seen as sustainable, it needs to meet criteria of balanced inputs and outputs, with the former tending to outweigh the latter. All industries have traditionally taken raw materials and processed them into consumable goods. Human endeavour and finance are important ingredients, with the focus on economic factors such as investment, returns and savings. The construction industry is no exception. At the end of their useful lives, most products are disposed of in landfill or are reused in new processes. Some of the latter have been very successful, with materials such as metals and glass being reclaimed for use in new products. These benefits include the following elements.

21.5.1 Capital costs

These costs need to be carefully established at the outset, since the decisions made at inception or during the feasibility stages affect the various design, construction and in-use phases. Designs that are made early on in a project are likely to result in less of an increase in capital costs than those which are made at a later stage requiring revisions to the original design. Increased capital costs may be required to meet sustainable objectives, but, with little effort, these can often be offset against latent costs in use. This emphasis is especially important to those carrying out whole life cost analyses.

Reduced costs are possible through the use of lean construction methods, the more efficient use of resources during design and construction, eliminating waste and reusing resources wherever possible. There is growing evidence that sustainable construction does not inevitably lead to higher capital and whole life costs. The pattern of expenditure or cost profile will differ when compared to a more traditional and conventional solution.

21.5.2 Operational costs

These costs are diverse, ranging from utilities such as energy, water and waste disposal to staff costs, rents and management costs. It has already been suggested that operational resource costs account for more than five times the initial building costs of a building over a sixty-year life horizon. Staff costs may frequently account for as much as 200 times the initial capital costs. Efficiencies in working practices can result in significant ongoing cost savings and, since this represents a high cost, it is clearly one to be targeted and where savings might be achievable.

21.5.3 Investment

Investment appeal can be influenced through the attractiveness of the building project or company. Investment values can be improved where increased income levels are achieved through rents or resale. Investment appeal will also be increased through a speedier realisation of those income levels through reduced procurement times. Greater flexibility in the design will increase investment appeal through the greater potential uses of the building's design. However, this increased flexibility may have the effect of increasing the initial construction costs through over specification.

21.5.4 Profitability

This remains the key item for business, for without this the other considerations remain largely theoretical. To remain profitable, it might be necessary to supplement income with grants, where these are available, in order to achieve sufficient profits at least for future investment. Profitability is affected by productivity, sales, market profile and competitiveness. The drive for profitability in design and construction can result in reduced innovation and experimentation.

21.5.5 Legislation

The need for compliance with current statutory requirements and obligations are well understood in the building procurement process. The planning and building control processes become major milestones for many projects. Both industry and investors are concerned with the need for future proofing of their projects to ensure that they fully meet any anticipated changes in legislation. With the average life of a procurement programme lasting several years, the current rapid changes and improvements to legislative requirements can make this a significant issue.

For some developers, the various legislation is seen as the standard to achieve. For those concerned about sustainable projects, the criteria outlined in the various Acts might be seen more as a minimum expectation that can be improved upon.

21.5.6 Liability

All stakeholders have liabilities towards the staff whom they employ, occupiers and third parties. In an increasingly litigious society, concerns over liabilities are increasing and it would appear that nothing can be taken for granted. This can provide some incentive to move from traditional to more innovative solutions. However, designers, constructors and their clients are all wary of innovations that are insufficiently tried and tested and which may include repercussions that no one envisaged at the time.

21.5.7 Productivity

This has been a focus of the construction industry for the past few decades. The UK construction industry has made significant strides in this direction. Buildings that have been carefully designed are likely to reflect improved buildability aspects and thus reduce both the time and the costs of construction on-site.

21.5.8 Staffing

Sustainable design and construction are likely to enhance the status and motivation for those who are engaged. There are now a great many construction companies that show the proper respect for people as outlined in the Egan Report. This results in less effort being placed on recruitment and training, and such companies will be able to attract and retain more conscientious personnel.

21.5.9 Management

The whole ethos of sustainability values will result in a more socially aware and concerned management approach when dealing with the appropriate use of site resources. This factor also makes good business sense.

21.5.10 Enhanced public relations

Since there is a focus and concern on sustainability across many sectors of industry now, companies who engage in such methodologies are likely to be preferred bidders for new projects in the industry.

21.6 Case study: the Swiss Re headquarters

The Swiss Re building (colloquially known as The Gherkin), designed by Sir Norman Foster and Partners, at 30 St Mary Axe, is claimed to be the first ecologically tall building to be constructed in England. It was designed and constructed between 1997 and 2004, is forty floors tall and was the first skyscraper building to be built in London for thirty years. Its distinctive form is an instantly recognisable addition to the London skyline and has become a landmark in Europe's leading financial centre. The majority of the structure is devoted to offices, although there are double height retail outlets at the ground level and a restaurant at the top. The restaurant offers a 360-degree panorama over the city and beyond. The tower occupies the historical site of the former Baltic Exchange that was damaged by an IRA bomb in 1992.

The Gherkin is an environmentally progressive building. Its uncompromising modernity is allied towards a sensitivity to the natural environment. A comprehensive range of sustainable measures means that the building will use up to 50 per cent less energy than the traditional prestige office block. Fresh air is drawn up through the spiralling light wells to naturally ventilate the office interiors and to minimise reliance on artificial cooling and heating. The light wells and the shape of the building maximise natural daylight, lessen the use of artificial lighting and allow views out from deep within the building. The balconies on the edge of each light well provide strong visual connections between floors and create a natural focus for communal office facilities. The interior atria are expressed on the exterior by the distinctive spiral bands of grey glazing.

A number of complex fluid dynamic studies of the local environment conditions suggested a strategy for integrating the building within its site and allowing it to use natural forces of ventilation. The 180-metre-high tower breaks with the conventions of traditional box-like buildings. Its circular plan is tapered at the base and the crown to move connections to the surrounding streets and allow the maximum amount of sunlight to the plaza level. The circular plan enables much of the site area to be used as a landscaped public plaza, with mature trees and low stone walls that subtly mark the boundary of the site and provide seating. Half of the tower's ground level is shops and a separate new building houses a restaurant serving an outdoor café spilling out to the plaza.

The exterior form explores a series of progressive curves with the aid of parametric computer modelling techniques. Indeed, without the use of the computer it would have been difficult to provide such a structure. The shape and geometry have affinities with forms that recur in nature; the pinecone, for example, has a natural spiral and, like the building's elevation, opens and closes in response to the weather. The building's smooth shape also directs air movement around the building and minimises the amount of wind at plaza level to improve pedestrian comfort. The external diagonal structure is, by virtue of its triangulated geometry, inherently strong and light, permitting flexible, column-free interior space.

The exterior cladding consists of 5,500 flat, triangular and diamond shaped glass panels, which vary at each level. The glazing to the office areas consists of a double-glazed outer layer and a single inner screen that sandwich a central ventilated cavity containing solar-controlled blinds. The cavities act as buffer zones to reduce the need for additional heating and cooling and are ventilated by exhaust air drawn from the offices. The glazing to the light wells that spiral up the tower consists of openable double-glazed panels with a combined grey tinted glass and high-performance coating that effectively reduces solar gain.

The structure has a floor area of 76,400 square metres and was built at a cost of £130 million.

21.7 Building resilience

It is important to understand the meaning of the word disaster in the context of the built environment. According to United National International Strategy for Disaster Reduction (UNISDR) (2009), disaster refers to a 'serious disruption of the functioning of a community or a society involving widespread human, material, economic or environmental losses and impacts, which exceeds the ability of the affected community or society to cope using its own resources'. It is a comprehensive definition, intertwining the impact a disaster could make on the built and natural environments. It relates to human habitations, and thus implies the built environment. Disasters are, essentially, natural (earthquake, tsunami) or otherwise (artificial: Chernobyl nuclear disaster). In recent years, an increase in the rate of disasters worldwide (including natural disasters) has been observed. Advancements in science and technology, coupled with awareness and better preparedness, have reduced the mortality levels from disasters. However, the converse has been true in terms of consequent economic losses. It is estimated that natural disasters around the globe have resulted in economic losses of roughly US$7 trillion (£5 trillion) since 1900, and this can no longer be considered as unavoidable. A strategic approach to managing economic losses resulting from disasters is required to help societies better cope with them.

The concept of resilience primarily refers to the ability to recover from change. UNISDR (2009) defines resilience in the context of disasters as 'the ability of a system, community or society exposed to hazards to resist, absorb, accommodate and recover from the effects of a hazard in a timely and efficient manner, including through the preservation and restoration of its essential basic structures and functions'. In the context of the built environment, this means the ability of the community to recover to its original status of the built environment (or better) upon being subject to a disaster. In this context, building resilience refers to enhancing and empowering societies to attain greater levels of resilience to disasters.

Solutions in the built environment are now required to consider building resilience as a key construct alongside aspects of sustainability, globalisation and digitalisation. Building and cities need to be socially, environmentally and economically resilient in a similar manner to achieving sustainability.

21.7.1 The Sendai Framework

The UNISDR have been in the forefront of leading and harmonising global efforts in managing disasters and developing strategies for disaster resilience. In 2005 the Hyogo Framework for Action (HFA) was launched, identifying priority areas of focus and pronouncing an overall strategy for dealing with disasters. It covered the period of 2005 to 2015. In 2015, the Sendai Framework for Disaster Risk Reduction was launched as the successor to the HFA. The Sendai Framework is for the period 2015–30 and comprises four priority areas, as indicated in Figure 21.3.

The priority actions 3 and 4 have direct implications on the way buildings are designed and managed. Improving disaster resilience means enabling new technologies, methods and processes to be integrated into building design that can enhance the levels of resilience of buildings and structures against most likely types of disasters that these structures will have to face. For example, building in a flood prone area may consider design improvements of buildings in terms of flood barriers at entrances to the building, use of flood resistant type of flooring materials, and so on. Priority 4 inculcates the idea of the need to 'build back better' in case of post disaster reconstruction. This means that buildings constructed as replacements to disaster damaged or destroyed buildings should incorporate improvements that can withstand similar type of disasters in future, reducing the level of vulnerability of built assets.

21.8 Sustainability and contractual procedures

It has sometimes been argued that the way to change behaviour is to offer incentives for those who are willing to comply and penalties for those who do not see the need for change.

Chart of the Sendai Framework for Disaster Risk Reduction

2015–2030

Scope and purpose

The present framework will apply to the risk of small-scale and large-scale, frequent and infrequent, sudden and slow-onset disasters, caused by natural or manmade hazards as well as related environmental, technological and biological hazards and risks. It aims to guide the multi-hazard management of disaster risk in development at all levels as well as within and across all sectors.

Expected outcome

The substantial reduction of disaster risk and losses in lives, livelihoods and health and in the economic, physical, social, cultural and environmental assets of persons, businesses, communities and countries

Goal

Prevent new and reduce existing disaster risk through the implementation of integrated and inclusive economic, structural, legal, social, health, cultural, educational, environmental, technological, political and institutional measures that prevent and reduce hazard exposure and vulnerability to disaster, increase preparedness for response and recovery, and thus strengthen resilience

Targets

- Substantially reduce global disaster mortality by 2030, aiming to lower average per 100,000 global mortality between 2020–2030 compared to 2005–2015
- Substantially reduce the number of affected people globally by 2030, aiming to lower the average global figure per 100,000 between 2020–2030 compared to 2005–2015
- Reduce direct disaster economic loss in relation to global gross domestic product (GDP) by 2030
- Substantially reduce disaster damage to critical infrastructure and disruption of basic services, among them health and educational facilities, including through developing their resilience by 2030
- Substantially increase the number of countries with national and local disaster risk reduction strategies by 2020
- Substantially enhance international cooperation to developing countries through adequate and sustainable support to complement their national actions for implementation of this framework by 2030
- Substantially increase the availability of and access to multi-hazard early warning systems and disaster risk information and assessments to people by 2030

Priorities for Action

There is a need for focused action within and across sectors by States at local, national, regional and global levels in the following four priority areas.

Priority 1
Understanding disaster risk

Priority 2
Strengthening disaster risk governance to manage disaster risk

Priority 3
Investing in disaster risk reduction for resilience

Priority 4
Enhancing disaster preparedness for effective response, and to «Build Back Better» in recovery, rehabilitation and reconstruction

Figure 21.3 Overview of the Sendai Framework 2015–30.

Source: www.unisdr.org/files/44983_sendaiframeworksimplifiedchart.pdf.

The various conditions of contract do not, at the present time, provide for these in respect of sustainable construction. Indeed, the contracts are silent about this concept.

One approach is to develop a culture change amongst developers, designers, constructors and others who are involved in the construction process. No new legislation or set of rules will be required, only a series of principles to guide those involved. Of course, projects that do not embrace sustainable principles may well be difficult to sell or lease, and this factor alone will assist with change.

There has been much progress made in focusing on sustainable development in the last twenty years. Evidence might suggest that the involvement of the larger organisations and major projects, especially in the public sector, has been relatively successful in this respect; however, it has proved to be more difficult to include others, working at different levels within the industry. Do we have sufficient time now available, and is action not now crucial before some processes become irreversible?

It is necessary to consider what the implications are on sustainability for a particular procurement option. Do the different methods of procurement now available to us at the beginning of the twenty-first century have different implications for the natural and the built environments? Chapters 8 and 9 have examined, to some degree, the different methods and criteria. Those methods that can be argued to support sustainable development should be adopted and the remainder rejected, unless they can change their attributes. What is especially clear is that the fundamental principles of sustainability should be accepted by every partner organisation who forms part of the supply chain management. It is desirable, or even necessary, to get such organisations to sign up to such principles at the outset of the project as a requirement of their being involved.

22 Information communication technologies in construction procurement

22.1 The fourth industrial revolution

The human race embarked on a journey of evolution from prehistoric periods, but recently on an accelerated journey that the world has never seen before. The first industrial revolution saw water and steam power transforming the production process. The second industrial revolution was based on the use of electric power, changing gear in the production processes to mass-scale production and considerable benefits of economies of scale. The third brought about the use of electronics and information technology, enabling automation of the production process. Now the fourth industrial revolution is unfolding with unprecedented velocity, disrupting all industries and processes around us. Klaus Schwab, the executive founding chair of the World Economic Forum, states that this revolution is characterised by the fusion of technologies blurring the boundaries between the physical, digital and biological spheres. Unlike the previous three industrial revolutions, the fourth is rapidly changing the way we operate, do business and govern, disrupting the status quo (see Figure 22.1). It is abundantly clear that the convergence of technology in developing solutions for the construction industry is already happening. The construction industry is reaping the benefits of the fusion of technologies; the integration of BIM (building information modelling) with VR (virtual reality) and 3D printing, and components supplied using drones, are becoming a reality.

The implications of this progress on construction procurement is manifold: the types of material used in buildings are rapidly changing, and the digitisation process is starting to generate a massive influence on procurement processes and the move towards digital smart contracts (discussed in detail in section 0). Construction is moving from on-site to off site, and greater degrees of automation are coming into force at both on-site and off site production lines. The use of robots on-site in robot-assisted construction, the use of drones for assembly, transport, progress monitoring and site layout mapping, and the use of mobile technologies in managing projects through use of the Internet of Things (IoT) are some examples indicating the disruptions in technology presently unfolding across construction industries worldwide.

22.2 E-procurement

Construction procurement is the process of obtaining construction goods (material, component or the whole building or structure) or services (construction or design related). The process of buildings procurement is complex and requires comprehensive effort. In procuring a building, a process of designing, preparation of documentation for procurement and some sort of bidding process or negotiation to obtain prices and construction are required. However, the term 'procurement' is not limited to construction procurement. It can be used for obtaining any other

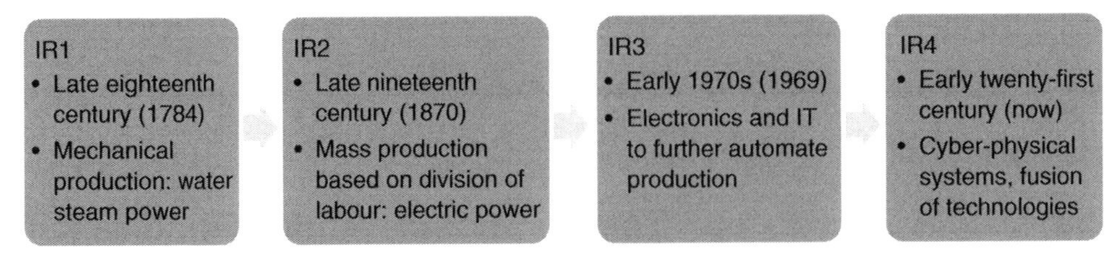

Figure 22.1 Chronology of industrial revolutions.

good or service. The procurement process includes preparation and processing of a demand or invoice as well as the end receipt and approval of payment.

The term 'e-procurement' refers to the use of electronic means for procurement of goods or services. The electronic processes of construction procurement originated with the use of disks, then CDs and subsequently memory devices to send electronic documents. With the advent and proliferation of the internet, most electronic procurement activities are now carried through the internet.

In the construction industry, construction project procurement can be categorised into three types:

1 Procurement of labour, material, plant and components required for a construction project.
2 Procurement of services required to deliver the construction project (such as consultancy services).
3 Procurement of works (the entire building or structure).

In each of these cases, e-procurement is now significantly utilised. E-procurement in the first type mostly occurs within the constructor's (builder's) organisation and normally follows general goods procurement processes. It involves e-sourcing and e-invoicing processes and construction companies may often have integrated web-based e-procurement systems. The second type of procurement does not often happen on e-procurement platforms and instead uses traditional processes; however, there is evidence that that is also changing. The classic type of e-procurement occurs in procurement of buildings and other construction projects. This is now often carried out through service providers (such as software as a service (SaaS) providers). There are many companies who offer e-tendering services along with extranets for procurement of construction projects (for more details, see Ashworth and Perera 2015, chapter 23). Further detailed discussion of e-procurement can be found in *Advances in Construction ICT and e-Business* (Perera et al., 2017).

22.3 Building information modelling (BIM)

BIM represents a digital presentation of the physical and functional characteristics of a building or structure. It creates an information resource that can be shared by everyone involved in a project, thus forming a reliable basis for all the decisions that need to be made throughout a project's life cycle, from inception to eventual demolition.

BIM typically uses three-dimensional, real-time, dynamic building modelling software to help increase the productivity and efficiency of building design and construction. The process relies upon, and uses a range of, information such as building geometry, spatial relationships, geographic information and quantities and properties of the building components. BIM is a process of demonstrating both the graphical and non-graphical aspects of the project. It uses the concepts of object orientation in computer information modelling. As such, all building elements, components and items are modelled as objects in the BIM. Each of these objects represents attributes and value pairs expressing the characteristics of the object whilst encompassing its relationship to other objects in the model. The modelling framework expresses these relationships of the objects both graphically and semantically, making it possible to make visual sense of the model and interact with it in a logical and true to life manner.

One of its chief advantages is that it offers a single repository of drawings, specifications, schedules and other information. Some define these as dimensions, where 4D represents the geometric three dimensions, with the fourth dimension as time, fifth as cost, sixth specifications, seventh facilities management information, and so on. In theory, BIM can achieve *n* number of dimensions, which led to the concept of *n*D modelling in the early 2000s. All of the design information, including constructional details for the project, is able to be interrogated by all members of the design and construction team in real time. This enables the design performance to increase considerably for any aspect of the project, whether aesthetical, structural or environmental. A building services engineer, for example, would be able to consult the architect or designer immediately about the project's energy consumption strategies in the provision of heating, air conditioning and insulation. In this way, BIM facilitates a much-improved coordination and collaboration amongst the design and construction team, and helps to reduce information loss experienced with some aspects of project management. This approach provides benefits in respect of construction time analysis and cost and value calculations, which can have an important bearing at both the design stage and after the project is brought into use by the client. Many BIM programmes efficiently and accurately integrate cost management and control along with programme planning. This allows cost to drive both the design and the programme and gives increased accuracy and confidence in the data derived. Most software does not store cost and programme information directly in the BIM model, but in add-on software that communicates with BIM. This is very much a legacy issue, where a plethora of software applications developed for cost control (estimating) and scheduling are being still used to store such data. It is expected that these will become mainstream BIM integrated over time as the industry becomes more accustomed to the use of BIM.

BIM covers a complete range of professional services, such as building geometry, spatial relationships, light analysis, geographical information, quantities and the properties of building components, such as manufacturers' details and data sheets. Information from a wide range of sources can be imported to enhance the model; however, this level of comprehensive BIM model is not yet a reality. It will take a long time (more than another decade) to see comprehensive BIM models that are similar to what exists in the manufacturing sector.

BIM can be used to demonstrate the entire building life cycle, including the processes of facility operation after completion of the construction work and handing over of the project to the client. BIM is therefore very valuable to facility managers as they manage and evolve the project during its lifetime use. However, it all depends on the state of knowledge that exists in terms of typical lifetimes of buildings and components embedded in buildings.

Quantities and shared properties of materials can be generated, but, like many other systems, it is important to ensure that the quantities calculated are those which are assumed and intended. This means that, when designers design, details of the design come after some time, but estimates are all encompassing. As such, design teams will have to understand estimating assumptions and vice versa. It is important to distinguish, for example, between floor areas that are calculated based on net or gross areas. Something that may be obvious within a manually calculated process may not be so obvious in a computer generated model, and may even represent something entirely different. This will depend on the rules that have been developed for the BIM. A professional using such a model would need to regard it as more than simply a black box approach.

BIM goes far beyond switching to a new type of software, the computing requirements can be considerable – a basic laptop would not have the capability to run a BIM programme today. BIM also requires changes to the definition of traditional design and construction phases, resulting in more data sharing between architects, engineers and quantity surveyors than occurs at the moment.

There have been some, limited, attempts at creating a BIM for existing built facilities. This process is not as easy to prepare as might at first be thought possible. The validity of such models requires numerous assumptions about the design standards, building codes, construction methods, materials, and so on, that were relevant at the time such a facility was constructed. Over time, the introduction of BIM documentation will make this more manageable for the users of new buildings.

22.3.1 Implications of BIM

The use of BIM in practice provides an array of issues that confront practitioners, consultants and contractors in the development of a fully integrated system. A fully integrated BIM project would be best supported by an entirely different approach to contractual procedures and practices. Paradigm shifts of this nature are difficult to manage. It requires a multi-party agreement, for which a standard would ultimately emerge. Liability policies towards a loss-based whole-project insurance will be too messy and counterproductive.

In the UK, the construction industry is expecting the government to eventually require some form of BIM to be adopted in the future, but mainly for major construction projects. The contractual and legal implications, as well as technical issues such as IT infrastructure and systems inter-operability, will thus become significant issues. Investment in new systems and training will become very important to ensure that expectations are properly achieved.

The Construction Products Information Centre (CPIC) in the UK is responsible for providing best practice guidance on construction production information. It has proposed a definition of BIM for adoption throughout the UK's construction industry. It believes that the proliferation of interpretations of BIM currently hampers the adoption of a working method to improve the construction industry and the quality and sustainability of the deliveries from the design and construction team to clients.

22.3.2 Benefits of BIM

Managing building information using a building information model can lead to substantial cost savings, from design and construction through to maintenance. The model saves time and waste on-site, and extra coordination checks are largely unnecessary. The information generated from the model leads to fewer on-site errors caused by inaccurate, and especially uncoordinated, information. When all members of the design and construction team work on the same model, from early design through to completion, changes are automatically coordinated across the project and the information generated is therefore of high quality. Information technology is an integral part of today's commerce, and transferring information from designers to the producers and constructors is an example where, with the availability of modelling software, the tools are in place.

The proponents claim that BIM offers:

- improved visualisation
- improved productivity due to easy retrieval of information
- increased coordination of construction documents
- embedding and linking of vital information, such as vendors for specific materials, location of details and quantities required for estimation and tendering
- increased speed of delivery
- reduced overall costs both in construction and design
- the use of BIM by facility managers can result in considerable savings in life cycle expenditures.

22.3.3 Future of BIM

BIM is the future of construction and long-term facility management, but there is still much confusion about what exactly it is and how it should be utilised and implemented. Although, more than thirty years have elapsed since its introduction, the level of BIM adoption varies significantly the world over. BIM is a relatively new technology in a construction industry typically slow to adopt change. The manufacturing industry has been much more proactive in utilising this technology. Yet, many early adopters are confident that BIM will grow to play an even more crucial role in building documentation, information integration and collaboration. BIM provides the potential for a virtual information model to be handed from design team (architects, surveyors, engineers and others) to contractor and subcontractors and then to the owner. Each add their own additional discipline-specific knowledge and tracking of changes to the single model. It would act as the information repository for a building, encapsulating all relevant data related to it. The result greatly reduces information losses in transfer. It can prevent errors being made at the different stages of development and construction by allowing the use of conflict detection, where the model actually informs the team about parts of the building clashing or in conflict. It also offers detailed computer-based visualisation of each part in relation to the total building. As computers and software become more capable of handling greater amounts of information, this will become even more pronounced than in current design and construction projects.

BIM is now considered as the most significant step of the construction digitalisation process. Countries, including the UK, USA, Singapore, Sweden and the EU, are progressing well with adoption, although these are not yet to the levels one would ideally expect. Progress in Australia is more limited and there is not much government stimulation for its adoption. As such, it is mostly commercially driven.

22.4 The Internet of Things (IoT)

IoT stems from the concept of connected ‘things’. Connected things could include devices, machines and equipment to people. The main thread of connectivity comes from the fact that these are connected to the internet (mostly through Wi-Fi networks) or to each other through a network. A ‘thing’ in IoT could be a person, whose movements are tracked through a connected GPS monitor, an internet connected refrigerator, where contents are monitored by its owner, a solar powered micro generation system, where power generation is regulated over the internet, or a home with a connected security system, where the owner can remotely monitor status. Simply, these are devices for which an internet protocol (IP) address can be assigned and connected to one another over the internet or other network. IoT is a product of convergence of technologies: wireless technologies, micro-electromechanical systems (such as sensors, actuators, etc.), micro services and the internet.

Commonly connected devices include smartphones, headphones, washing machines, lights and lamps, power switches, solar inverters, smart meters, wearable devices and many other devices considered as smart devices. By definition, a smart device is an electronic device generally connected to other devices or networks via different wireless protocols, such as Bluetooth, NFC, Wi-Fi, 4G and 5G, that can operate to some extent interactively and autonomously. As such, smart devices are inherently part of IoT in most instances. Businessinsider.com defines IoT as a 'network of internet-connected objects able to collect and exchange data using embedded sensors'.

Most analysts predict that by 2020 there will be more than 20–30 billion IoT devices around the globe. Many industries, such as manufacturing, transport, defence, agriculture, food, banking and health care, amongst others, benefit from IoT implementations. Industries directly connected to the construction industry also benefit from IoT. These include smart buildings and smart cities, smart infrastructure (such as smart roads and railways), smart homes and automation, amongst others. The HVAC (heating, ventilation and air conditioning) industry has embraced smart technologies and IoT in the commercial building sector, according to a report recently published by British Iron and Steel Research Association (BSRIA) (2017).

22.4.1 Implications for the construction industry and procurement

Building automation and management systems are becoming ever more common and seen as the way forward for the future. This is mainly due to the fact that there are many potential benefits for users, building owners and managers:

- IoT devices enable communication between devices as well as remote access. This allows buildings to be remotely accessed and managed.
- There would be major cost savings for facility managers. They can use IoT devices to report issues with buildings and dynamically monitor buildings for system performance. For example, building HVAC systems can be run, monitored and controlled remotely through the internet, enabling optimum use of the systems and on demand service provision. Sensors could be used to monitor occupant levels, and air conditioning load requirements calibrated accordingly. The potential here is unlimited and future connectivity will revolutionise facilities management operations.
- Building maintenance requirement detection and system performance monitoring enables continuous and breakdown free performance. Facility managers could monitor performance and schedule maintenance operations based on actual condition as opposed to periodic standard service.
- Building users can remotely control building service systems as required. This includes monitoring the security status of their homes or offices and controlling lighting, air conditioning and sound systems, amongst others. This helps to reduce energy consumption and improve comfort and usability of homes and offices.
- Efficient building management systems will save energy consumption in overall terms, which makes a direct contribution towards reduction of operation carbon emissions from buildings.
- Smart meters are another case in point. These enable building users to dynamically monitor energy use and make corrective actions where required, providing greater controllability and dynamic feedback to help make lifestyle adjustments where required.

Smart Energy:	**Smart Security:**	**Smart Lifestyle:**
• Smart sensors • Smart power points • Smart light bulbs • Smart batteries • Other smart energy components (incl. other smart heating and cooling systems) • Smart energy services (subscriptions and managed services) • Smart energy installation services	• Smart alarms • Smart cameras (including smart baby and pet monitors) • Smart motion sensors • Smart locks • Other smart security components (incl. smart smoke detectors and smart intercom systems) • Smart security services (subscriptions and managed services) • Smart security installation services	• Smart appliances (e.g. smart fridges, washing machines, air conditioner, robot vacuum, etc.) • Smart gardening • Other smart lifestyle components (incl. smart garage doors, smart blinds and curtains and smart pool maintenance equipment) • Smart lifestyle loT@Home services (subscriptions and managed services) • Smart lifestyle installation services
Smart Hubs: (e.g. dedicated control hubs such as Samsung SmartThing Hub, WINK Hub 2, smart speakers such as Amazon Echo, Google Home and security hubs such as Trend Micro Security Box)	Google home Samsung SmartThings Hub	Amazon Echo Wink Hub 2

Figure 22.2 The scope of IoT @home.

Source: www.telsyte.com.au/announcements/.

- Smart hubs are becoming popular with households to enable them to control connected devices through the home Wi-Fi network. The hubs help in monitoring and controlling the household environment using voice commands or through smartphone connected apps. Figure 22.2 from the Telsyte Australian IoT @home study 2017 indicates the current and potential use of IoT devices.

In the future, construction product and component procurement is expected to follow the digital revolution of its general procurement counterparts. It is expected that there will be quantum changes in procurement processes, with rapid adoption of blockchain technology for procurement (see section 0). Increasingly, building products are tagged with radio-frequency identification (RFID) chips or other wireless communication methods, making components or equipment installed traceable and/or controllable. This makes a great difference in product compliance and performance monitoring.

22.5 Big data analytics

The construction industry is traditionally data driven and characterised by masses of data produced in the context of its projects. Quantity surveyors, estimators and construction managers the world over use construction data for understanding their projects and predicting costs and time related issues. However, data within the construction industry is rarely shared or open sourced. Project related data is considered confidential and often available for access only within the immediate stakeholders of the project. That limits the amount of data to process and conforms to traditional data processing techniques.

Big data represents mass volumes of data. Although there is no specific amount defined, it is often in the range of terabytes, petabytes and even exabytes of data captured over time. Data can be captured instantaneously in real time or near time, periodically or in batches. Data can also be presented in variety of forms, such as databases, audio, video, social media, structured to unstructured, mobile or IoT device-generated, amongst others. These massive volumes of data often cannot be analysed using traditional data processing software.

The large volumes of information generated by buildings and components attached to IoT devices provides a rich source of data. However, if the data is not analysed it does not become useful. Figure 22.3 indicates how data can be analysed to make sense and maps traditional data with a typical big data scenario. The only major difference here would be the volume, velocity and variety of data generated in the case of big data. The distinction between knowledge and wisdom is debatable, and many authors have not considered the category of wisdom in recent publications, whilst others have and further classifications of data as well.

It is clear that if big data generated or collected is not analysed, it will not have any impact. Thus, big data analytics becomes important for the construction sector as a way of making behavioural and or trend predictions. The *Digital Built Britain* report (2015) anticipates that there will be significant use of big data analytics in analysing data generated from smart cities, smart buildings and smart homes. These will be related to:

- health care and assisted living, patient monitoring, digital records and administration
- smart energy grids, demand management and renewable source integration
- transport, traffic and congestion management, road charging, emergency response, public information, managed motorways and smart parking
- water management, consumption measurement, wastewater treatment
- waste management, collection and processing of all types of waste including carbon.

The *Digital Built Britain* report (2015) predicts that commercial applications based on the use of big data derived from the internet will require new forms of contract to establish commercial

Construction Big Data		**Traditional Construction Data**
Understanding of the thermal profile of the building and be able to predict and modify usage	Wisdom	Understanding of which types are more expensive and where and when to procure
Mapping of the temperature zones of the building	Knowledge	Average material prices
Classification of data by location and time	Information	Material prices classified by material type
Thermostat sensor data for the whole building from multiple sensors	Data	Building Material Prices

Figure 22.3 Data hierarchy in the context of construction.

models for 'semantic' transactions and their associated benefits, obligations and liabilities. The fact that big data analytics enable prediction of behaviours of people and buildings means that such knowledge generated would lead to key decisions regarding design, construction and use of the buildings. It would have contractual implications. Existing traditional forms of contracts will not be able to deal with decisions based on data analytics.

22.6 Blockchain technology and crypto currencies

The construction industry is a gigantic market place where multiple transactions occur between contractors, subcontractors, suppliers, manufacturers, designers, construction managers, clients and many others. Central to all these transactions are contracts, currencies and trust. Often, these factors are compromised, resulting in disputes, litigation and transaction (or even project) failures. There are issues of breaches of trust between parties or supply of non-compliant (or substandard) building materials and components. As a result, construction supply chains are notoriously expensive and inefficient.

Blockchains provide a technology that can help resolve most of these issues by bringing back trust and swift performance for digitally secure transactions. Blockchain is an open and distributed digital ledger that records efficiently verified transactions between parties. It is a distributed series of records of peer to peer transactions recorded as a permanent set of transactions that cannot be altered by individuals, which makes it secure and trustworthy.

Technologies that underpin Blockchains are not new, as they have existed in many domains before. Ledgers, receipts and invoices are from the accounting domain and digital transactions from e-business (ICT domains). What is key here is bringing together these different technologies into one platform to create blockchains.

A block is a record of valid network activity of a transaction performed, and all blocks formed in chronological sequence form a blockchain. Each block in the chain can be defined as a cryptographically coded set of data pertaining to a transaction. All blocks in the chain are recorded as a database of digital ledger distributed to all parties in the network. This makes the entire blockchain transparent because the blocks cannot be altered or changed. Any tamper of a record in the ledger will result in an invalidation, which stops further processing of transactions, making the blockchain secure. Anyone in the network can add more blocks to the chain, expanding the scope of transactions. This process allows blockchains to contain a complete, immutable record of all transactions shared between all in the network. It allows two or more secure transactions to occur over the internet between parties who did not know or trust each other without a mutually trusted intermediator, such as a bank, government or cash. The blockchain therefore acts as a shared single source of truth. The most common form of crypto currency used is a Bitcoin, which can represent anything from a physical currency unit (£ or $), a barrel of oil, a bag of cement, a standard door, a share in a company, a digital certificate of ownership or even a vote of a citizen (See Figure 22.4). It can contain up to 100 million units.

22.6.1 Advantages and benefits of blockchains

There are many purported benefits from using blockchains in general, and in the construction industry specifically:

- It provides a single shared digital ledger that is tamper evident to all participants of the network. Once a transaction is recorded, it cannot be modified or altered. It works as a series of shared records amongst all participants of the business network, eliminating the need to reconcile disparate ledgers. These could simply be transactions carried out between a contractor and a subcontractor in a construction supply chain where the main contractor purchases aluminium windows from a subcontractor.

- All parties in the network need to give consensus for any transaction to occur, but this happens in a non-invasive way, without individual interventions, through Bitcoin mining. Transactions that are recorded cannot be deleted, even by system administrators. This makes it a secure and trustworthy network – a single source of truth.
- The decentralised network eliminates single points of failure that could happen in a centralised system. Risks are decentralised to each of the nodes of the blockchain, making the whole blockchain more durable.
- Each member in the network has levels of access rights enabling them to maintain confidentiality of information shared on a need to know basis. It means in a series of transactions that, although there is a shared ledger, privacy can be maintained to the business network concerned to make sure appropriate levels of details are available to the relevant participants of the network.
- It eliminates manual and paper-based processes, making transactions cost and time efficient. All transactions occur in the digital platform. A contractor will not be raising paper invoices or the supplier providing paper receipts signed by various authority levels. All of these happen digitally.

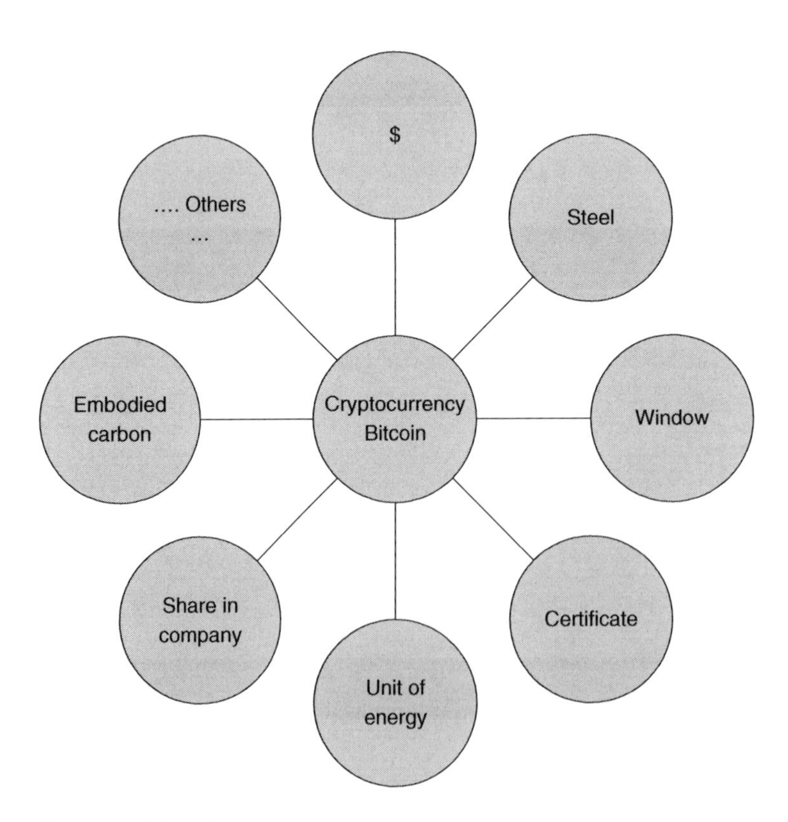

Figure 22.4 Bitcoin can represent any property.

22.6.2 Applications of blockchains in the built environment

There are many construction-based applications of blockchains. The complicated and multi-level transactions that occur in construction supply chains can utilise blockchains as the platform for transactions. This will make sure all goods finally delivered to site are duly paid for in time, whilst products comply with relevant standards and specifications as designed by the building/facility design team. Construction payment and progress payment systems can be implemented in blockchains. Payments can be automated on compliance.

Construction projects are highly data/information rich. There are multiple sources of project information and multiple types of standards, building codes and regulations to which designs should comply. These can be effectively implemented in blockchains, where sources of information generated indicate the level of compliance achieved before it is passed on to the next stage in the chain. A typical construction industry blockchain network is illustrated in Figure 22.5.

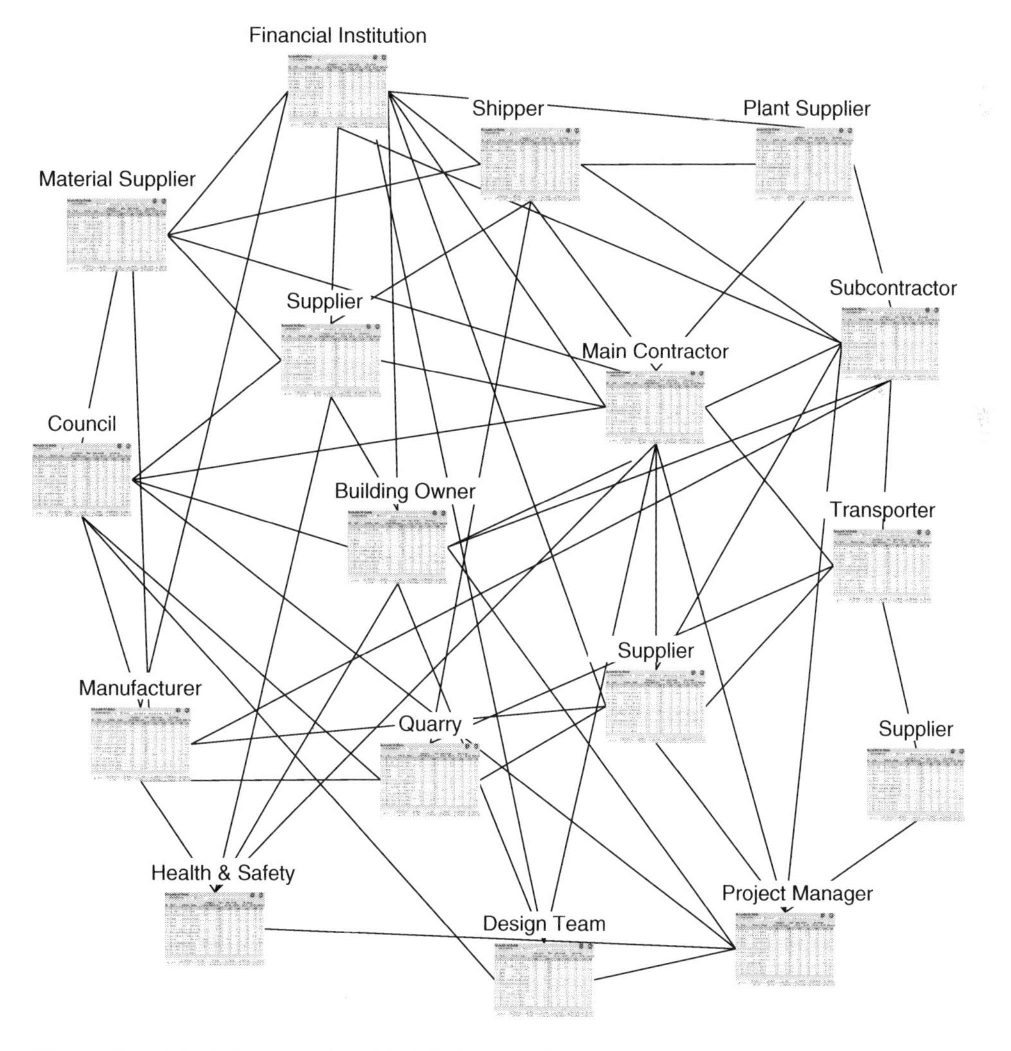

Figure 22.5 A typical construction blockchain network.

There is huge potential for the implantation of IoT in buildings and facilities in the build environment (as discussed in section 0). Data generated from IoT devices can be integrated to a blockchain where automated or semi-automated processes can follow through. For example, maintenance monitoring devices in HVAC systems can provide a signal indicating imminent failure of a component, which can then be procured on a blockchain-enabled procurement system, supplied, installed and progress monitored. So, there are endless applications of complementary blockchain systems that can be implemented in smart buildings and smart cities, facilitating smart construction.

The property world is also likely to be digitalised and captured in blockchains. 'Smart property' is real life physical property for which a digital identity is created. This then allows the ownership of properties to be transferred digitally in blockchains. This will harmonise with the associated smart contracts that would be developed to facilitate the legality of transactions. Such technologies will see future property transactions in the built environment operating in fully digital platforms. It would make the operations of property developers much easier, leaner and more efficient, with minimal transaction costs whilst ensuring security of transactions. Estate agencies selling property will adopt these technologies, and it can then be integrated with financial institutions who provide finance and mortgages and land registries.

22.6.3 Implications for the construction industry

It is expected that the next decade will bring about quantum changes in the way the business world makes transactions and operates. Blockchains will play a central part in this transformation, driven by globalisation and digitalisation of supply chains. The construction industry will not be immune to these changes and will follow the same destiny. But the construction industry stands to benefit even more than other industries as such digitalisation approaches will push greater efficiency gains, reducing overall costs of construction and reducing waste.

22.7 Smart construction and smart contracts

Digital Built Britain (2015) envisages that there will be performance-based contracts supported through digitally procured data. It states: 'Contracts will focus on the capture of performance intelligence and project feedback and the deployment of a data-based briefing process'. In Level 3 of its Construction and Infrastructure Delivery Model, it foresees more effective data exchange and data-enabled collaborative working based on transactional contracts. The advancement of technologies and scope of blockchains indicates that this will happen within implementation of blockchains for the construction industry.

A blockchain-enabled contract is referred to as a 'smart contract'. Therefore, smart contracts can be defined as computer protocols intended to facilitate, verify or enforce the negotiation or performance of a contract. The term 'smart contract' was coined by Nick Szabo in 1996 in his book, *Smart Contracts: Building Blocks for Digital Free Markets*, and subsequently further developed.

The expectation of smart contracts is the possibility of having partially or fully self-executing and/or self-enforcing contracts. Such would be secure and have very low transaction costs. The most prominent smart contract is implemented in the Ethereum Blockchain platform. It has a smart contract scripting functionality implemented as an open source public blockchain-based distributed computing platform. As such, it allows companies to build their own blockchains for the network of companies with which they transact. Implementing such a smart contract in a construction project within a blockchain platform is quite plausible. It will allow execution of contracts to be automated to a great extent; for example, if a construction contract states that payments to the main contractor are linked to a fuel price index where each fluctuation in fuel price will result in automatic payment adjustment to the contractor.

There are many benefits expected from smart contracts implemented in the construction industry:

- It will significantly reduce the administrative processes involved in construction contracts. Therefore, billing and checking and subsequent calculation of costs can be simplified and automated where possible. Linking with smart devices (IoT) will enable achievement of a greater degree of automation.
- It will minimise cash flow issues rampant in the construction industry as blockchain payments would be instantaneous, based on actual delivery or component or performance of task. Payments down the supply chain from contractor to subcontractor to supplier (and many other intermediates) will be carried out with minimum delays as smart contract and securing of blockchain will ensure that the full transactions related to each level is accurately and fully performed.
- The one-off nature of construction projects means that, often, the supply chains involved in projects are different from one another. There is less chance to build trust over a longer period of business engagement. This is further exacerbated because the construction industry is often cyclical, and boom and busts means that there is a high turnover of companies in business. Trust in construction contracts can be significantly increased with the use of blockchains and smart contracts for execution of transactions. The mathematical algorithms and coding in the software platform ensures unbiased and secure transactions between parties, eliminating the need for human trust to do business. The very fact that the supply chains are increasingly becoming globalised means that there is, anyway, less opportunity for traditional processes of trust based business. In the blockchain, this is taken care of by the coding and the blockchain structure and process itself.
- Traditional construction contracts require third parties to enforce (lawyers) and administer (quantity surveyors or project managers). Use of such authorities will be minimised with the use of smart contracts in a digital environment.

22.8 Future directions

The construction industry and the whole of the built environment is in an irreversible process of globalisation and digitalisation. This means that there is greater need to remove barriers when making transactions that span the globe. Administrative processes and transaction costs need to be lowered in order to make these processes lean and efficient. BIM provides the platform for digitalising the construction industry's product, i.e. the building or facility being built. The way business transactions are carried out will be shaped in future through the use of blockchains and smart contracts. This means that construction procurement in the future is likely to be carried out in blockchain platforms supported by smart contracts. The process is compatible with the requirement for globalisation of supply chains and global sourcing of material and components. This will enable the achievement of smart construction that is central to an efficient, high-performaing and sustainable built environment.

The entire commercial legal domain is currently in a state of flux. The Decentralised Arbitration and Mediation Network (DAMN) is a recent start-up providing an opt-in justice system for commercial transactions targeting the cross-border dispute resolution market. Major banks in the UK and Australia are fast adopting blockchain technology and smart contracts for their business purposes. These include Barclays Bank, Commonwealth Bank of Australia, National Australia Bank and Westpac Banking Corp, all testing Ethereum-based smart contracts, amongst others. The traditionally slow to adopt construction industry will have to move in the digitalisation direction fairly rapidly now.

Achieving greater levels of sustainability in the built environment is another global objective requiring the industry to reduce waste and carbon emissions. Smart precincts, smart cities and smart buildings would provide a sustainable future for the built environment. IoT is expanding rapidly, with a high degree of potential application in the construction industry and the built environment (see section 0). Big data analytics will provide the platform required to analyse the vast amount of data generated from IoT devices utilised in smart buildings. Together, these technologies will enable us to understand and predict trends and behaviour patterns that would in turn shape the built environment in which we aspire to live.

There will be a great impact on the labour market across industries and worldwide. Many will see that their jobs are now redundant, yet others will see that their jobs have changed significantly. Technologies such as artificial intelligence (AI) and IoT are highly invasive in making this change happen. Technologies such as blockchain will change the entire global business process architecture, creating new ways of procurement and supply chain behaviour. The advancements in nanotechnology, innovative materials, cyber-physical systems and automation have a significant impact on all professions. These are not only going to change the skilled and unskilled labour markets, but also the highly specialised professions of law, medicine, accounting, creative industries and education, amongst others. The construction industry will not be immune to these changes. All construction professions will see the impact of these changes significantly changing the role they perform. More than just the demise of professions, we will see the radical change in the way in which professions do operate. AI is infiltrating the architecture profession, structural design is becoming standard and susceptible to AI, quantity surveying is subjected to BIM and blockchains, mechanical, electrical and building services design will become standardised and taken over partially by AI. Construction management and project management will have less to do on-site and will be aided by IoT, blockchains and similar technologies. The story is long and covers the breadth of construction professions as well as trades; but, rather than mass social chaos, it will result in new types of jobs and even new types of professions. The universities and the tertiary institutions will change and respond to the new dynamics of the labour markets in creating new jobs and professions. These will initially come as high-end masters' programmes, but will gradually filter into the undergraduate themes. Already this trend is seen in universities where, for example, nanotechnology started as masters' content and filtered down into undergraduate content. At a technical level, we could see more cobots (collaborative robots) aiding skilled tradesman on-site, taking over mundane, repetitive and complex tasks, thereby making construction more safe, efficient and cost-effective. Technical training will move from using electrical tools to how to use the cobots for their trade related work.

We are still at the dawn of the forth industrial revolution, which is seeking fusion of digital, physical and biological spheres. The built environment is no doubt at the centre of this revolution, affected by the advancement and fusion of technologies, processes and societies, the fusion of biological, physical and digital spheres. There are many other technologies upon which this chapter did not focus, yet would create significant impact on the built environment and construction industry. These primarily include the advancement of AI. The technological advancements in AI are already having a significant affect in the built environment. Our day-to-day lives are now either directly or indirectly influenced by AI. These are in the forms of smartphone applications, word processors and other software or computers that we use. Robots are significantly changing construction technology and the way construction activities are performed. Advancements in nanotechnology and material science is bringing in new and revolutionary building materials to the construction industry. The impact of these means that the next decade will crucially transform the construction industry and its processes, influencing the contractual procedures and procurement systems that it utilises.

Part 4

Principles of the conditions of contract

23 Introduction, articles of agreement, appendix

The various forms of contract used for building and civil engineering works of construction are discussed in Chapter 6. Changes in procurement methods will have an impact on exactly what the different conditions of contract describe. All new forms of contract examine what has gone before, so that revisions and changes are made in the light and knowledge of previous best practices. These revisions are necessary because of changes in the law and legal principles, changes brought about through case law and changes resulting from the way the industry is now being organised. The industry has observed wide-ranging changes in the way in which it procures construction projects, from the smallest to the largest schemes. Chapter 7 referred to the RICS *Contracts in Use Survey* series (see RICS 2010) and the new *National Construction Contracts and Law Survey* (NBS 2015) that helped to record changes in procurement practice.

The forms of contract appear to have become more complex than some of their predecessors. In some cases, the forms in use today are, for example, more than double the size of those that were in use fifty years ago. This may just go to show how much more litigious society and the construction industry has become over recent years. Some forms of contract used for a single project, including their additional supplements, amount to more than 800,000 words. This has sometimes been compared to the Lord's Prayer, but that achieves its aim successfully in a mere seventy words!

The different forms of contract used in the construction industry have been developed on the general presumption of fairness in the allocation of risk to the party that is more likely to be able to control it. Whether this allocation of risk is fair depends upon its use and interpretation in practice. Some of the newer forms of contract do, however, aim to place more risk away from the employer; but this is done at a cost. The forms have been written in an attempt to cover every possible eventuality, so that should a dispute arise there are principles, processes and procedures that can be followed and applied. However, several surveys of users in the construction industry indicate that they are often unaware of the full implications of the different forms of contract, consider them unnecessarily complex and describe them as ambiguous. The forms are also written in a legalese language that is not easily understood. There remains continued criticism of subcontracting nomination procedures in some of the forms.

The different forms of contract contain:

- Articles of agreement – which the parties need to sign.
- Conditions of contract.
- Appendix – which needs to be completed for each project.

There may also be a series of annexes to the forms, such as alternative methods that might be used to deal with price fluctuations, bonds and guarantees and VAT agreements. The different forms of contract, whilst they cover a common range of issues, differ in detail on how these issues should be dealt with.

The basic principle of all the forms is that they are entire contracts which are modified by a number of their own provisions. Separate editions may be available for use by private and public sector clients. In practice, local authorities have traditionally used different forms of contract from those of central government departments and even the latter may not use the same set of procedures. There are usually separate forms for the different variants: with quantities, without quantities and with approximate quantities – although some forms of contract do seem able to deal with these alternatives in a single form since the differences between them are often minimal.

The two parties to the contract are the employer (client) and the contractor. The employer is responsible for making payments to the contractor, who in turn has obligations to complete the works in accordance with the contract provisions. Neither the architect or engineer nor the quantity surveyor are parties to this contract, although they figure strongly within its clauses. The architect or engineer acts on behalf of the employer, but also sometimes in an impartial way, giving benefit to the contractor where this is appropriate. The architect or engineer often have different roles to fulfil, emphasising the different natures of the building and civil engineering industries. The quantity surveyor is impartial in accounting for the financial transactions that take place before, during and after the project has been completed. In some of the forms of contract the name 'supervising officer' is used, which might mean either architect or engineer.

23.1 Articles of agreement

The articles of agreement precede the conditions of contract and they include information that must be completed for a particular contract. The articles define the various persons referred to in the contract and they describe the procedure to be followed should a dispute arise between the employer and the principal contractor. The following information therefore needs to be written into the articles of agreement:

- *The date*: this is the date when the agreement is made. It is not the date of tender, which may have some significance in respect of fluctuation contracts; nor is it the date when the project will start on-site. This date is defined, in the appendix to the form of contract, as the date of possession. The date in the articles of agreement will therefore have little significance as far as the contract is concerned.
- *The parties*: the contract is between the employer and the contractor, and their names and addresses must be inserted in the articles of agreement. These names, employer and contractor, are used consistently throughout the conditions of contract. Their registered office addresses are also included. Some forms of contract may use a different name for the employer, such as promoter.
- *The works*: the employer requires the contractor to complete the works in accordance with the contract documents. The works are essentially the construction project, but, by implication in the conditions of contract, may also mean the site. A description of the project and its location (address) should be given to avoid any doubt of where it will be constructed. The contractor is given possession of the site on the date of possession as stated and detailed in the appendix.
- *The drawings and bills of quantities*: these, together with the form of contract, constitute the contract documents. In some cases, a specification is used in lieu of bills of quantities. On some types of project, typically civil engineering works, there may be both bills of

quantities and a specification. The contract drawings are formally registered by their numbers and should be signed by both parties. The contract bills, which would be a priced copy of the bills of quantities, are also to be signed by the parties.

- *The designer*: the drawings and bills of quantities are prepared under the direction of the pre-contract architect or engineer. This emphasises that the design is the responsibility of the architect or engineer rather than of the contractor. Where the contract is for design and build, then different arrangements will apply. On some projects the designer may not be the person who supervises the work on-site. The architect or engineer referred to in the conditions comes within this latter definition. In the event of the architect's or engineer's death, or when no longer fulfilling this function, the employer must nominate a replacement. If the contractor has a reasonable objection, then the employer must appoint someone with whom the contractor approves. Such an appointee must not revoke decisions already acted upon by the contractors. The new architect or engineer can, of course, ask the contractor to rebuild part of the completed works, but, if these have previously been accepted, then the contractor is entitled to a variation along with the appropriate payment. In some cases, the supervisor of the works may not be an architect within the definition of the Architects (Registration) Act 1938. In this case, the term 'supervising officer' is used to refer, perhaps, to an engineer or surveyor.
- *The quantity surveyor*: this refers to the client's quantity surveyor. In the event of death, or when replaced by the architect or engineer, this must be done with the approval of the contractor. Any objection by the contractor must be sufficient to invoke the appointment of the arbitrator. The term quantity surveyor is not used in all of the forms of contract, some prefer to use the more ambiguous term of contract administrator.
- *The contract sum*: the employer's main obligation is towards the payment of the contractor for the work executed. The total amount is described in the contract as the contract sum and this will usually be paid by instalment, although there are several ways in which payment may be made. Although this sum is agreed upon by the two parties, it will be subject to adjustment within the terms of the contract as the project progresses, for example through change orders or variations to the design, materials or methods of construction. It is very unusual if the contract sum remains unaltered by the time the project is finished.
- *The employer*: the status of the employer for the purpose of the statutory tax deduction scheme must be stated in the appendix.
- *Contractor's obligations*: the contractor's obligations under the terms of the agreement are to carry out and complete the works in accordance with the contract documents. These obligations are further reinforced throughout the conditions of contract.
- *The planning supervisor*: this is the person, either the architect, engineer or other person, appointed as a result of regulation 6(5) of the CDM Regulations.
- *The principal contractor*: this refers to the main contractor in pursuance of regulation 6(5) of the CDM Regulations.

23.2 Settlement of disputes – adjudication

Where disputes or differences of opinion occur between the employer (or the architect or engineer) and the principal contractor, the parties agree to attempt to settle the matter first of all, usually through adjudication. Where the entry in the appendix has not been deleted, then it can then be referred to arbitration. The provisions of the Arbitration Act of 1996 or any subsequent amendments then apply. Where these do not resolve the differences, the dispute will be referred to legal proceedings.

23.3 Attestation

The attestation page is for the various parties to sign the contract and for the signatures to be witnessed. The signatories must have the necessary authority to sign for this purpose to provide validity to the contract. The contract may be executed under seal or under hand, and the Limitation Act of 1980 states that actions become barred after twelve years and six years, respectively. The current ruling suggests that the cause of action begins to accrue when the damage is discovered, or when it should reasonably have been discovered. The actual time of discovery may therefore not be relevant. There is no contract between the parties until the articles of agreement have been executed, unless the facts show otherwise. The carrying out of work prior to contract may indicate an intention on both parties, and therefore, if the project does not proceed, a claim for *quantum meruit* may result.

23.4 The appendix

The appendix to each form of contract provides a considerable amount of detailed information about the contract that is of particular importance to the project. The purpose is to set out in one place a schedule of these details, which are then referred to through the conditions of contract. The completion of the appendix is necessary to give the conditions their full meaning. For example, if the date used for the possession of the site is not stated, then this will have a knock-on effect to several different terms and items in the conditions of each contract. These different terms are discussed later in the following chapters. There are, however, some differences in principle between the treatment of items in the appendix. Some items will always need to be completed to make sense, whereas others can be left blank. In other circumstances, the appendix makes a recommendation and, unless anything is entered to the contrary, this suggestion will apply.

24 Quality of work during construction

The combination of ensuring that the contractor complies with the statutory regulations and the specification of the works that have been described should result in a construction project that achieves the desired standards and quality that were expected. In addition, there is provision in the contract for regular inspection by a clerk or inspector of works, or someone similar, and intermittent inspection by the architect or engineer. Most of the forms of contract include an added provision for inspection of goods and materials in the workshops off-site, should this be so desired. With much more work being prefabricated, this can be very important to the successful outcome of the project. The contractor must also seek to ensure that competent tradespeople are employed to do the work required. It is expected that such people will hold National Vocational Qualifications (NVQs) or similar, especially for those who are carrying out the craft work. Someone on-site must be constantly available to accept instructions from the architect or engineer. Such individuals will also be responsible for the daily site management of the project and the people working on it. On large projects this will be a site manager, or site agent, with several assistants fulfilling the different roles. In addition, there are clauses covering levels and setting out, which make sure the project gets off to a correct start from the outset. There is also provision for uncovering work for inspection if the architect or engineer feels this is necessary. The different forms of contract specify the outcome of this depending upon whether the work complies with the specification requirements as intended.

24.1 Statutory requirements

Contractors must comply with, and give all notices required by: any Act of Parliament; any instrument, rule or order made under any Act of Parliament; any regulation or by-law of any local authority or statutory undertaker which has any jurisdiction with regard to the construction works. However, the contractor will not be liable for compliance with this requirement where the works themselves do not properly comply with these requirements. This is the designer's responsibility. Nevertheless, the contractor should not deliberately and knowingly execute work that will contravene such regulations, and will then require future modification for compliance purposes. On the other hand, the contractor is not a watchdog to ensure that designers carry out their duties properly and efficiently.

Where the principal contractor finds any divergence between the statutory requirements and any of the contract documents or an instruction, then the person responsible for designing the works should immediately be informed in writing, pointing out the discrepancy. Within a reasonable period of time, which will be specified in the contract (typically about a week), of receiving such a written notice, the designer must then issue instructions in relation to the divergence. If the instruction requires the work to be varied, then this will be treated as a variation.

It is sometimes necessary for a contractor to comply urgently with a statutory regulation. For example, where an existing structure on the site is in danger of collapsing or where health and safety are in danger. This may require the contractor to supply materials or execute work prior to receiving instructions. The contractor should do what is reasonably necessary to secure the immediate compliance with the statutory requirement and inform those who are responsible of this action. The materials and work executed in these circumstances will then be treated as a variation.

The principal contractor will not be liable to the employer for work carried out in accordance with the contract that is subsequently found not to be in compliance with any statutory requirements.

It is normally the principal contractor's responsibility to pay and indemnify the employer against liability in respect of any fees, charges, rates, taxes, etc. These may be required under an Act of Parliament, regulation or by-law. The amount of such fees and charges will be added to the contract sum unless:

- they arise in respect of goods or work done by a local authority or statutory undertaker
- they are included in the contract sum by the contractor
- they are included as a provisional sum in the contract bills.

24.1.1 Construction (Design and Management) Regulations

The Construction (Design and Management) Regulations 2015 (formerly 2007, see Chapter 20) are more commonly known as the CDM Regulations. The employer shall ensure:

- that the planning supervisor carries out the relevant duties under the CDM Regulations
- the principal contractor carries out their duties under these regulations.

The principal contractor must ensure that the health and safety plan has the features required by CDM Regulations. Any amendment to this plan by the principal contractor must be notified to the employer, who in turn will notify the planning supervisor and the designer. The principal contractor should also ensure the compliance with this provision of the subcontractors working on the site. The principal contractor should have a health and safety plan and maintain a health and safety file for the project (see Chapter 20).

24.2 Setting out of the works

The architect or engineer will be responsible for providing the principal contractor with all the information necessary in order that they can set out the works at ground level. This information should comprise properly dimensioned drawings and levels. The actual setting out of the works from this information is entirely the responsibility of the principal contractor. If mistakes are made during this process, then these must be corrected without any additional cost or charge to the employer. If the architect or engineer delays the presentation of this information, this could result in an extension of time to the contract. The inaccurate setting out of the works could result in trespassing on the land of an adjoining owner. If this is due to inaccurate information provided, then this is the responsibility of the employer.

24.3 Work, materials, goods and components

A number of clauses in any contract conditions will refer to the quality of materials and standards of workmanship. Such clauses will reinforce the information contained in the bills of quantities, schedules or specification. This may include the performance-specified work given to the principal contractor. All the work must be carried out in a proper and workmanlike manner and in accordance with the health and safety plan.

Circumstances may arise where the quality specified becomes no longer possible to obtain. Before proceeding with some alternative, the principal contractor should seek instructions rather than proceeding with their own ideas. It may be necessary on some occasions to prove that the materials being used are in accordance with the specified requirements. The architect may require documentary proof that the materials are in accordance with those specified. The principal contractor may need to produce vouchers, such as invoices, for this purpose. The contractor must not substitute any materials or goods without consent.

24.3.1 Construction Skills Certification Scheme

It is important that the principal contractor and subcontractors use appropriately qualified labour. The workforce should be registered card holders under the Construction Skills Certification Scheme (CSCS, www.cscs.uk.com). This recognises that the construction industry and the carrying out of construction work should only be done by those who hold recognised qualifications (see also Chapter 17).

24.3.2 Inspection and testing

Work that does not comply with the contract will be rejected. This can include an incorrect choice of materials by the principal contractor or standards of workmanship that do not comply with the specified standards stated in the contract documents. It is important that such dissatisfaction is expressed within a reasonable time from the execution of the unsatisfactory work. From time to time, instructions will be given to the principal contractor regarding tests on materials and workmanship. Such routine tests will be those that have been described in the contract documents. This may involve the contractor in testing the materials, goods or workmanship. Other tests may be required to be performed that have not been specified. Unless these are unreasonable, they must be carried out by the principal contractor, but may form an additional charge to the contract. In some circumstances, it may be necessary to include a provisional sum for any unusual testing if this is to be expected.

Inspection or testing of any part of the works can usually be carried out up to the issue of the final certificate. The purpose of this is to ensure that the works are in accordance with that specified in the contract. The costs of opening up the works for this purpose will be borne by the contractor if the inspection or tests show that it is not in accordance with the contract. Where works conform with that specified, then the costs of opening up and making good will be added to the contract sum. This may result in an extension of time or a claim for loss and expense.

Where the materials or workmanship do not conform to the contract, instructions may be given to remove these from site at the principal contractor's own expense. However, it should be noted that in most construction contracts permission is usually required to remove materials once they arrive on-site. There is no limit in time during the contract when such defective materials might be discovered. However, reasonable inspection should take place during the progress of the works. If a failure in the foundations did not occur until a building was suitably loaded, and such a failure was as a result of the use of defective materials, then the principal contractor would be liable fully to rectify the problem.

The principal contractor (and subcontractors) must carry out the works in a proper and workmanlike manner and in accordance with the health and safety plan. Where this does not occur, then instructions will be given to ensure compliance. Such instructions will not result in any addition to the contract sum and no extension of time will be given.

If any of the work, materials or goods is not in accordance with the contract, then the following may apply:

- Issue instructions for its removal.
- Allow such work to remain, and confirm this in writing to the principal contractor. This will probably result in an appropriate deduction from the contract sum.
- Issue instructions for opening-up inspection using the following code of practice.

24.3.3 Code of practice

- The event of the non-compliance is unique and is unlikely to happen again.
- The need to discover whether any non-compliance in a primary structural element is a failure of work standards and/or materials such that rigorous testing of similar elements must take place. Where the non-compliance is less significant, then it can simply be repaired.
- The significance of the non-compliance, with regard to the nature of the work in which it has occurred.
- The consequence of any similar non-compliance on the safety of buildings, its effect on users, adjoining property and the public and compliance with any statutory requirements.
- The level and standard of supervision and control of the works by the contractor.
- The relevant records of the contractor and any relevant subcontractor resulting from the supervision and control referred to above.
- Any codes of practice or similar advice issued by a responsible body which are applicable to the non-complying work, materials or goods.
- Any failure by the contractor to carry out, or to secure the carrying out of, any tests specified in the contract documents or in an instruction to the architect.
- The reason for the non-compliance when this has been established.
- Any technical advice that the contractor has obtained in respect of the non-complying work, materials or goods.
- Currently recognised testing procedures.
- The practicability of progressive testing in establishing whether any similar non-compliance is reasonably likely.
- If alternative testing methods are available, the required time and the consequential costs for such alternative testing methods.
- Any proposals of the principal contractor or any other relevant matters.

24.3.4 Exclusion from the works of persons employed thereon

Instructions may be given to exclude anyone from the site. This must not be done unreasonably or to specifically annoy the principal contractor. A previous warning of this impending situation is likely to have been given in order that appropriate corrective action can be taken. It is not a common occurrence, but it may arise in circumstances where standards are persistently bad, when instructions have been ignored or where someone is having a negative effect on the progress and quality of the works.

24.4 Royalties and patent rights

All royalties and other similar sums included for work described in the contract are deemed to be included in the contract sum.

24.5 Person in charge

The principal contractor must keep a person in charge or site manager on the site. This is the person who is responsible for the daily running of the site, who will receive instructions, and to whom a clerk or inspector of works can give directions. Different construction firms will use different names for this person, such as project manager or general overseer. A contracts manager, however, who is often responsible for several sites, will not easily fit within this description since such a person is not 'kept up on the works'. The instructions given to the person in charge will be deemed to have been issued to the principal contractor. In order to satisfy this last requirement, the person in charge may delegate an assistant who can receive instructions in their absence. Some contracts use the term 'constantly on-site', which really means at all reasonable times, to ensure that the work is correctly executed and to receive any instructions or directions from the superintending architect or engineer.

24.6 Access to the works

In most construction contracts, the architect or engineer, or indeed one of their representatives, has the right of access at all reasonable times to the construction site and the contractor's workshops. This will also cover the workshops of domestic and nominated subcontractors. In some circumstances, for example, it may be important for the structural engineer to be allowed access to the fabrication process of the steelworker. It is important to be able to inspect the quality control of any of the firms involved in the construction of the works. It will be noted that this does not sometimes include suppliers, since their relationship is different. However, different forms of contract will sometimes include such firms and organisations.

Due to the increasing amount of specialist work included in buildings, provision is now necessary and is provided to protect the proprietary interests of the contractor or subcontractor, whilst not detracting from the right of access. It might be achieved through confidentiality agreements or limiting the inspections of specialist manufacturing processes where some form of indemnity could be provided.

24.7 Clerk or inspector of works

Most forms of contract include the provision for inspecting works as they are being constructed. On small projects, this might be solely by the architect or engineer themselves. On larger projects, it is often necessary to appoint a clerk or inspector of works. Their duty is to act solely as an inspector on behalf of the employer under the directions of the architect or engineer. The principal contractor will provide them with all reasonable facilities in order that their designated duties can be performed properly. They are not employed in the capacity of the architect's or engineer's representative, since employment is solely for inspection purposes. If they choose to give the principal contractor instructions, even in writing, they will have only little contractual effect unless they are ratified by the architect or engineer. This must be done within a reasonable time frame (normally within a week) in order that the contractor can comply with such instructions. If such instructions are only confirmed orally by the architect or engineer, then the appropriate procedure is to put this in writing. The clerk or inspector of works is usually resident on the construction site and therefore in constant touch with the progress of the works. In some circumstances they can make a quick decision. In practice, therefore, the principal contractor must receive many such statements or instructions from the clerk or inspector of works in good faith.

25 Costs of construction

An important factor in any construction contract is the calculation and payment of the sum of money by the employer to the building contractor. The importance of the contract sum and its relationship to the final account, together with the matters relating to the timing of payments, are discussed in this chapter.

There are many ways of calculating construction costs and making payments to a contractor. In respect of the latter, the usual way is to make interim payments throughout the contract period according to an assessment of the work completed. Payment may then be done through an agreed stage payment structure. Relatively few contracts only allow for a single payment to be made on completion. Such projects are likely to be of only a minor nature, since cash is the lifeblood of the contractor. There are many examples where a contractor has gone into liquidation, not because the work was unprofitable but because of negative cash flows that their banks would not support. In some cases, there is provision for advance payments, although very few people or organisations would enter agreements to pay the full amount of the contract costs up front. This would be the case in circumstances where such contracting firms were financially secure.

25.1 Contract sum – adjustments to the contract

The project costs often begin with the contract sum, and the integrity of this amount is clarified within the contract conditions. There are, of course, projects where there is no agreed contract sum (see Chapter 8). It is extremely unusual for the contract sum to remain unchanged and be paid as the amount of the final account upon completion. Changes beyond the contractor's control will be made, and these must eventually be costed to the appropriate amount. Changes to the contract sum can occur in respect of some of the following items:

- substructure items, which often cannot be fully determined until the contractor starts work on-site and are usually labelled as a provisional amount
- remeasurement of provisional sums
- adjustments to the accounts of nominated subcontractors and suppliers
- variations to the design instigated by the employer or architect, and occasionally by the principal contractor
- daywork accounts
- contractual claims or *ex gratia* payments
- liquidated and ascertained damages.

The different forms of contract set out rules on how these changes are to be evaluated in terms of their costs. In those circumstances where contractors consider that they have been insufficiently reimbursed, there is provision in the contract to rectify this, should their case be proven. In other circumstances, the employer may have suffered damages due to a late completion by the contractor. Adjustments to the final account for this, whilst relatively uncommon, are also provided for within the terms of the contract. It is uncommon, because the contractor is usually able to identify mitigating circumstances for late completion of the works. On the majority of contracts of any size, contractors are unable to complete the works without some form of interim payment or payments on account. The different contracts have procedures for calculating the amounts owing to the principal contractor both during the period of construction and also upon final completion of the contract.

25.2 Variations, change orders and provisional sums

The vast majority of the forms of contract used in the UK and elsewhere around the world allow for variations or change orders arising at some stages during the project. The absence of such a clause within the contract conditions would necessitate a new contract being drafted if variations did arise. The disadvantage of such a clause is that it frequently allows the designer to delay making some decisions almost until the last possible moment. The assumption that a project can be fully designed prior to the contractor starting work on-site is false. During the process of construction, new ideas will emerge and changes will be made. However, these can have serious repercussions on the planning of the project and in executing the works efficiently on-site. The employer will be bound by any variation given by the architect or engineer as long as they do not exceed the powers given to them under the terms of the contract.

The question sometimes arises as to what constitutes a variation. The different forms of contract aptly describe those circumstances where a variation can arise, but it must be presumed and implied that a limit on these change orders will apply in practice. If the nature of the works described in the contract documents can be shown to be 'substantially different' from that carried out, then the main contractor may be entitled to re-price the work in a different way. The phrase 'substantially different', however, may require the courts to formulate a decision in individual cases. A variation resulting in a change in the size of a window will be admissible. The increase in the size of a proposed extension from 200 to 2,000 square metres of floor space will almost certainly not be described as a variation, since the nature of the works has changed considerably from what was originally envisaged. The contractor may be willing to carry out the work, but on a different contractual basis.

25.2.1 Definition of variation

The different forms of contract attempt to define a variation. In simplistic terms, a variation arises where the:

- actual work to be carried out changes
- circumstances in which the work is carried out are different from those originally envisaged.

The former is described as the alteration or modification of the design, quality or quantity of works and may include the:

- addition, omission or substitution of any work
- alteration of the kind or standard of materials or goods
- removal from site of work, materials or goods that were formerly in accordance with the contract, but which have now been changed.

A change in the circumstances in which the work is carried out might include the following:

- access and use of the site
- limitations of working space
- limitations of working hours
- changes made to the sequencing of work.

Some forms of contract place restrictions on the way in which variations can be issued. For example, work that has been defined in the contract bills as the principal contractor's (or one of the domestic subcontractors, should the intention have been to sublet it) cannot then be omitted and then awarded to another firm within the contract period. It can, of course, be omitted entirely and carried out after the contract has been completed. It can also be awarded to a firm employed directly by the building owner, assuming that the main contractor will allow access to the works for this purpose during the contract period. Of course, if the main contractor agrees to such changes, then this will override the condition.

25.2.2 Instructions requiring a variation

Most of the forms of contract envisage the issue instructions to cause a variation. The principal contractor has the right to the reasonable objection to these, as set out in the different contracts. The architect or engineer may sanction, in writing, any variation made by the contractor. This may be done for expediency. In practice, the principal contractor should not vary the works without first obtaining express permission. All instructions requiring a variation from the contract must be in writing. This is usually stated in one of the conditions of the contract. The quantity surveyor has no power to measure and value varied work without such a written instruction. Any variation made by either the architect, engineer or principal contractor can usually be confirmed in writing at any time prior to the issue of the final certificate.

25.2.3 Instructions on provisional sums

Instructions will be required from the architect or engineer on the expenditure of provisional sums in the contract and in nominated subcontracts. Where no such instruction is given, then the amounts included will be deducted from the final account for the project.

25.2.4 Valuation of variations and provisional sum work and work covered by an approximate quantity

These rules apply to each of the following circumstances:

- All variations required by an instruction of the architect or engineer or subsequently sanctioned in writing.
- All work which under the contract is to be treated as a variation.
- All work executed by the contractor in accordance with instructions regarding the expenditure of provisional sums.
- All work executed by the contractor for which an approximate quantity has been included in the contract.

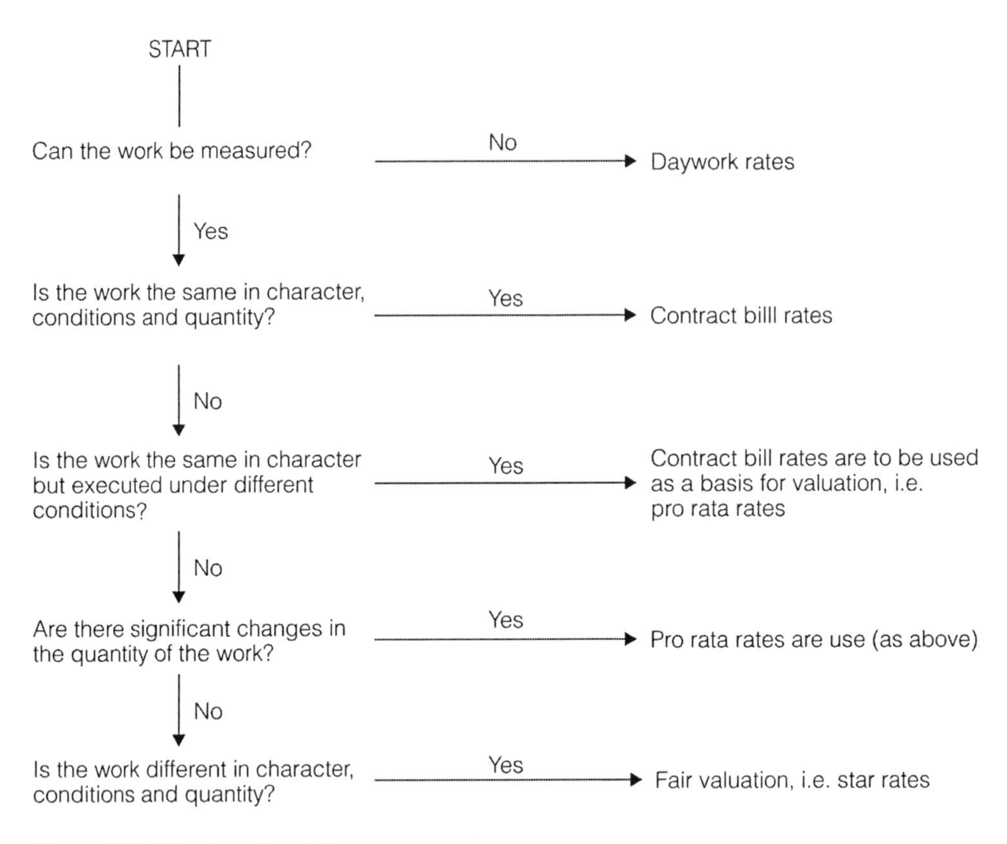

Figure 25.1 Valuation of variations.

Variations usually necessitate the remeasurement of the work on-site or from drawings. The principal contractor has the right to be present at the time that such measurements are taking place. It is important that such measurements are agreed by both of the parties prior to the work being priced.

25.2.5 General principles for pricing variations

The different forms of contract explain how such variations will be valued, normally against a set of principles that are set out in the different contracts, and these are illustrated in Figure 25.1:

- Where the additional or substituted work is the same in character, conditions and quantity to items in the contract bills, then bill rates or prices are to be used to value the variation.
- Where the additional or substituted work is the same in character, but is executed under different conditions or results in significant changes in quantity, then the bill rates or prices are to be used as a basis for valuing the variation. These are known as pro rata rates.
- If the additional or substituted work is different from those items in bills, then a fair method of valuation is to be used, often known as star rates.

- Where the approximate quantity is a reasonably accurate forecast of the quantity of work, then the rate or price quoted for the approximate quantity will be used.
- Where the approximate quantity is not a reasonably accurate forecast of the quantity of work, then the rate or price quoted for the approximate quantity will be used as a basis and a fair allowance for such differences will be added.
- Omissions are valued at bill rates unless the remaining quantities are substantially changed, in which case a revaluation of these items will become necessary.

Omitted work is at contract bill rates unless the remaining work changes substantially in quantity, in which case a revaluation of these items becomes necessary.

It is sensible to adopt the same rules and principles for measurement that have been used for the preparation of the contract bills. If any percentage or lump sum adjustments have been made in the contract bills, perhaps for the correction of errors, then they will need to be adjusted within the variation account. Also, if it can be shown that the value of any preliminary items has changed as a result of variations, then these also will require adjustment. Preliminary items are frequently priced as either lump sums, time-related or method-related charges. Their adjustment will generally follow this pattern of recalculation.

The same principles also apply to approximate quantities in contract bills. If the approximate quantities are not a reasonable forecast, then the rates used in the bills will only be used as a basis for revaluation. At the extremes, i.e. where the approximate quantity does not change or where it differs by, say, 100 per cent, then the above rules can easily be applied. In the grey areas between these extremes, the test of reasonableness will need to be applied by the parties concerned. This principle only applies where the work as executed is not altered or modified other than by quantity alone. Specific instructions in respect of the execution of work for which an approximate quantity is included in the contract bills do not need to be given.

25.2.6 Daywork

If the varied work cannot be properly measured or valued, then the work is to be paid for on a daywork basis. The documents to use might include the schedule *Definition of Prime Cost of Daywork Carried out Under a Building Contract* issued by the Royal Institution of Chartered Surveyors (RICS 2007), together with the percentage additions determined by the principal contractor in the contract documents. If the work is within the province of any specialist trade, defined as either electrical or heating and ventilating, then their appropriate definition for daywork, together with the percentages inserted by the contractor in the bills, should be used. Vouchers showing the daily time spent on the work, the workers' names and the plant and materials employed should be verified. This is recommended to be done weekly and then checked by, for example, the clerk or inspector of works. The reason for it being done on a weekly basis is to avoid forgetfulness by either party. It should be noted that the verified voucher is only an agreed record. It does not mean that the method of valuation adopted will be daywork. This is a matter for the quantity surveyor to determine.

25.2.7 Provisional sums

The valuation of a provisional sum follows a similar process to the valuing of the principal contractor's work above. For example, if a provisional sum had been included in the contract bills for 'repairs to timber staircase', and it was agreed that daywork should apply, then the above rules for valuing daywork would be used.

25.2.8 Performance-specified work

The valuation of performance-specified works follows the similar procedure of valuing construction work generally. The valuation may need to take into account the preparation and production of drawings, schedules or other documents.

25.2.9 Principal contractor's price statement

An alternative approach to the above is to request the contractor to provide an all-inclusive price for any variations that are authorised under the terms of the contract. This can be given before the work is authorised as a lump sum amount.

25.3 Variation instruction – contractor's quotation in compliance with the instruction

25.3.1 Contractor to submit quotation

The valuation of variations will normally follow the procedures set down in the form unless the architect's instruction states that the treatment and valuation of the variation should be dealt with differently.

25.3.2 Principal contractor's quotation

Some forms of contract allow the principal contractor (or subcontractor) to provide a written quotation for a variation in advance of the work being carried out. The quotation should be based upon:

- where relevant, the rates and prices contained in the contract bills
- where appropriate, the adjustment of preliminary items
- any adjustment in time required for completion of the works
- any amount paid in lieu of direct loss and expense
- a fair and reasonable amount in respect of preparing the quotation.

The instruction from the architect or engineer may also require the principal contractor to identify any additional resources that may be required and the method to be adopted for the carrying out of the work. The quotation should provide sufficient information to allow the work to be properly evaluated on behalf of the employer.

25.4 Contract sum

25.4.1 Quality and quantity of work included in contract sum

The contract sum is assumed to have been calculated on the basis of the quality and quantity of work as described in the contract bills. A part of the quantity surveying process on any project is to establish the correctness of the contract sum, both arithmetically and technically. This means that the bills should have been checked to see that the item costs have been extended correctly, that page totals add up and that the various collection sheets and summaries have been correctly completed. The adequacy of the principal contractor's rates should also have been examined in order to avoid difficulties that might arise during the agreement of the final account. However, failure on the part of the quantity surveyor to perform these duties efficiently will not provide the principal contractor with any redress against the building owner. It is the contractor's responsibility to determine the sufficiency of the tender, with no alternative but to honour the pricing should any errors occur at a future date during the contract. Any errors discovered prior to the signing of the contract can be corrected by using one of the alternative methods described in the *Code of Procedure for Selective Tendering* (see Chapter 8). It is prudent, therefore, on the part of the quantity surveyor to be satisfied on the nature of the prices prior to the signing of the contract.

The only errors that can be corrected are those made by the quantity surveyor during the preparation of the contract bills. These may have occurred because of errors in the quantities or descriptions or because items have been omitted. If the bills depart from the method of measurement stated, and this has not been brought to the principal contractor's notice, then this will also be deemed an error that will need to be corrected.

25.5 Materials and goods unfixed or off site

25.5.1 Unfixed materials and goods – on-site

Materials and goods which have been delivered to the site or placed on or adjacent to works, and are intended for the works, must not be removed from the site unless written consent is received from the architect or engineer. Also, once the value of these goods and materials has been included in an interim certificate and paid for by the employer, they automatically become the property of the employer. There has been considerable case law on this point. The principal contractor continues to remain responsible for any loss or damage they may suffer and usually for their insurance. This will depend upon the specific terms of the particular contract conditions under which the project is being constructed. Whilst the employer becomes the true owner of these items, if the principal contractor becomes insolvent, then this may raise questions of the reputed ownership of such materials and goods. The employer would need to ensure that the contractor's title to the goods was thus not defective.

25.5.2 Unfixed materials and goods – off site

The value of listed items of materials or goods intended for the works which are stored off site can also be included in an interim certificate and paid for by the employer. These listed items of materials and goods then become the property of the employer. The principal contractor must not remove these items from the premises, usually the place of manufacture, except to the project for which they are intended. The principal contractor continues to be responsible for any loss or damage that may occur, and usually for their insurance in accordance with the particular contract.

This clause has particular relevance and importance to factory-manufactured components, which have become much more common across many trades in recent years. Understandably, the principal contractor may be loath to bring to site items of manufactured joinery until their incorporation in the works is imminent, due to the possibility of damage or theft. The option of paying for such items will assist the principal contractor's cash flow, and should cause no problems if the provisions of the contract are fully complied with. These items will only refer to manufactured items that are specific to the project concerned that require only to be fixed or placed in position. The contract provisions will not usually consider items of a standard nature that can be used on any type of project.

25.6 Damages for non-completion

If the principal contractor fails to complete the works by the completion date, the architect or engineer will usually issue a certificate accordingly. Such a certificate will allow an employer to deduct from monies due to a principal contractor the liquidated and ascertained damages at the rate specified and agreed in the appendix to the contract. The amount of damages claimed is based upon the period between the completion date and the date of practical completion. If there is an insufficient amount owing to the contractor by which to offset the damages, the employer is able to recover the balance as a debt. The amount stated in the appendix for liquidated and ascertained damages must be a realistic sum related to possible actual damages suffered by the employer. If the amount stated is shown to be a penalty, then the courts will set a fair, but smaller amount as damages payable by the contractor.

25.7 Loss and expense claims

The main purpose of these clauses in the conditions of contract is to reimburse the principal contractor in those circumstances where loss and expense have been suffered, and which will not be reimbursed elsewhere under the terms of the contract. The loss and expense claim must usually be directly attributable to matters that have substantially affected the regular progress of the construction of the project (see Chapter 5).

In order for the principal contractor to claim for reimbursement, the matter will be raised, followed by a written application to the architect or engineer. The correspondence will state that such losses are likely to be incurred during the execution of the works, and that adequate reimbursement will not be made under any of the provisions of the contract. The basis of claim is that the regular progress of the works has been disturbed and this has resulted in loss and expense. Loss in this context means that the principal contractor has been inadequately reimbursed for work carried out. Expense implies that additional resources were necessary to complete the works. The contractor should:

- make the application as soon as it becomes apparent that the regular progress of the works has been or will be affected
- support the application with sufficient information to allow an opinion to be formed
- submit details of the financial loss or expense.

25.7.1 List of matters

The different conditions of contract list various matters that could affect the regular progress of works. The principal contractor must usually cite one or more of these items in support of a claim for loss and expense:

- Delay in the receipt of instructions, such as drawings, details or levels or the expenditure of provisional sums. The principal contractor usually must have requested this information in writing, although the different conditions of contract vary on this.
- Opening up of work for inspection or testing of materials and for their consequential making good, providing the work was in accordance with what was specified.
- Discrepancy or divergence between any of the contract documents.
- Work being carried out by firms employed directly by the employer, including the supply of materials and goods by the employer for the project.

- Postponement of any of the work to be executed under the provisions of the contract.
- Failure on the part of the employer to give ingress to or egress from the works by the appropriate time.
- Variations or the expenditure of provisional sums for performance-specified work.
- The execution of work for which an approximate quantity in the contract bills is not a reasonably accurate forecast of the quantity of work required.
- Impediment, prevention or default, whether by act or omission, by the employer or any person for whom the employer is responsible.

25.7.2 Relevance of certain extensions of completion date

Upon receipt of the correspondence from the principal contractor, an assessment must be made whether the claim is justified. It will be necessary first of all to assess whether the regular progress of the works have been disturbed by one of the events listed above. It will then be necessary to refer to the contractor's master programme for this purpose. The mere fact that the project is running behind schedule may be due in part to the contractor's own inability to perform the works adequately. Some reduction on the period of delay will therefore need to be taken if this assumption is correct. Secondly, the quantity surveyor will need to ascertain the amount of loss and expense actually incurred by the contractor.

The contractor's written application should be made as soon as it is apparent that the regular progress of the works has been affected. Relevant information should be submitted in support of the claim that will help form an opinion. The principal contractor should provide some indication of the amount of loss and expense that has been suffered. If any extension of time has already been agreed under the contract, then the principal contractor should be informed accordingly. This may affect the calculation and assessment of the contractor's claim.

25.7.3 Nominated subcontractors

Nominated subcontractors can also make a claim for the loss and expense caused by the disturbance of the regular progress of the subcontract works. A claim for this will be first submitted to the principal contractor. If this is supported, then the contractor will forward it to the architect or engineer for proper consideration. This claim will then follow a similar process, as described in section 25.7.2.

25.7.4 Reservation of rights and remedies of contractor

The inclusion of a loss and expense clause within a contract provides only one remedy for the principal contractor. Other actions that the principal contractor may wish to take must not be too remote in the eyes of the law if they are to be successful. The loss of profit may be a reasonable claim to be made under this heading.

25.8 Valuations, certificates and payments

The appendix to the form of contract, which is completed for each project, will recommend the period of time when interim payments are to be made to the principal contractor. Once this is received, the principal contractor will make payments to the various subcontractors involved with the project. The period of one month is common in practice, although the parties do have the option of stating any period. Some conditions of contract also stipulate that the valuation must achieve a minimum amount, that is stated in the contract, before any monthly payment will be made. The purpose of such a recommendation is to attempt to ensure that the contractor makes regular progress of the works. On very large contracts it is often necessary to make payments on a weekly basis on the basis of a very approximate valuation in order to aid the principal contractor's cash flow. These would be stated in the appendix to the specific contract. More accurate payments would continue to be made at the monthly interval. Such practices may be required at peak construction periods, for example during the summer months on mass earth-moving contracts. In some contracts, a stage payment system may be used as a preference, where the value of the contract sum is allocated work stages. This is a common method used on large housing projects. The certificate will state:

- the amount due to the principal contractor from the employer
- to what the amount relates
- the basis on which the amount was calculated
- the date for payment (typically fourteen days from the issue date of each interim certificate).

Within five days following the issue of an interim certificate (this period will vary with the form of contract being used), the employer must give written notice to the principal contractor specifying the amount of the payment, to what the payment relates and the basis on which the amount is calculated. Not later than five days before the final date for payment, the employer may give a written notice to the principal contractor specifying any monies that are to be withheld or deducted from the due amount and the grounds for deducting those monies. Where the employer does not provide any written notices to the principal contractor, then the amount stated in the interim certificate must be paid to the contractor.

25.8.1 Delays in payment

Where an employer fails to pay the approved amount within the stipulated period of time for honouring the certificate, then interest will normally be added to the amount outstanding. The interest is calculated on a simple interest basis above the Bank of England base rate. Contracts will state the percentage to be applied, which will typically be in the order of 5 per cent. The rate of interest is not construed as a waiver by the contractor in lieu of the correct payment at the correct time. Delaying the payment of interim certificates is a breach of the contract conditions and will give the principal contractor the right to suspend the performance of the contract or to terminate the employment.

25.8.2 Advance payment

In some cases, it may be appropriate to make an advance payment to the principal contractor. This may be necessary where the project start-up costs are excessive and would not normally be recouped quickly by the contractor. The terms will be stated in the appendix to the form of contract. The principal contractor will be required to provide an advance payment bond as a surety. This will be approved by the employer and on terms suggested by the British Bankers Association. The advantage to the employer is that such an arrangement will have the effect of reducing the employer's overall contract costs, rather than the principal contractor borrowing finance at a high rate from a third party.

25.8.3 Interim valuations

Interim valuations are usually prepared by the client's quantity surveyor in agreement with the contractor's surveyor. The valuations will be prepared on a monthly basis (or on whatever basis has been stated in the appendix) until the certificate of practical completion is issued. If the work has been accurately valued at this point, then only retention monies will be outstanding and these can therefore be released with the final certificate. There will therefore be no requirements to issue a certificate between these two certificates. The certificate of practical completion has therefore often been referred to as the penultimate certificate, although there is nothing in the contract to prohibit the issue of further certificates beyond this one. However, it is often not possible to value accurately the works at completion, and this therefore results in further certificates being issued.

Where the formula adjustment method is used to calculate either increased or decreased costs due to inflation, the effect of the timing of this on interim valuations needs to be carefully considered. Using the traditional method of price fluctuation, i.e. increased cost sheets, these are independent from the value of the measured works. The interim valuation of the contract works was also looked upon as a means to an end, and as long as it was reasonably correct overall, that was all that mattered. The principal contractor needed to be paid a reasonable sum for the work carried out, and the employer needed to be satisfied that certificates represented a reasonable amount. The introduction of the New Energy and Industrial Technology Development Organization (NEDO) formulae has meant that interim valuations need to be rather more precise than was previously the case, since the appropriate percentages used to calculate the increases (or decreases) use the valuation amounts in their computation. Although at interim valuation time the indices may only be provisional, the work valued must be actual, to closely resemble the work completed. Using the traditional method for calculating price increases, it was generally accepted that it was in the best interest of the principal contractor to overvalue the works, since this helped to improve cash flow. The adoption of the formula method will usually result in a higher overall payment should the contractor delay as long as possible the inclusion of work in a valuation.

In practice, it is often the contractor's surveyor who often prepares the valuation, since they are more frequently on-site and therefore more familiar with the progress of the works. The quantity surveyor will visit the site at the appropriate date to examine, approve and agree the amount of the valuation. Should there be any dispute between these parties on the amount to be authorised, the client's quantity surveyor's assessment will be adopted. The valuation is then forwarded for certification with the provision that the work included is in accordance with the contract. The quantity surveyor should not knowingly include in a valuation work that does not conform to the contract. Quality control, however, is not the prerogative of the quantity surveyor. Where there are any doubts over some aspect of work, then this should be brought to the attention of the architect or engineer, otherwise the valuation will form the basis of the certificate. The certificate is sent to the employer with a copy to the principal contractor, and this should be paid within the time (often fourteen days) stipulated in the contract. Nominated subcontractors may also be informed that the certificate has been issued with the amounts that they should receive.

25.8.4 Issue of interim certificates

Interim certificates are issued at the period specified in the appendix to the form of contract. This period is up to and including when the certificate of practical completion is issued. The final certificate is referred to as the certificate of completion of making good defects, and is issued either at the end of the defects liability period or once the defects have all been made good, whichever date is the later. The certificate of practical completion of the works includes the release of one moiety (one-half) of the retention monies, the remainder being released with the certificate of making good defects.

25.8.5 Suspension of operations by the principal contractor

Interim payments have often been described as the lifeblood of the principal contractor's business. The failure by an employer to honour the agreed certificate by making payments to the contractor at the appropriate time is therefore a potentially serious situation for the contractor. Most forms of contract recognise this by allowing the principal contractor to suspend the progress of the works. This is the ultimate sanction by the contractor. However, the principal contractor cannot suspend the works immediately a payment becomes due. They must first issue a written notice to the employer, with a copy to the architect or engineer, of their intention to suspend construction activities. The contractor must then allow time for the employer to respond by making the appropriate payment within a period of time, usually one week. This recognises the importance and seriousness of not honouring certificates when these have been properly prepared.

25.8.6 Calculation of the amounts due in interim certificates

The amount stated as due in an interim certificate is the gross valuation agreed between the parties and then issued in the interim certificate. The amount calculated is then paid to the principal contractor, but with deductions for retention monies (the percentage will vary depending on what has been stated in the appendix to the form of contract), payments already made in previous certificates and any advance payments. The gross valuation includes the following items:

- Work properly executed by the principal contractor (including subcontractors), including variations carried out and agreed in terms of their financial consequences.
- Adjustments to these values arising as a result of the application of the price adjustment formulae or other methods that might be used to calculate increased costs.
- Materials, goods and components delivered to the site. This, in practice, often only includes those items of a major financial importance. The materials on-site will only include those items that are reasonably, properly and not prematurely delivered to site. The inclusion, in the valuation, of sanitary wares that are already on-site at the start of the contract may be refused because they have been prematurely brought to the site. The employer, when paying for materials on-site, will need to be satisfied in every respect regarding their safety from damage or theft. Materials on-site for long periods are more susceptible to these occurrences. The provision for payment may also require adequate protection from the weather as a prerequisite condition. This clause also applies in respect of materials from suppliers.
- Materials, goods or components off site as long as they conform to the requirements of the contract.
- Nominated subcontractor's work including materials on or off site as appropriate and agreed. These amounts are only normally included where a subcontractor supplies an invoice. It may also be necessary to show that previous payments have been made to

any nominated subcontractors. In some contracts there is a provision that a final payment to nominated subcontractors can be made.

- The profit of the contractor in respect of nominated subcontractors' or suppliers' invoices. Attendance items will be added where appropriate.
- Loss and expense payments, although in practice these are often not agreed until the project is completed.
- Any increased payments in respect of contributions, levies and tax fluctuations.

25.8.7 Off site materials or goods

The amount stated as due in an interim certificate can include the value of any listed items of materials, goods or components before delivery or adjacent to the works, provided that the following conditions have been fulfilled:

- The contractor has provided reasonable proof that the property is uniquely identified and vested in the contractor. Immediately upon payment by the employer, these become the property of the employer.
- If the appendix to the form of contract requires a surety bond in favour of the employer, then this must also be provided by the contractor prior to payment.
- The listed items must be in accordance with the contract.
- The listed items at premises where they have been manufactured, assembled or stored must be set apart or clearly marked to identify the employer and their destination as the works.
- The contractor must also provide the employer with reasonable proof that the items are insured against loss or damage for their full value under a policy of insurance protecting the interests of the employer and contractor in respect of specified perils. This insurance must cover the listed items from the time of transfer of property ownership until they are delivered to the works, when they will then be covered by the insurance of the works.

In practice it is usual to interpret these conditions to mean materials or goods that are special to this contract. Common bricks or quantities of standard-sized timber which could be used on any project would not normally be considered as allowable items to be included within a certificate, although there may always be exceptions. Nothing should remain to be done to these items prior to their incorporation within the works. The typical sorts of items covered by this clause envisage some form of prefabrication for the project concerned, although other materials and goods can also be included. The listed items may therefore be at the contractor's workshops or the workshops of a subcontractor.

25.8.8 Retention

The contractor is not paid the full amount of the certificate, and this is subject to a deduction in the form of retention. Retention is applied to work that has not reached practical completion and to materials and goods referred to above, including nominated works. The purpose of this retention is to provide some incentive for the contractor to complete the works, and it also provides some security should the contractor default in construction.

The amount retained by the employer under this clause will vary depending upon the form of contract that is used. The typical figure is usually between 5 and 10 per cent, although a lower percentage can be agreed between the parties. Higher rates of retention than these are not normally used. On large projects, the percentage may be limited to 3 per cent. Some contract conditions indicate a limit to the retention fund, so that, once this is reached, all of the remaining work certified will be paid in full. The retention is released to the contractor as follows: one-half of that which is retained by the employer at the certificate of practical completion of the works, and the remainder with the issue of the certificate of making good defects. The percentage rate of retention is indicated in the appendix. Some contracts allow for the principal contractor to provide a bond in lieu of retention.

25.8.9 Employer's interest in retention

The employer has only a fiduciary interest in the retention. In this respect, the retention is held on trust only and need not therefore be invested to accrue interest. The principal contractor or a nominated subcontractor can, under some circumstances, require the employer to put the retention monies into a separate bank account. Any interest accruing from this account will be only for the employer's benefit.

Where the principal contractor defaults in payment to a subcontractor, some contract conditions provide for the employer to pay a nominated subcontractor direct. The principal contractor must be informed of this action and its subsequent effect upon the retention. In this case, the principal contractor will forfeit the cash discount to that subcontractor.

25.8.10 Final account

Generally, contracts require that either during the contract period or within six months after practical completion of the works, the contractor shall provide all the documents necessary for the preparation of the final account. These documents should include the appropriate details from nominated subcontractors and nominated suppliers. The following items are to be deducted from the contract sum:

- Prime cost sums, which include the contractor's profit and attendance items.
- Provisional sums and provisional work (approximate quantities) in the contract bills.
- Amounts omitted by reason of variations caused by architect's instructions.
- Any other amount that is required by this contract to be deducted from the contract sum, for example, liquidated damages.

The following items are to be added:

- Nominated subcontract sums as finally adjusted or ascertained under the relevant provisions of subcontract.
- Prime cost sum items that have been undertaken by the principal contractor.
- Nominated suppliers' accounts adjusted within the terms of the contract.
- Profit and attendance items on nominated subcontractors' and nominated suppliers' items.
- Variations authorised or approved by the architect or engineer.
- Adjustments for provisional sums and approximate quantities in the contract bills.
- Loss and expense claims.
- Increases in contributions, levies and taxes.
- Any other amount that is required to be added to the contract, such as the agreement of the contractor's claims.

25.8.11 Issue of final certificate

The final certificate should be issued within two months of the last of the following events:

- The end of defects liability period.
- The date of issue of certificate of making good defects.
- The agreement of the final account.

The final payment must be made to the contractor within twenty-eight days of the issue of the final certificate. If the employer fails to make this payment, simple interest is added at an agreed rate above the base rate of the Bank of England. The final certificate must state:

- the amounts already paid to the contractor under interim certificates
- the contract sum adjusted within the terms of the contract
- to what the amount relates and the basis on which the statement in the final certificate has been calculated
- the difference expressed as a debt to either the employer or the contractor.

25.8.12 Effect of the final certificate

The final certificate is often considered to be the final and conclusive evidence that:

- the quality of materials and standards of work are as described on the contract drawings and in the contract bills and to the reasonable satisfaction of the architect
- the contract sum has been properly adjusted within the terms of the contract
- relevant extensions of time have been given
- reimbursement of direct loss and expense is in final settlement of any contractor claims.

However, if there has been any accidental inclusion or exclusion of work, or arithmetical errors in computation, these are able to be corrected. The same principle applies in the case of fraud.

If any adjudication, arbitration or other proceedings have been commenced by either party before the issue of the final certificate, then the final certificate shall have effect as conclusive evidence if provided after the earlier of these two outcomes:

- Such proceedings have been concluded and the final certificate is subject to the award or judgment that has been made.
- After a period of twelve months in which neither party has taken further steps to solve the problem.

If any adjudication, arbitration or other proceedings are commenced within twenty-eight days after the final certificate has been issued, the final certificate shall still remain conclusive evidence other than in respect of the matters relating to the proceedings.

26 Time factor of construction

There are certain clauses in the forms of contract that describe the start and finish dates of the project and the contractual implications associated with these dates. The difference between the start and finish dates is the contract period. The distinction between any completion date and date for completion should be carefully noted. Should a principal contractor fail to complete the works by the completion date, then it is usual for a certificate of non-completion of the works, or something similar, to be issued. Liquidated damages may then be payable by the principal contractor to the employer for not finishing the project on time. Upon the practical completion of the project, the principal contractor is relieved of most of the contractual obligations under the particular contract that was signed. The principal contractor will still remain responsible for defects that may arise, due to poor standards of work that need to be rectified. This responsibility continues until the end of the defects liability period. All forms of contract will specify this period of time, which is typically at least six months after practical completion. Debate continues on whether this responsibility should extend further for some aspects of work, up to ten years (see Chapter 11). Normal guarantees for other products have been greatly extended in recent years, for example, with motor vehicles. In some circumstances, the principal contractor may be able to secure an extension of the contract period for one of the reasons listed in the contract. The advantage of this course of action to the principal contractor is either the reduction or elimination of liquidated damages. The date for completion is stated in the appendix and any approved extension of time that is added to this then provides the later completion date. The completion date cannot occur prior to the date of completion stated in the appendix, unless the contractor agrees to such a revision. This does not mean that a project cannot be completed earlier. In some extreme circumstances, the works may be postponed, perhaps indefinitely. All contracts must consider this eventuality. The works may also offer partial possession in the case of a phased project awarded as a single contract. In this case, phased completion is treated almost as a separate project as far as time is concerned. Finally, circumstances can, and sometimes do, arise where either the employer or the principal contractor feels that the contract should be terminated. This usually results because of a breach by the other party. Although it is an unusual occurrence, the contract must allow for such a possibility.

26.1 Practical completion

When the works become practically complete, in the opinion of the architect or engineer, a certificate of practical completion of the works must be issued. In practice, this certificate will be issued even though minor items of work still need to be carried out within the terms of the contract. This does not presume that these items are either irrelevant or need not be completed.

The date of practical completion often coincides with the handover date to the employer, and this latter consideration may have resulted in this certificate being issued even though the works are only 'almost' complete. Prior to the issue of this certificate, the architect or engineer needs to be satisfied that:

- the work has been carried out in accordance with the contract documents and the relevant instructions
- the construction project is in an appropriate state to be taken over and used by the employer.

Usually, when the employer wants to occupy the building urgently, the principal contractor will request that such a certificate should be issued.

The date of practical completion is important, since the following take effect automatically:

- The start of the defects liability period.
- The beginning of the period of final measurement.
- The release of the first moiety (half) of the retention fund.
- The ending of the insurance of the works.
- The end of any liability to liquidated damages by the employer.
- The opening of matters referred to arbitration.
- The removal of the liability to frost damage.

Many of these items will also automatically occur to those parts of a project affected by partial or sectional completion where this is a part of the contract conditions for a project. Upon achieving practical completion, the principal contractor is no longer obliged to accept any further instructions in respect of additional work, since the works are now complete. The principal contractor may choose to do this work on the basis of a fresh agreement, and where good relationships have been maintained this is likely to take place. If such additional work is of a minor nature, then a new informal agreement will be drawn up. The existing rates from the original contract for this work will no longer apply.

26.1.1 Defects liability

The principal contractor is responsible for the making good of defects, shrinkages or other faults that appear within the defects liability period. These defects must be due to materials and work standards that are not in accordance with the contract, or from frost damage occurring before the practical completion of the works. When defects occur because of the inadequacy of the design, the principal contractor is not obliged to make these good at their own expense. These situations are often a cause of disagreement, especially when it cannot be easily decided which party is really responsible. Variations cannot be issued of the certificate of practical completion in order to remedy design faults. Items which 'appear' during the defects liability include those that were presumably unnoticed during the contract period. These items must be included on a schedule of defects not later than fourteen days after the expiration of the defects liability period. The length of the defects liability period is inserted in the appendix to the conditions of contract. Six months is usually the minimum period, other than for small works projects. The making good of the principal contractor's defects should be done within a reasonable time and entirely at the principal contractor's own expense.

26.1.2 Defects, etc.

Instructions can be issued at any time for the making good of any defect, shrinkage or other fault within the defects liability period. Such defects must be due to the use of faulty materials that were not in accordance with the conditions of contract or workmanship that was substandard. It will also usually cover defects arising because of frost action occurring before the practical completion of the works. The principal contractor must comply with such instructions within a reasonable period of time and at no expense to the employer. Faults not made good by the principal contractor will result in a deduction from the contract sum. In practice, a schedule of defects will be issued not later than about fourteen days from the expiration of the defects liability period.

26.1.3 Certificate of completion of making good defects

When the defects requested have been made good, and the defects liability period has expired, a certificate of making good defects will be issued. The issue of this certificate releases the remainder of the retention monies due to the principal contractor, and removes the defects obstacle that would prevent the issue of the final certificate. However, principal contractors are not totally released from the responsibility of defective work, since they will still be affected by the statute of limitations in common law.

26.2 Partial possession by employer

The employer, with the agreement of the principal contractor, may decide to take possession of a section of the works. This is particularly appropriate where the contract concerned may be in two distinct parts; for example, alteration work and a new extension to an existing building, two extensions to an existing building, a housing scheme where numbers of units are released at different time intervals, or a road and bridgeworks project where the bridges might be fully completed ahead of the road works. A written statement identifying the parts of the works taken into possession by the employer and the date when this occurred will be issued to the principal contractor.

This certificate has the same effect on the part completed as the certificate of practical completion has on the whole of the works. The items listed under 'Practical completion' apply equally to this part of the works that are now complete, and have received the aforementioned certificate. Partial possession by the employer clearly has contractual advantages to the principal contractor, and also earlier use or occupation by the employer.

26.3 Date for possession, completion and postponement

26.3.1 Date of possession – progress to completion date

The date for possession of the site by the principal contractor is stated in the appendix to the conditions of contract. Failure on the part of the employer to provide the site by this date will result in a breach of contract. However, there are provisions in most forms of contract, without the employer being in breach of contract, to defer giving the contractor possession of the site for a limited period of time, typically up to six weeks. The deferment of the date of possession is referred to in the appendix. The principal contractor must start the project and then proceed 'regularly and diligently' with the project up to the date of completion or the completion date if this has been extended. Failure on the part of the principal contractor to make regular progress with the works will also result in a breach of contract. The employer is usually able to determine the principal contractor's employment on the works for this reason.

The completion date is the date at the start of the defects liability period. It is also the date, should the principal contractor fail to complete by this time, of the starting point of the provisions relating to liquidated damages. The date for completion is this date amended to take into account any extra time allowed under an extension of time.

26.3.2 Postponement

The project, in exceptional circumstances, can be postponed for a short period of time or even indefinitely. Where postponement exceeds the period of delay stated in the appendix, the principal contractor can determine the contract. In circumstances where the postponement is a relatively short period of time, the usual redress for the contractor is to claim an extension of time and loss and expense incurred.

26.3.3 Possession by contractor – use or occupation by employer

The principal contractor retains possession of the site and the works up to and including the date of the issue of the certificate of practical completion. The employer, with the consent of the principal contractor in writing, may use or occupy the site or the works for the purpose of storage of goods or otherwise before the date of practical completion. The provisions of the insurers should be obtained before this consent is given. If any additional insurance premium is required, then this will be borne by the employer.

26.4 Extension of time

Any reference to delay, notice or extension of time will include further delays, further notices or further extensions of time.

Once the principal contractor realises that the progress of the works is likely to be delayed, then the architect or engineer should be informed in writing, giving the reasons for the cause of the delay. Where the delay includes reference to a particular nominated subcontractor, this firm should also be given details of the delay. The effect of this is to reduce or eliminate any damages that the principal contractor may suffer due to non-completion of the works by the agreed date. The initiative to be taken in seeking either an extension of time or a further extension of time must come from the principal contractor.

The contractor, in giving notice, should, where possible, provide particulars of the expected effects of the delay and also estimate the extent in respect of completion of the works. Where these cannot be assessed, this should not delay the notification of this to the architect or engineer, and they should still be informed of a probable delay and be provided with the detailed information at a later date.

26.4.1 Fixing completion date

Upon receipt of the notice of delay, the architect or engineer will decide whether any of the events listed are relevant. If, in their opinion, these are not considered to be a reasonable basis for an extension of time, the extension to the contract should be refused.

If the reasons listed by the principal contractor are accepted as being a relevant event, then the architect or engineer should write to the contractor giving details of:

- The events that are relevant in these circumstances for an extension of time.
- A new completion date that has been estimated to be fair and reasonable. (If variations or omissions have been issued since the fixing of a previous completion date, then they can be taken into account in the fixing of a new date.)

The principal contractor must always, using the best endeavours available, try to prevent any delay in the progress of the works. The principal contractor must also proceed with the regular progress of the works.

26.4.2 Relevant events

The following represent relevant events that can result in an extension of time being given. These events will vary depending upon the form of contract being used.

Force majeure

The meaning of *force majeure* is imprecise, but it is generally accepted as 'exceptional circumstances beyond the control of either of the parties of the contract'.

Exceptionally adverse weather conditions

Exceptionally adverse weather conditions are weather conditions that could not normally be expected at the time of the year in the location of the project. Heavy rainfall during the winter months would not fit within the definition of 'exceptional' unless it was abnormal. In the winter of 1978/79 the total rainfall was 579 mm compared to the average for that period over the preceding years of 452 mm (Manchester Weather Centre). This may have been considered exceptional, but such a decision will rest with the courts of law if the parties cannot finally agree upon an interpretation. This application will also be influenced by the type of project and its completeness. A road project involving heavy earth-moving will result in a different outcome from an office building nearing completion when all of the work to be carried out was inside.

Loss or damage occasioned by any one or more of the specified perils

The perils referred to in this clause are generally those items that are an insurable risk, such as fire, lightning, explosion, storm, tempest, etc. The different forms of contract compile their own lists, and these will be defined in the respective conditions.

Civil commotion, etc.

Civil commotion includes both local and national strikes and lockouts by management of any of the trades employed upon the works. It is also extended to similar actions in the preparation, manufacture or transportation of the goods and materials.

Compliance with instructions

These can include the following:

- discrepancies in or divergences between documents
- an instruction requiring a variation
- the expenditure of provisional sums
- a provisional sum for performance-specified work
- the postponement of any work
- the finding of antiquities on the site, since this may delay construction activities
- nominated subcontractors' work
- nominated suppliers.

Opening up of work for inspection

If the principal contractor is required to open up work for inspection or testing, and if the work complies with the contract, then this becomes an admissible event. If the work is not in accordance with the contract, any delays ensuing will not be grounds for an extension of time.

Failure to comply with an information release schedule where this has been provided

This may include delays in the receipt of instructions, drawings, details or levels from the architect or engineer. If this is to be accepted as a relevant event, the contractor must have already requested the information beforehand. The principal contractor must also use some foresight and cooperation in respect of when the information might be required. Reasonable notice should be given so that the information can be prepared and provided by the requisite time. In practice, the principal contractor, by reference to the programme of works, should be able to request such information well in advance of when it is needed. It must not be assumed that the architect or engineer will only work at the *direction* of the principal contractor, but that they will be diligently preparing the information that will be required anyway. In the event of this assumption being invalid, the principal contractor could be involved in preparing a checklist, and then be penalised for doing this inadequately.

Delay on the part of works or persons engaged by the employer

If the employer decides to undertake some of the work direct, outside the scope of this contract, then such work should be carried out in order not to cause a delay to the main contract. The principal contractor in these circumstances will have only minimal control. It is up to the employer, therefore, to take the necessary steps to ensure as little inconvenience as may be expected. This also covers the supply, by the employer, of materials and goods that the employer has agreed to provide for the works.

Government intervention

If, once the contract has been signed, the government restricts the availability or use of labour essential to the carrying out of the works, then the principal contractor may apply for an extension of time. In the case of war, for example, the government may call up all 'able-bodied' men to the armed services, thus removing tradespeople from the labour market. In other circumstances, government may commandeer the essential building materials, or it may restrict the use of fuel or energy resulting in only limited working, as occurred in 1973 with the three-day week.

Two points are worth noting. Firstly, the government intervention must restrict the essential supplies of labour, materials, fuel or energy. If the work can be easily performed in some other way, without detriment to the project, then no claim for an extension of time will be permitted. Secondly, the government intervention must occur after the signing of the contract. If the principal contractor knew or should have known before entering into the contract that the government would introduce certain restrictive measures, this cannot be claimed as a basis of defence.

Contractor's inability to secure labour and material

The foregoing comments on government intervention may apply equally to this relevant event. For example, a sudden unexpected upturn of the construction industry in a certain area may starve some sites of the essential supplies of labour and materials.

Work by local authorities or statutory undertakings

Although these may be part of the contract, the principal contractor may have severe difficulties in attempting to control them. In some circumstances, there may be delays awaiting statutory inspections by local authorities. The principal contractor must, however, have taken all reasonable steps to avoid difficulties occurring in the first place.

Failure on the part of the employer to give access

If the employer is unable to give to the principal contractor, by the dates stipulated in the contract, proper ingress to or egress from the works as defined in the contract documents, then the principal contractor may be entitled to an extension of the contract period.

Inaccurate forecast of an approximate quantity

Where such work is executed by the contractor, then it must be a reasonably accurate forecast.

Delays which the principal contractor has taken all practical steps to avoid

This may be due to changes in statutory requirements that necessitate some alteration or modification to performance-specified work.

The use or threat of terrorism

The authorities may take action to deal with terrorist threats.

Delays arising from a suspension by the contractor

This might be as a result of the employer failing to honour certificates.

The above matters relate only to an extension of the contract period. The events listed may form a good reason, but it must not be assumed to be an automatic result following any one of the above events. In any case, the agreement for an extension of time could sometimes result in only a few days. Another important consideration will be in respect of any financial adjustment to the contract sum. This might be in regard to damages to the employer for non-completion or the loss and expense to the principal contractor resulting from such delays.

An extension of the contract period, for any reason, may be unacceptable to the employer. A date for opening the project may have been fixed and can involve important dignitaries attending or considerable inconvenience if the project is not complete. The architect or engineer will therefore reasonably expect the principal contractor to take appropriate action to make up any lost time. It is reasonable to expect that the principal contractor will be paid for the costs involved in such action. There is, however, no obligation under the contract for the contractor to take such action.

26.5 Termination by employer

In certain circumstances, the employer is rightfully allowed to terminate the employment of the principal contractor. Where a default occurs, then the architect or engineer may issue a notice in writing to the principal contractor. This notice must be delivered by actual, special or recorded delivery specifying the default. The reason why the notice must be delivered in this way is to avoid any claim by the principal contractor that it was never received. The notice will be deemed to have been received by the contractor forty-eight hours after the date of posting.

26.5.1 Default by principal contractor

The following defaults must occur before the date of practical completion:

- Unreasonable suspension in carrying out of the works, either wholly or substantially.
- Failure to carry out the works in a regular and diligent manner.
- Failure to comply with an instruction to remedy defective work, resulting in further damage to the works. The failure to comply with other instructions merely results in instructions being issued to another firm and recouping the necessary amount from the principal contractor.
- Assignment or subletting a portion of the works without permission.
- Failure to comply with the requirements of the CDM Regulations.

In the case of one of the above defaults, the principal contractor should be required to remedy the contract, perhaps informally at first. If the principal contractor continues to ignore the warning, then the notice of termination will be issued. The principal contractor, upon receipt of such a notice, usually has a period of time (possibly fourteen days) in which to rectify the problem. If, after this date, the principal contractor is still in default, the employer may then issue a notice that effectively terminates the employment. This must be done reasonably and should not be done solely to annoy the contractor. Most of the contracts would describe this as vexatious or unreasonable behaviour on the part of the employer.

26.5.2 Insolvency of the contractor

If a principal contractor becomes insolvent, or arranges with creditors for voluntary winding up (Companies Act 2006 and Insolvency Act 1986), then the employment of the principal contractor will be automatically terminated (see Table 26.1). Most contracts describe this as:

- making a proposal for a voluntary arrangement for a composition of debts
- having a provisional liquidator appointed
- having a winding-up order made
- passing a resolution for voluntary winding up
- under the Insolvency Act 1986, having an administrative receiver appointed.

However, there is provision in most contracts for the employment to be reinstated where both the employer and the principal contractor, or liquidator, agree. In practice, there are immense disadvantages to the employer where bankruptcy of the principal contractor occurs. It is therefore in the best interests to retain the services of the principal contractor where this has been previously satisfactory. The liquidator is also in a position of wishing to continue with those contracts that are likely to be profitable and quick to execute. Contracts that are nearing completion are therefore likely to be retained, since access to retention monies might be soon available. Those contracts in their earlier stages of construction are most likely to be dispensed with.

26.5.3 Corruption

The employer may terminate a principal contractor's employment in cases of corruption. In some circumstances this can be extended to all contracts between the employer and the

Table 26.1 Example calculation of contractor's insolvency

Position as it should have been		£
Contract sum		500,000
Value of variations		25,000
Theoretical final account		525,000
Position at termination		
Value of work executed at time of liquidation		340,000
Interim payments	300,000	
Less retention	–15,000	
	285,000	
Nominated subcontractors paid after termination	30,000	
	315,000	315,000
Amount outstanding at termination		25,000
Position at final account		
Paid as above, at termination		315,000
New completion contract		250,000
Employer's loss and expense		25,000
		590,000
Less as it should have been		–525,000
Cost of termination/bankruptcy		£65,000

principal contractor where an irregularity of this nature has occurred on only one contract. The principal contractor may, for example, be already undertaking several different contracts on behalf of the employer. A bribe might have been used to secure further work. Where this is proven, then all the contractor's work may be terminated. Corruption comes in different guises. It may result from the method of obtaining work or from the execution of the contract. Corruption can also occur where one of the contractor's employees offers favours that are neither known nor sanctioned by the principal contractor. The offences would usually come within the remit of the Prevention of Corruption Acts 1889 to 1916. Although corruption is generally well understood, it is often difficult to draw a line between the actions that may be classified as such. For example, many would regard the offer of a bottle of spirits at Christmas time as quite acceptable, and not as a bribe or offered in return for a favour. A crate of spirits would probably prove to be a different matter!

26.5.4 Insolvency of the principal contractor

The employer's first action is to give notice to terminate the employment of the principal contractor. The employer may choose to make an agreement with the principal contractor on the continuation or novation of the contract or the contract may be simply terminated. The employer and the principal contractor may make interim arrangements for work to be carried out. The employer may take reasonable measures to ensure that site materials, the site and the works are adequately protected, and also goods and materials that are adjacent to the site are not removed. The employer may calculate and deduct the reasonable costs of taking such action.

26.5.5 Consequence of termination

The respective rights of the employer and the principal contractor, upon the termination of the principal contractor's employment, are as follows:

- The employer may engage another firm to complete the project. This firm is to be allowed to use all the temporary buildings, plant and equipment until completion, where they are owned by the principal contractor. Where they are not owned by the principal contractor, the consent of the owner must be obtained by the employer.
- Unless the termination occurs because of bankruptcy, the contractor must, without payment, assign to the employer any benefits for the supply of goods and materials or for the execution of subcontracts. In practice, the principal contractor could encourage bad relations to develop in order to deprive the employer of these benefits.
- Where insolvency has not occurred, the employer may pay any supplier for materials or goods, or a subcontractor for work executed, who have not previously received payment from the principal contractor. In practice, those who should have been paid, but have not been because of a default on the part of the principal contractor, may be unwilling to carry on their work unless they receive a payment. The employer may therefore be forced into this situation, in order to see progress in the works, and then try to recover the amounts from the principal contractor.
- The principal contractor must remove temporary works and other items within a reasonable time after receiving an instruction. Where the contractor fails to do this, the employer can remove and sell the items, and after deducting costs, pass on the proceeds for the benefit of the principal contractor.

- If the employer suffers any loss or expense because of this termination, then this will become a debt of the principal contractor.
- The employer is not bound to make any further payments to the principal contractor until the project has been properly completed and the final accounts agreed for the project as a whole.

26.5.6 Employer decides not to complete the works

If, after the termination of the contractor, the employer chooses not to proceed with the works, then the principal contractor must be notified, usually within six months of the date of termination. The total value of work that has been properly executed is calculated, less any reasonable costs and expenses borne by the employer. The difference is then expressed as a debt or credit to the principal contractor. These provisions are without any prejudice to other rights and remedies which the employer may possess.

26.6 Termination by contractor

A number of situations can arise that can give the principal contractor the legal right to terminate the contract. If the contractor decides this is an appropriate course of action, then a notice in writing must be sent to the employer specifying the default that has caused this to take place. This notice must be delivered by actual, special or recorded delivery specifying the default. The reason why the notice must be delivered in this manner is to avoid any claim by the employer that it was not received. The notice will be deemed to have been received by the employer usually within forty-eight hours after the date of posting. This does not imply that a principal contractor should take into account the peculiar state of the employer in respect of, for example, the employer's temporary shortage of funds or finance due to a cash flow crisis.

In a majority of cases, this course of action suits no one. It is therefore introduced as a last resort, and there will already have been an attempt to resolve the dispute that caused the termination in other ways. Also, whilst a principal contractor may be completely innocent, damage might nevertheless be suffered in terms of reputation or financial loss. Termination usually results from a series of actions rather than a single misdemeanour on the part of the employer. This might, for example, result from a repetition of one of the events listed below.

26.6.1 Default by employer – suspension of uncompleted works

Financial reasons

Perhaps the employer does not discharge in accordance with the contract the amount properly due in respect of a certificate or VAT in pursuant of the VAT agreement. Or perhaps the employer interferes or obstructs the issue of any certificate. If the employer fails to pay a certificate, usually within fourteen days, and continues this default, typically for a further seven days, then the principal contractor can issue a notice of intention of termination. The first time there is a delay in payment, the principal contractor is more likely to approach the architect or engineer regarding the absence of the payment. The certification of monies due is not a matter of debate by the employer. If an employer has misgivings about the performance of the contractor, this should be brought to the attention of the architect or engineer. Also, the employer cannot deduct sums of money from a certificate that has already been authorised.

Delays in honouring a payment is the most usual circumstance that results in termination by the principal contractor.

Assignment and subcontracts

If the employer attempts to assign a part of the contract to another, without the principal contractor's consent, then this can give rise to termination by the contractor.

CDM Regulations

If the employer does not comply with the requirements of the CDM Regulations, this can give rise to termination by the contractor.

Suspension of the works

The project has been suspended by the architect, engineer, employer or by someone else such as a local authority. The period for such a delay is normally entered in the appendix to the form of contract, and is typically one month. Works may be suspended for a variety of reasons:

- Instructions, drawings, details or levels were not received in due time after the principal contractor made a specific request in writing. This would normally be done through the information release schedule. Such requests must allow sufficient time to provide this information.
- There was a failure to provide further drawings and details.
- There was a problem relating to compliance with instructions issued in respect of discrepancies between documents, variations or postponement, unless caused by the principal contractor.
- There was delay in the execution of work caused by persons employed directly by the employer.
- The employer failed to give ingress to or egress from the site of the works through or over any land or buildings in due time.

The employer should be given sufficient time (typically fourteen days) to rectify any of these defaults. The principal contractor must then, within a ten-day period, issue a further and final notice of terminating employment under the contract. These notices must not be given vexatiously or unreasonably by the principal contractor.

Insolvency of the employer

The employer can make a composition or arrangement with creditors, become bankrupt or cease being a company in the following ways:

- Make a proposal for a voluntary arrangement for a composition of debts.
- Have a provisional liquidator appointed.
- Have a winding-up order made.
- Pass a resolution for voluntary winding up.
- Have an administrative receiver appointed under the Insolvency Act 1986.

The employer must inform the principal contractor in writing when any of the above situations occur. This is in accordance with the Companies Act 2006 or the Insolvency Act 1986. Termination will take effect upon the date of receipt of such notice.

26.6.2 Consequences of termination

The effects of termination by the principal contractor result in the following financial arrangements, with the relevant sums being paid by the employer:

- The reasonable costs of the removal of any temporary buildings, plant, tools, equipment and goods, including subcontractors. This must be done within a reasonable period of time, taking precautions to prevent possible injury, death or damage.
- The payment of any retention monies within twenty-eight days.

Within a reasonable time, the principal contractor should prepare an account setting out the following:

- The value of work already completed and the value of work under construction, values that have been properly authorised.
- Any sum resulting from loss and expenses.
- The reasonable cost of removal in pursuing this course of action.
- The direct loss and expense resulting directly from this termination.
- The costs of other materials and goods already ordered and provided for the works.

The respective procedure, therefore, following termination by the principal contractor is as follows. The principal contractor will clear the site, and the employer will pay for the work completed to date, including materials and goods for the works and retention amounts. In addition, payments for loss and expense will be made to cover this premature action by the employer. The above provisions are without any prejudice to other rights and remedies that the employer may possess.

26.7 Termination by employer or principal contractor

These are in effect neutral events caused by neither party. They should thus be given as grounds for termination by either party. The events listed relate to the suspension of the works:

- *force majeure*
- loss or damage to the works occasioned by any one or more of the specified perils
- civil commotion
- instructions in respect of:
    - discrepancies or divergencies between documents
    - variations
    - postponements.
- hostilities in the UK, whether or not war has been declared
- terrorist activity.

Either the employer or principal contractor may give notice in writing to the other party by registered post or recorded delivery in the way described previously. In this situation, presumably neither party wants the contract to end, but this has occurred as a result of events beyond the control of either party.

26.7.1 Consequences of determination

The consequences of determination are as follows. The removal from the site of the principal contractor's temporary buildings, plant, tools, equipment, materials and goods and site materials. This must be done within a reasonable length of time and with precautions to prevent injury, death or damage. This applies also to subcontractors working on the site. The employer will release half of the retention monies within twenty-eight days of the notice.

Within a further two months, the principal contractor must provide the necessary documents to prepare the final account for the project. This is to be calculated as follows:

- Total value of works at determination.
- Sum ascertained in respect of direct loss and expense under:
    - loss and expense
    - antiquities.
- Reasonable cost of removal:
    - cost of all materials and goods ordered for the works, which then become the property of the employer.
- Any direct loss or damage caused by the termination.

This amount, together with any outstanding retention sums, should be paid to the principal contractor within twenty-eight days. This could include loss resulting from one of the specified perils that might have been caused by the negligence of the employer.

27 Works by other parties

This chapter relates to those parts of the works that are done by firms other than the principal contractor. In some cases, the principal contractor has no choice in their selection, other than a reasonable objection to such firms or persons on the works. The different forms of contract deal with these in different ways:

- assignment and subcontracts
- works by employer or persons employed or engaged by employer
- nominated or named firms.

The first of these groups allows a principal contractor to appoint domestic or trade subcontractors. The number and type of subcontractors employed will depend on the overall trades and facilities provided by the principal contractor. On some projects, the principal contractor may work almost solely in a coordinating and organising role and choose to subcontract the whole of the construction work and activities. On some contracts the architect or engineer must approve all these firms, although approval should not be unreasonably withheld. However, this does not in any way diminish the principal contractor's responsibilities under the terms of the contract.

The second group are those firms who are the employer's own directly employed contractors or workers and their right to be engaged upon the works during the normal contract period. These firms may already be employees of the employer or be of a very specialised nature, such as sculptors. The employer may also wish to retain a more direct control over their work. These firms need to be careful not to interfere with the work of the principal contractor in the execution of their work.

The third group include named or nominated subcontractors who, once appointed, will work in very much the same way as the principal contractor's own subcontractors. These might also include statutory undertakings. These are a special type of subcontractor that sometimes none of the parties – employer, architect, engineer or principal contractor – has any choice about in their appointment and involvement in the project.

27.1 Assignment and subcontracts

Without the written consent of the other party, neither the employer nor contractor should assign the contract. Assignment is described as transferring the interests of one party to another.

Subletting occurs where the principal contractor chooses to enter into a subcontract with other firms for some part, or even the whole, of the contract, whilst still maintaining the existing relationship to the employer in all respects. Assignment occurs where another firm takes over the contractual rights of the contractor, and for all purposes the contractor to whom the project was contracted ceases to exist. Although subletting occurs to some degree on all construction projects, assignment is extremely unusual.

27.1.1 Subletting

Usually, the principal contractor must obtain the written consent of the architect prior to subletting any portion of the works. The consent must not be unreasonably withheld. The principal contractor will not undertake all the trades, and sometimes none at all, but some of the work will be done by subcontracting firms. In order that an architect or engineer can have some influence over these subcontractors, it is necessary to know who they will be in order that they can be approved. This approval would only be withheld where past experience or knowledge considered them to be unreliable or where they previously performed a poor standard of work.

27.1.2 Conditions of any subletting

If for any reason the principal contractor's employment is determined under the contract, then the employment of the subcontractors will also cease, although they may be re-engaged under a new agreement.

Where subletting occurs, then all the conditions in the contract between the principal contractor and the employer must apply equally to the subcontract. The principal contractor will have to ensure these agreements are written into the subcontracts. Otherwise, conditions will apply to the principal contractor that will not be able to be enforced on the subcontractor. These conditions also apply to materials and goods and, for example, access to the subcontractor's premises, such as workshops, for any required inspection purposes.

The principal contractor's tender will be based upon subcontractors' prices and quotations. However, since the subcontractors are subject to approval, this adds a risk to the principal contractor if they are not subsequently approved. Other subcontractors will then have to be invited to tender for the subcontract works, possibly at a higher price. This higher price is borne by the principal contractor and is not normally passed on to the employer. The approval of such subcontractors should therefore be sought prior to the signing of the contract.

The principal contractor must pay the subcontractors the correct amount and at the appropriate time. When payments are not received, then the subcontractor is allowed to add simple interest until they are. The interest rate charged is 5 per cent above the base rate of the Bank of England. A subcontractor may determine the subcontract under these circumstances.

27.2 Works by employer or persons employed or engaged by employer

The employer may choose to employ some of the building or specialist trades directly. This work may be of a relatively minor nature, or may be concerned with items that are perhaps only loosely related to the building industry. However, it is important to secure the right to the construction site for such persons or firms. The principal contractor, knowing that these will require entry to the site, can organise and allow for any possible inconvenience or disruption ahead of the work being carried out. This knowledge will also give the principal contractor the opportunity to include in the tender any expense that may thus be incurred. If there is any delay on the part of their work, the contractor has some redress by an extension of time, loss and expense or determination.

If the execution of work under this heading is an afterthought on the part of the employer, the principal contractor must be prepared to give the same reasonable access to the works. Note that the employer is responsible for their proper organisation and control. They will not fall within the definition of a subcontractor or the jurisdiction of the principal contractor.

27.3 Named or nominated subcontractors

It should be noted that the various forms of contract deal with these types of firms in often very different ways. These firms are effectively preferred firms of the designer or a recommendation of the employer. Where they are described as such, then the principal contractor usually has no choice in their selection or appointment. Traditionally, in bills of quantities, they would be included as a prime cost sum. Some forms of contract differentiate between those who supply goods and materials from those who also execute the work.

It is the architect's or engineer's prerogative to nominate firms to undertake some of the construction work. The type of work that these firms carry out is often specialist in nature and character and may include, for example, lift installations, piling or security systems. The firms selected should meet the approval of the contractor. The nomination, however, may still proceed in certain cases where the principal contractor believes there are reasonable grounds for refusing such an approval. In these circumstances, it may be unwise to proceed with the nomination of such a firm. If the principal contractor has a reasonable objection to a proposed nominated subcontractor, then this should be put in writing as soon as possible. This obviously needs to be done prior to actual nomination, but also in sufficient time to enable other firms to be selected without causing any delay to the contract programme.

27.3.1 Contractor's tender for works otherwise reserved for a named or nominated subcontractor

The principal contractor may also wish to carry out work for which a prime cost sum has been included in the contract documents. It is usual to supply with the tender form both a list of the main contractor's domestic subcontractors and a list of prime cost sums for which the contractor desires to tender. Both of these forms are often required to be returned with the contractor's tender. The following conditions will apply regarding the expenditure of prime cost sums:

- The principal contractor, during the ordinary course of business, must directly carry out this work. This does not preclude subletting.
- The principal contractor must give notice of this intention as soon as possible. This is in order to pre-empt a contract being awarded to another firm.
- The architect or engineer must be willing to allow the principal contractor to tender. Refusal to tender, or to accept the principal contractor's lowest tender, will usually mean a preference for a specialist firm to do the work.

If the principal contractor is allowed to submit a tender for this work, then the invitation to submit a price must make it absolutely clear how discounts, profit and attendances will be dealt with. Where the price is submitted in competition, then the importance of these items may influence the choice of the successful tender. An obvious method is to treat the principal contractor's quotation in the same way as each of the other subcontractors. However, practical difficulties associated with these aspects may favour a composite price being provided from the principal contractor.

27.3.2 Procedure for naming or nomination of a subcontractor

The different forms of contract deal with these in different ways. A comprehensive range of documents may be used, which will typically comprise the following:

- the employer's invitation to tender
- tender by a subcontractor
- particular conditions
- the conditions of contract for nominated subcontractors.

27.3.3 Contractor's right of reasonable objection

Architects or engineers should not nominate a subcontractor to whom the principal contractor has a reasonable objection. A reasonable objection should be made within seven days of receiving the instruction. If possible, the objection should be removed so that naming or nomination can then proceed. Alternatively, a different firm should be named or nominated.

27.3.4 Interim payment for a named or nominated subcontractor

Invoices must normally be provided before the subcontractor's work can be included in a valuation. These will be checked by the quantity surveyor against the accepted quotation. The rates, the prices and the arithmetic should be checked, and a check should also be made to ensure that the quantity of work has been properly executed. Although it is not the quantity surveyor's duty to inspect the quality of materials and the standard of work, work should not be included that is obviously not meeting the requirements of the specification. Where there are any doubts, the matter should be brought to the attention of the architect or engineer. The approved invoices are then included in the quantity surveyor's valuation of the work, and then subsequently the certificate.

It is usual and good practice, upon the issue of each interim certificate, to state the amount that has been included for each named or nominated subcontractor. It must also be clearly stated whether these amounts are interim or final payments. The individual subcontractors should be informed accordingly. The principal contractor must then make these payments as directed. Deductions cannot be made at this stage for any contra-charges unless of course these have been agreed in writing by the subcontractor concerned.

27.3.5 Direct payment to nominated subcontractor

Before the issue of each certificate, the principal contractor must provide some evidence that the appropriate amounts shown on previous certificates have been paid. Such evidence might include a receipt for the amount listed. The absence of this evidence will not automatically assume that the subcontractor has not been paid, but before proceeding to issue another certificate there must be reasonable satisfaction that this is not the case.

In those cases where a principal contractor defaults in payment, many forms of contract provide a remedy for paying a named or nominated subcontractor direct. In circumstances where adequate proof is unavailable to the satisfaction of the architect or engineer, then a certificate must be issued accordingly, together with a copy to the subcontractor. Any amounts paid by the employer in this way will be deducted from future payments to the principal contractor. However, if it is known that the principal contractor is about to go into liquidation, such a direct payment will not be made by the employer. In no circumstances does the employer want to be placed in a position of having to pay any firm or subcontractor twice. Unfortunately, situations do occur where this may be necessary in order to secure the services of a particular subcontractor, without which the project would not be completed.

27.3.6 Prenomination payments to named and nominated subcontractors by employer

There are provisions in some forms of contract to allow an employer to pay a named or nominated subcontractor for design work, materials and goods prior to the issue of an instruction of nomination. Such direct payments are ignored in computing interim and final certificate payments.

27.3.7 Extension of period or periods for completion of named or nominated subcontract works

The main contractor is not able to give a named or nominated subcontractor an extension of time without the written consent of the architect or engineer.

27.3.8 Practical completion of named or nominated subcontract works

When a named or nominated subcontractor achieves practical completion of the works, a certificate should be sent to the principal contractor to this effect, with a duplicate copy to the named or nominated subcontractor.

27.3.9 Early final payment

Final payment can be made to a named or nominated subcontractor prior to that of the main contract. Before this payment can be made, twelve months must elapse from the date of the subcontractor's certificate of practical completion. However, the subcontractor concerned must have remedied any defective work and must have sent, through the principal contractor to the quantity surveyor, the necessary documents for the final adjustment of the contract sum.

27.3.10 Position of the principal contractor

The principal contractor is not responsible to the employer, in respect of named or nominated subcontract works, for:

- design, where items have been designed by a named or nominated subcontractor
- selection of materials and goods that have been selected by a named or nominated subcontractor
- satisfaction of any performance specification.

However, the principal contractor remains responsible for carrying out the work in order to ensure that it complies with the design and that the materials and work standards are in accordance with the specification.

It is sometimes necessary to have to renominate a named or nominated subcontractor where the named or nominated subcontractor makes a default under the terms of the subcontract, becomes insolvent, terminates the contract, has the contract terminated by the employer or where there is a failure to comply with instructions. The contractor cannot determine the contract of any named or nominated subcontractor without an appropriate instruction from the architect or engineer.

28 Injury and insurance of the works

The definition of insurance is that one party (the insurer) undertakes to make payments to or for the benefit of the other party (the assured) upon the occurrence of certain specified events. The insurance contract between these parties is generally contained in a document called an insurance policy. The consideration, which is necessary to make such a contract binding, is provided by the assured in the way of a premium.

An insurance contract is said to be *uberrimae fidei*, 'based upon good faith between the parties'. The assured must therefore make a full disclosure of every material fact that is known. A material fact is information which, if disclosed, would influence the judgment of the insurer. Filling in the proposal form incorrectly can make the policy voidable by the insurer, even where an innocent mistake occurs. The insurance policies are usually printed on standard forms by the company issuing its own terms. Each policy must, however, be clear on the terms of insurance, and be accompanied by certain exclusions of liability.

If a situation occurs that may result in a claim by the assured, this must be notified to the company within a reasonable time of the event. The claim is often forwarded by the insurance company to a loss adjuster. This is the general procedure involving sums of even a relatively minor amount. The loss adjuster will then assess the amount of damage and the sum payable in respect of the insurance policy. Although the loss adjusters work on behalf of the insurance company, they will try to achieve an equitable settlement between the parties. In some circumstances, the employer or contractor may employ a loss adjuster on their behalf in order to negotiate with the insurance company's own loss adjuster.

The construction industry has a poor performance record in respect of death and injury, especially to workers during construction operations. The overall numbers of deaths continue to decline, but accidents and fatalities remain at an unacceptable level. The introduction of the CDM Regulations has assisted in attempting to make construction sites safer in respect of working practices (see Chapter 20).

28.1 Injury to persons and property and indemnity to the employer

28.1.1 Liability of contractor – personal injury or death – indemnity to employer

The principal contractor is liable for, and must indemnify the employer against, any expense, liability, loss, claim or proceedings of any nature. These may arise because of statute legislation or at common law. They are in respect of personal injury to, or the death of, anyone caused by the carrying out of the works by the principal contractor or others employed, such as sub-contractors. The principal contractor's liability does not extend to any act of negligence either by the employer or persons directly under the employer's control, such as the employer's own contractors. Indemnity in this circumstance means the protection that one party gives to another in respect of claims by a third party. Mainly because the principal contractor may be faced with liquidation, contracts provide for joint insurance. However, if the insurance fails for some reason, this does not relieve the contractor of the indemnity.

28.1.2 Liability of contractor – injury or damage to property – indemnity to employer

The principal contractor is also liable for, and must indemnify the employer against, injury to property. The injury applies to any type of property and must be as a result of carrying out the works. The two exceptions to this liability are the same as for personal injury, namely the employer's negligence and the employer's own directly employed contractors.

28.1.3 Injury or damage to property – exclusion of the works and site materials

The employer has the option of accepting some of the risks where the contract provides for this. Also, the principal contractor will not be responsible should damage occur because of an inappropriate or false design, but must ensure that even in these circumstances all reasonable precautions during the execution of the works have been taken. Where loss or damage does occur, the principal contractor's main area of defence is to point any negligence in the direction of the employer, architect, engineer or other agents working on behalf of the employer.

In the case of partial completion and the issue of the appropriate certificate, then the responsibility for insurance of the property automatically becomes that of the employer.

28.2 Insurance against injury to persons and property

28.2.1 Contractor's insurance – personal injury or death – injury or damage to property

Principal contractors have a liability to indemnify the employer against injury to persons and property. They are responsible for the whole works, including that of subcontractors. The principal contractor should therefore ensure that all subcontractors maintain proper insurances to cover the liability of the contractor. If, however, a subcontractor fails to provide the appropriate insurance cover, this will not remove the overall liability of the principal contractor to the employer. The principal contractor is only exempt in the case of negligence caused by the employer or persons for whom the employer is directly responsible.

Insurance is required against claims submitted by third parties for injury to persons and property. These third parties will include the owners, the occupiers of adjoining properties and the general public. It will also include all those who visit the site or who have lawful entry to the works. A public liability policy will protect the principal contractor against claims arising through negligence or mistake. There may also be occasions when a third party will be able to substantiate a claim even when negligence cannot be proved. The policy must therefore include clauses to guard against such claims. A major risk with constructional work is the danger of subsidence to adjoining property. It is equally important to ensure that the principal contractor's public liability policy covers this event.

The principal contractor will also be required to provide an employer's liability insurance for workers employed. Although National Insurance provides compensation to employees for injury, the principal contractor will still be liable at common law to compensate the employee for injury due to negligence, whether through the fault of the individual concerned or that of another employee. The premiums paid are often calculated on the basis of the annual wage bill.

The principal contractor must also indemnify the employer against all possible claims. The amount of the insurance cover required is usually stated in the appendix to the conditions of contract. Whilst the sum insured must not be less than the amount stated, the principal contractor can insure for a greater sum than required, as explained in a footnote to some of the conditions of contract.

The employer can at any time request the principal contractor, or any subcontractor, to produce for inspection documentary evidence that the insurances are properly maintained. This will usually require the appropriate policies to be produced at least some time during the early stages of the contract. The employer can, however, require them to be produced at other reasonable times.

If the principal contractor or a subcontractor fails to properly insure or continue to insure the works as requested, then the employer may insure on their behalf. The employer will pay the necessary premiums and deduct these amounts from sums that are due. The amounts recoverable from the principal contractor are those amounts that the employer actually pays, rather than any amount inserted by the contractor in the bills. Because this problem is more likely to occur at the beginning of the contract, the employer can deduct the appropriate amounts from interim certificates.

28.2.2 Insurance – liability of employer

In some circumstances a provisional sum may be included in the contract bills for insurances. The principal contractor should then maintain, in the joint names of the employer and contractor, insurances to the appropriate amount as stated in the contract bills. Such indemnity shall cover damage to property other than the works caused by collapse, subsidence, vibration, weakening or removal of support, or lowering of groundwater, unless the damage occurs because of:

- the contractor's own liability
- errors in the design of the works
- damage that is likely to be inevitable owing to the nature of the works, which is almost an uninsurable item
- the employer's acceptance of responsibility for the insurance of the works
- parts that are subject to the issue of a certificate of practical completion
- war, invasion, act of foreign enemy, hostilities, etc.
- direct or indirect causes by or contributions to or arising from the excepted risks
- direct or indirect causes arising out of pollution or contamination of buildings, structures, water, land or atmosphere
- issues in respect of damages for a breach of contract.

Any insurance required for the above should be placed with insurers approved by the employer. The principal contractor must also provide proof of the policy and that the premiums have been paid. The inclusion of a provisional sum may mean that some of these items are difficult to insure.

28.2.3 Excepted risks

By the effect of an excepted risk, in no circumstances will the contractor be liable either to indemnify the employer or to insure against: personal injury or death of any person; or damage, loss or injury caused to the works or site materials; work executed; the site or any property. The excepted risks will be defined in the contract and are typically:

- ionising radiations
- contamination by radioactivity from any nuclear fuel or from any nuclear waste from the combustion of nuclear fuel
- radioactive toxic explosive
- other hazardous properties of any explosive nuclear assembly or nuclear component
- pressure waves caused by aircraft
- other aerial devices travelling at sonic or supersonic speeds.

The amount of insurance cover required for any one occurrence or series of occurrences arising out of one event, in respect of injury to persons and property, will be stated in the appendix to the form of contract. The employer may need to seek expert advice in connection with this sum, but in any event the contractor may well wish to maintain third-party policies at a much higher level.

28.3 Insurance of the works

28.3.1 Insurance of the works – alternative clauses

The responsibility for the cost of making good any damage to the works during the period of construction can usually allow for one of three choices to be made by the employer. Two of these should therefore be deleted. The conditions do not envisage, in any circumstances, the principal contractor being responsible for insurance beyond the period of practical completion of the works. They assume that the date of handover of the project will become the entire responsibility of the employer as far as the insurance items are concerned. The three choices available for the insurance of the works are:

- Erection of new buildings: all-risks insurance of the works by the contractor.
- Erection of new buildings: all-risks insurance of the works by the employer.
- Insurance of existing structures: insurance of the works in or extensions to existing structures.

Where the project comprises two separate parts, one being new works and the other alterations, it is possible to use the first two options. However, some confusion might occur regarding the common stock of materials and the apportionment of the preliminaries costs. From a procedural viewpoint, the events would be governed by the alternative invoked when a particular incident occurred.

Essentially, two kinds of policy are available for property insurance. The first type is known as an indemnity policy. To indemnify means not only to protect against harm or loss and make financial compensations, but also to secure legal responsibility. The payments made by the insurance company are based upon the damage that occurs less any depreciation in the value of the property. The alternative policy has a reinstatement clause that commits the insurers to paying the full cost of replacement, no account being taken of depreciation. If property is insured for less than its true value, the policyholder then will be paid an average sum based upon a ratio of insured value to true value. For example, if the property is only insured at half the true amount, then, whatever the size of the claim, only half of it will be reimbursed. Some property owners choose to carry a part of the risk themselves, and in these circumstances the insurance company will then only be responsible for a proportion of the loss. In most circumstances, an excess clause will be included to avoid claims for trivial amounts of damage. In addition to the cost of reinstatement, the insured party could be put to further expense as a result of the incident, such as the provision of alternative accommodation. This is known as consequential loss and a clause to cover this should be included in the policy.

28.3.2 Definitions

All-risks insurance is defined as insurance that provides cover against any physical loss or damage to work executed and site materials, but excluding the cost necessary to repair, replace or rectify:

- property which is defective due to wear and tear, obsolescence, deterioration, rust or mildew
- work or materials lost or damaged as a result of its own defect in design, plan, specification, material or work standards or which relied upon the support or stability or work which was defective
- loss or damage caused by or arising from:
    - any consequence of war, invasion, act of foreign enemy, hostilities (whether or not war has been declared), civil war, rebellion, revolution, insurrection, military or usurped power, confiscation, commandeering, nationalisation or requisition or loss or destruction of or damage to any property by or under the order of any government *de jure* or *de facto* or public, municipal or local authority
    - disappearance or shortage if such disappearance or shortage is only revealed when an inventory is made or is not traceable to an identifiable event
    - an excepted risk.

The contract has been amended from terrorism in Northern Ireland to terrorism anywhere where the contract applies. This reflects the growing threat of terrorism on a worldwide scale. The contract refers to activities directed towards the overthrowing or influencing of any government *de jure* or *de facto* by force or violence. An all-risks policy insurance in the joint names of the employer and contractor must be provided in respect of physical loss or damage to the work executed, site materials and existing structures.

28.4 Erection of new buildings – all-risks insurance of the works by the contractor

28.4.1 New buildings – principal contractor to take out and maintain a joint names policy for all-risks insurance

The principal contractor should take out and maintain a joint names policy for all-risks insurance. The insurance to be provided by the principal contractor is to be for the full reinstatement value including professional fees, the percentage for which will be stated in the appendix to the form of contract. The policy will be maintained up to the issue of the certificate of practical completion or up to any date of termination. This is regardless of whether the validity of termination has been contested.

28.4.2 Single policies – insurers approved by employer – failure by contractor to insure

The joint names policy must be taken out with insurers approved by the employer. The principal contractor must send a copy of the policy and the receipt for the premiums paid to the architect or engineer. Where the principal contractor fails to provide adequate insurance cover in respect of the contract, the employer may insure against the risks described. The employer cannot extend the policy to cover other risks, but can deduct the premiums from amounts owed to the contractor under interim payments.

28.4.3 Use of annual policy

If, however, the principal contractor maintains an all-risks policy as the normal pattern of business efficiency, then this may be accepted as an alternative to the joint names policy outlined above. It will be subject to the endorsement of the employer's interest on the policy, and evidence that it is appropriate and being maintained and that the necessary premiums have been paid. The provisions outlined previously in regard to any default by the contractor in taking out and maintaining insurances apply equally in this case.

28.4.4 Loss or damage to works – insurance claims – contractor's obligations – use of insurance monies

If the necessity for an insurance claim arises, in respect of loss or damage affecting work executed or site materials, then the contractor shall proceed as follows:

- Inform the architect or engineer and employer in writing of the nature, extent and location of the claim.
- Allow for inspection by the insurers.
- Diligently restore any work which has been damaged, replace or repair any site materials which have been lost or damaged, remove and dispose of any debris and proceed with the carrying out and completion of the works.

The insurance monies received for this work will be paid to the employer. The principal contractor will be paid for the rectification of this work by instalments under certificates on a similar basis to that of interim certificates. The sums received for professional fees will be paid to parties concerned. The principal contractor will not be paid anything extra to cover the restoration of the damaged work other than that received from the insurers. Before proceeding 'with due diligence', the principal contractor should make sure that the insurers have accepted the claim. Unless it is necessary for safety reasons, this part of the works should not be disturbed until the company's loss adjusters have inspected the damage. The final amount received by the principal contractor and the employer for professional fees may be inadequate since the nature of the damages may not be pro rata in either situation.

28.4.5 Terrorism cover

If the insurers named in the joint names policy notify the principal contractor or employer that terrorism cover will cease from a specified future date, and will be no longer available, then the respective parties will need to decide what action they need to take.

The employer has few options available should this occur. A new insurer could be found, but this might be unlikely. The employer can terminate the project and terminate the principal contractor's employment. This will result in an uncompleted building and payments, such as release of retention monies and payments associated with termination being made to the contractor.

The alternative approach is to proceed with the works in the absence of any insurance. The employer will need to consider carefully how serious the terrorism threat is and to what extent the project might be damaged under these threats. If this alternative is adopted and terrorism damage occurs, then the principal contractor will, with due diligence, restore such work under an instruction. If the employer feels that the principal contractor or subcontractor has contributed in some way towards this damage, perhaps through a negligent act, any amounts due cannot be reduced without the principal contractor's agreement.

28.5 Erection of new buildings – all-risks insurance of the works by the employer

Under this clause, the employer is responsible for insuring the works, and the procedure is similar to the one above. However, the insured roles of the two parties are reversed and the employer is the person who takes out and maintains a joint names policy for all-risks, for no less than the cover agreed.

The principles of providing documentary evidence are with the employer. A subsequent failure to insure the works can be rectified by the principal contractor. In this case, the amount of any premiums involved are then added to the contract sum.

28.5.1 Loss or damage to works

The procedure to be followed by the contractor in the event of an insured claim follows this outlined pattern:

- The principal contractor must give notice, in writing, if any loss or damage arises in respect of the risks covered by the policy.
- The occurrence of such loss or damage is disregarded in calculating any amounts due to the principal contractor, under the contract.
- The work is inspected by the insurers, probably a loss adjuster.
- The principal contractor, with due diligence, restores the damaged work by replacement or repair.
- If materials have been lost or damaged, they should be removed from the site and replaced with the specified materials.
- The principal contractor and any named or nominated subcontractor involved will authorise the insurers to pay the money to the employer.
- The restoration, replacement or repair will then be treated as a variation.

The value of rectifying the damaged work will usually be on the basis of contract bill rates or quotations. Agreement with the insurance company's loss adjusters may also be necessary to determine the actual amount to be paid to the employer. Whether the insurer's basis of payment is the same as that of the contract basis is a matter which must remain at the employer's own risk.

28.6 Insurance of existing structures – insurance of the works in or extensions to existing structures

The employer is responsible for insuring the existing works under this clause in a joint names policy, with the same contractual risks occurring.

The employer will usually take out a joint names policy following a similar procedure to that in section 28.5. The amount of cover should include the full reinstatement value, including professional fees. This is to ensure that, should considerable damage occur, there remain sufficient funds to reinstate and complete the project. The full reinstatement value should include VAT.

If the employer fails to insure as requested by the contract, the principal contractor can insure the works and the existing building. This may involve the contractor in carrying out a survey and inventory of the existing building. The costs of the premiums involved and any associated costs are then added to the contract sum.

28.6.1 Loss or damage to works – insurance claims – contractor's obligations – payment by employer

A similar pattern of reinstatement by the principal contractor and payment by the insurance company occurs as previously discussed. One assumes that in this case the principal contractor will be paid the full cost of reinstatement, in accordance with the contract and valuation rules. Any shortfall between that received by the employer from the insurance company and that paid to the principal contractor would thus be borne entirely by the employer. The insurer may pay for reinstatement on the basis of the principal contractor's estimate for making good. Under this section, however, the occurrence of loss or damage by a specified peril may result in termination of the contract by the employer or the principal contractor. Presumably the damage could be so extensive as to make reinstatement of the existing building unwise as far as the employer is concerned. The principal contractor may also wish to terminate because the damage could result in a project that has become too complex to contemplate.

28.7 Insurance for the employer's loss of liquidated damages

The appendix to the forms of contract identify to the principal contractor whether or not this clause will be applied. Once the contract has been signed, the contractor may be requested to obtain a quotation for such insurance. This will be on an agreed basis of a genuine pre-estimate of the damages that may occur as the result of any delay and at the rate stated for liquidated and ascertained damages included in the appendix. The amounts expended by the principal contractor to take out and maintain this insurance are added to the contract sum. If the principal contractor defaults in taking out such an insurance, then the employer may insure against any of the risks involved.

28.8 Joint fire code

In recent years, insurers in the construction industry have become concerned with the high level of claims for fire loss and damage on construction sites. They have produced the fire code that imposes obligations on all those involved with construction projects, including employers, principal contractors, subcontractors and the professions. The main objective of the code is to ensure that adequate protection and detection systems are in place in respect of fire safety.

The code ensures that the employer and anyone who is involved in the project complies with its requirements. The principal contractor should ensure that the code's requirements are enforced by all those brought on to the site, including subcontractors, local authorities and statutory undertakers.

If a breach of the joint fire code occurs, the insurers under the joint names policy should specify the measures that are required and the time when they are to be completed. The principal contractor must ensure the remedial measures are carried out in accordance with the contract and by the remedial measures completion date.

29 Fluctuations in costs

Contracts are often described as either fixed price or fluctuating price. The essential difference is that fixed-price contracts expect the contractor to have allowed for any changes to the contract sum for items such as inflation, whereas fluctuating-price contracts do not. A fixed-price contract means that only the rates and prices will remain unchanged. It does not mean that the contract sum will not be revised for other items, such as variations. Most forms of contract include provisions for the following items:

- fluctuations
- contributions, levy and tax fluctuations
- labour and materials costs and tax fluctuations
- use of price adjustment formulae.

29.1 Fluctuations

Fluctuations in rates and prices are normally dealt with in the following three ways in a form of contract. Only one of these options may apply, and this will usually be stated in the appendix to the form of contract:

- Adjustments in the price of contributions, levies or taxes; this is technically known as a 'firm-price' contract.
- The assessment of labour and materials costs and tax fluctuations, using the traditional method of calculation.
- The calculation of price fluctuations by the formulae method.

29.2 Contributions, levy and tax fluctuations

The contract sum is deemed to have been calculated as detailed by this clause. The adjustments that may therefore be necessary will occur where differences eventually arise because of the following provisions. The contract sum is based upon the types and rates of contribution, levy and tax payable by a person in the capacity of employer.

The prices in the contract bills are based upon those contributions that are current at the base date. Should these change, or should the principal contractor's status be revised, then adjustment will become necessary. Whilst such contract clauses are often referred to as 'increased costs', money may sometimes need to be repaid to the employer where the contributions actually decrease. The levies payable under the Industrial Training Act 1982 are often expressly excluded from any adjustment.

To enable the principal contractor to claim fluctuations in the costs of labour, the different forms of contract usually specify in detail how the provisions are to be applied. Normally, each employee must have worked on the project for a minimum of two working days in any week for which the claim is applicable. The aggregation of days and parts of a day do not usually apply. The highest properly fixed tradesperson's rate must be used, provided that such a tradesperson is employed by either the principal contractor or subcontractor. The Income Tax (Employment) Regulations 1993 (the PAYE regulations) under section 204 of the Income and Corporation Taxes Act 2010 will apply.

Where any of the tender types or tender rates change, then the net actual amount of an adjustment must be calculated. This sum is then paid to or allowed by the principal contractor. The change is measured as the difference between monies actually expended and those that were appropriate at the base date used for these purposes. It has already been noted that some changes are based upon a theoretical adjustment. For example, where an overseer works as a craftsperson, any adjustments that may be due are based upon the craftsperson rate. In other circumstances where the workpeople are contracted out within the meaning of the Social Security Pensions Act 1975, the method of adjustment is on the basis that they are not contracted out employees.

The contributions, levies and taxes that are subject to adjustment are those which result because of an Act of Parliament or because of a change in an Act of Parliament. They are the amounts that a principal contractor must pay as a result of being an employer of labour, and therefore include sums such as the statutory insurances against personal injury or death.

29.2.1 Materials – duties and taxes

The contract sum is deemed to have been calculated in the following manner in respect of materials, goods and fuels:

- The prices contained in the contract bills are based upon the types and rates of duty and tax applicable at the base date (excluding VAT). These include amounts payable on the import, purchase, sale appropriation, processing or use of the materials, goods, electricity and, where specifically mentioned in the bills, fuels.
- If any types or rates of duty, in the context of the above, are altered, then the net difference between that actually paid by the contractor and what was presumed to have been allowed in the bills must be calculated. This sum will then be either paid to the contractor or refunded to the employer.

29.2.2 Subcontractors

Where the principal contractor has chosen to subcontract some of the works, then the following will apply. The subcontract must incorporate the same provisions as that of the main contract together with any percentage that may be appropriate. These adjustments will be paid to the principal contractor or refunded to the employer, whichever is appropriate.

29.2.3 Principles

The principal contractor must provide the quantity surveyor with the necessary evidence in order that the amounts of fluctuations may be calculated. This should also be done within a reasonable time. The incentive necessary for the principal contractor to do this is that payment will be received within the interim certificates once they are agreed. When a fluctuation is claimed in respect of employees other than workpeople (see section 29.2.4), the necessary evidence must include a certificate signed on behalf of the principal contractor each week, certifying the validity of the evidence. Fluctuations are usually net and exclusive of any profit.

Where the principal contractor is in default over completion, the fluctuations will not be adjusted in line with actual expenditure. The amount of fluctuations will be technically frozen at the level applicable at completion date. The principal contractor will therefore still be eligible for increases, but not at the true level of expenditure.

These fluctuation provisions are not applicable in respect of:

- dayworks, because these are usually valued at current rates and therefore take into account the appropriate adjustments
- specialist works carried out by named or nominated subcontractors, as all fluctuations will be dealt with by their invoices
- a principal contractor acting as a named or nominated subcontractor, where the above provision will apply
- changes in VAT, because these are exempt and are dealt with separately anyway.

29.2.4 Definitions

The different forms of contract will usually include a set of definitions in order that fluctuations can be applied within a set of rules:

- *Base date*: an agreed date is written into the appendix and called the base date, which brings the terminology in line with the formula rules for price adjustment. The necessity for the date is to avoid principal contractors having to make last-minute adjustments to their tenders.
- *Materials and goods*: this excludes consumable stores, plant and machinery, but can include timber used in formwork, electricity and fuels, if this has been specifically stated.
- *Workpeople*: persons whose rate of wages are governed by rules, decisions or agreements of the Construction Industry Joint Council or some other wage-fixing body. Overseers, for example, who work as tradespeople will be assumed to be tradespeople for the purpose and application of this clause. Note that workpeople not only include those employed by the principal contractor on the project, but also those who may work off site in workshops, e.g. a joinery shop.
- *Wage-fixing body*: a body which lays down recognised terms and conditions of workers under the Employment Protection Act 1975 or the Employment Rights Act 1996.

29.3 Labour and materials costs and tax fluctuations

This is the traditional full fluctuation clause. The principal contractor is able to claim for increases in the costs of employing labour and the costs of purchasing materials in addition to the other statutory increases. The different forms of contract set out how the rules need to be applied.

29.3.1 Labour

Contract sums are deemed to have been calculated under the following assumptions:

- The adjustment is in respect of workpeople on-site and of those employed directly upon the works.

- It also includes workpeople directly employed by the principal contractor who are engaged on the production of materials or goods in the workshops.
- The rules or decisions of the Construction Industry Joint Council or other appropriate wage-fixing body is used.
- Any incentive scheme or productivity agreement under the provisions of rule 1.16 of the Construction Industry Joint Council.
- The rates of wages also include other emoluments and expenses, such as holiday credits and insurances.
- Only wage changes unknown at the base date are eligible for adjustment. Promulgated changes are those that have already been agreed, and must therefore have been allowed for in the tender, even though they may not come into operation until some time after the project has begun. The actual amount of the changes must, however, be known. Proposed unagreed rates are therefore not promulgated within this definition.
- If the terms of the public holiday agreements change during the duration of the project, the appropriate costs of these changes can also be included. If such changes were promulgated at the time of tender, then these are assumed to have been covered by the contractor.
- The contributions and levies of the Construction Industry Training Board (CITB), for example, are excluded from adjustment and the principal contractor must therefore bear the costs of all increases. In practice, considerable notification time is generally given in respect of the future increases in these payments.
- Promulgated wage rate increases known before the base date would also mean that the principal contractor should have allowed for increases in the respective liability insurance. Only increases that are unknown at the base date can be claimed. Because the employer's liability insurance is often based upon the wages bill of the firm, this would be an accepted adjustable item in other circumstances. However, should the employer choose to provide a more expensive insurance, then this claim would not be valid. Increases because of changes in the wages bill or because of inflation are therefore allowable where they were unknown at the time of tender.
- Increases in the cost of productivity bonuses resulting from increases in standard rates of wages are recoverable. Thus, where the appropriate wage-fixing body agrees upon an increase in the basic rate, the principal contractor will be able to reclaim this similar proportion paid by way of their productivity agreement.
- The principal contractor's employees, who are outside the scope of workpeople but who are engaged on the works, are subject to fluctuations as if they were craft operatives. An overseer working as a bricklayer would be reimbursed as a bricklayer.
- The prices in the bills are based upon basic transport charges as submitted by the principal contractor and attached to the contract bills. Provision is therefore made for increases in travel allowances payable to workpeople under a joint agreement, including increased costs of employer's transport where used, or public transport fares.

29.3.2 Materials

The contract sum is deemed to have been calculated in the following manner:

- The prices contained in the contract bills are based upon the market prices of materials, goods, electricity and, where specifically stated in the bills, fuels that were current at the base date.
- The above market prices are referred to as basic prices, and are included on a basic price list for these items. In practice, only the major cost items are listed. This may be done by the

quantity surveyor, in which case the list is known at the time of tender. Alternatively, the principal contractor is asked to prepare a list together with prices.

- Any changes in the market prices of these items are then subject to fluctuations and adjustments of the net changes. Only the items included on the basic price are subject to adjustment.
- Market price changes include any changes in duty or tax (other than VAT) on the import, purchase, sale appropriation, processing or use of goods, materials, electricity or fuels as specified.
- In order to avoid contention later, the quantity surveyor must approve the basic prices shown on the list. These are usually supported by invoices from a single supplier. It is usual then for the principal contractor to obtain the actual materials from this supplier. Any changes between quotation and invoice can therefore be calculated easily. Where it is necessary for the principal contractor to use an alternative supplier, the change in price may be calculated between the invoice and the price on the basic price list and adjusted for any changes due to the use of a different supplier.

29.3.3 Subcontractors

The provisions relating to subcontractors correspond generally to those for labour and materials in sections 29.3.2 and 29.3.3.

29.3.4 Excluded work

The following are usually excluded from the reimbursement:

- work for which the contractor is paid daywork rates, since these will be current rates
- work executed by named or nominated subcontractors
- changes in the rate of VAT, since this is not dealt with under the contract.

29.4 Use of price adjustment formulae

The traditional method of calculating fluctuations has many drawbacks, which can include:

- disagreements over what is allowable
- shortfall in recovery
- delays in payment
- extensive clerical work in documenting and checking claims.

An alternative procedure was therefore developed in the 1970s, based upon the idea of using index numbers. (It should be noted that in the 1970s inflation in the UK was running at high levels of over 20 per cent per annum.) This method is known as the formula method of price adjustment and is based upon a set of rules. Separate sets of indices have been developed for building and civil engineering works, although the principles of application are similar. Under the Joint Contracts Tribunal (JCT), the rules are given contractual effect by incorporation in the contract. Two series have been devised:

- *Series 1*: thirty-four work category indices (March 1975).
- *Series 2*: forty-eight work category indices (April 1977, amended 1980).

The series 2 rates are in three sections:

- definitions, exclusions, correction of errors
- operation of work category and workgroup methods
- application of formulae to the main contractor's specialist work.

These have now been consolidated in the latest publication, which was issued in 2016.

The date of tender is chosen as the base date to which the indices will apply. Since the contract sum is exclusive of VAT, its operation will in no way affect the VAT agreement and its subsequent adjustments.

The following definitions typically apply:

- *Base date*: this is defined in the same way as above.
- *Index numbers*: obtained from the monthly *Bulletin of Construction Indices* published by The Stationery Office.
- *Base month*: normally the calendar month prior to the date when the tenders are returned.
- *Valuation period*: the date of the valuation and the midpoint of the period when the fluctuations are calculated.
- *Work categories*: the thirty-four (series 1) or forty-eight (series 2) classifications of the contract work.
- *Workgroups*: the less detailed classification that may be used in preference to work categories. It operates by aggregating work categories into larger units. It reduces the number of calculations, but may make the results untypical of the work being measured.
- *Balance of adjustable work*: some sections of the contract bills, e.g. preliminaries, are excluded from the work categories and are valued on the basis of averaging the index numbers used.
- *Non-adjustable element*: a proportion of the value of work may be excluded from the operation of the formulae.

The formula calculation adjustment is added to each valuation on the basis of the work carried out that month. Unlike the calculation of increased costs by the traditional method, the amounts calculated are subject to retention adjustments. Initially, the amount will be calculated using provisional indices, and three to four months later these will be firmed up when the final indices can be determined.

Some goods and components may be manufactured outside the UK, and it would therefore not be realistic to apply the formula rules in such cases. If any variation in the market price occurs, they will be subject to fluctuation adjustment using similar provisions to those of the traditional method.

If the contractor decides to sublet any portion of the works, these provisions must be incorporated into the subcontract as appropriate.

The quantity surveyor and the principal contractor have the power to agree any alteration to the methods and procedures for the formula for fluctuations recovery. This is presumably to help reduce the number of tedious calculations where an agreeable amount can be ascertained. The amounts calculated are then deemed to be the formula adjustment amounts. This is provided that such sums are a reasonable approximation to those that would have been calculated by the operation of the formula rules.

Where the principal contractor does not complete the works by the completion date, the following procedure is to be adopted: the value of work completed after that date is subject to formula adjustment on the basis of the indices applicable at the relevant completion date. Where a contract is running behind schedule, the employer will consistently be paying higher amounts month by month than would otherwise be the case, but, if the architect or engineer has granted an extension of time and fixed a later completion date, then formula adjustment will continue to run at the current dates.

30 Financial matters

Two clauses have been included under this heading, dealing with matters relating to financial legislation:

- Valued added tax – supplemental provisions.
- Construction industry scheme.

30.1 Value added tax (VAT)

30.1.1 Definitions – VAT agreement

VAT was introduced to the construction industry through the Finance Act 1972 (updated 2010). It is now administered through HM Revenue & Customs (HMRC). Via the Chancellor's annual Budget Statement there is the opportunity to amend both the extent and the rate of this tax. This has been done many times since its introduction.

Many types of work done by contractors and similar trades are standard-rated for VAT. However, certain types of work can sometimes be charged at a reduced rate of 5 per cent, or at the zero rate. It is important to charge the right VAT rate. The reduced or zero rates can only be applied if certain conditions are met. The conditions will relate to different aspects of the work, including:

- the type of construction work or project carried out
- the type of work done and the equipment installed
- when the work is carried out
- who the work is carried out for.

Other aspects of the work can also affect the VAT rate. The VAT guide can be accessed through the HMRC website, which explains in detail the application of the VAT rules (see www.customs.hmrc.gov.uk).

The correct application of VAT on construction contracts is not always straightforward, and expert advice should be sought whenever there is doubt. Quantity surveyors have the best knowledge to be able to offer this advice to employers and principal contractors.

30.1.2 Contract sum

The contract sum, on most forms of contract, is exclusive of VAT. Adjustments to the contract sum will also be exclusive of VAT, since the conditions specifically state that VAT will not be dealt with under the terms of the contract.

30.1.3 Possible exemption from VAT

If, after the date of tender, the goods and services become exempt from VAT, then the employer must pay the principal contractor an equal amount to the loss of the principal contractor's input tax. This will then equate with the amount that the principal contractor would otherwise have recovered.

The employer agrees to pay the appropriate tax to the principal contractor that is chargeable by HMRC. The employer must therefore pay to the principal contractor the appropriate amount in respect of any positively rated items in the contract. The procedure for this payment is as follows:

- The principal contractor gives to the employer a written provisional assessment of the respective values of any goods and materials which have been included in a payment, and are subject to VAT.
- This should be done not later than the date for the issue of each interim certificate.
- The principal contractor must specify the rate of tax chargeable on these items. The rate is fixed by HMRC, but could change, and several different rates could also be introduced.
- The employer, on receipt of the provisional assessment from the principal contractor, must calculate the amount of tax due.
- This amount is then included with the amount of the interim certificate and paid to the principal contractor.
- If the employer has reasonable grounds for objection to the provisional assessment, the principal contractor must be informed in writing within three working days of receipt.
- The principal contractor must then either withdraw the assessment, and thereby release the employer from the obligations, or confirm the assessment.
- When the certificate of making good defects has been issued, the principal contractor must then prepare a written final statement of the respective values of all the supplies of goods and materials. The principal contractor must specify the respective rates of tax on these items.
- This final statement can be issued either prior to or after the issue of the final certificate. For practical purposes, it is advantageous to issue the statement after the final certificate has been accepted.
- The employer, on receipt of the final statement, must calculate the tax that is due. The balance to the contractor is usually paid within twenty-eight days of receipt of the statement. In the unlikely circumstance where the employer may have overpaid the principal contractor in respect of VAT, a refund of the appropriate amount will be due to the employer.
- The principal contractor must issue to the employer receipts under the certificates for the appropriate amount of tax that has been paid.
- Any set-off in respect of liquidated damages when calculating and paying the amounts of VAT between the employer and the principal contractor must be ignored.
- If the employer and the principal contractor disagree, it may require the decision to be made by the VAT commissioners. This should be done before the payment becomes due. The employer can also request the principal contractor to appeal further to the commissioners should there still be disagreement with their findings. In these circumstances, the principal contractor would be able to claim any costs or expenses involved from the employer. Before the appeal can proceed, the employer must pay to the contractor the full amount of the tax that has been charged.

- The commissioners' appeal decision is final and binding, unless they subsequently introduce a correction to the tax that has been charged.
- Arbitration is not applicable to VAT assessments by the commissioners.
- If the principal contractor fails to provide a receipt for the tax paid by the employer, then the employer is not obliged to make any further payments. This applies only if the employer requires a validated receipt for tax purposes, or where the employer has paid tax in accordance with the provisional assessments.
- If the employer has terminated the principal contractor's employment, any additional tax which the employer may have to pay as a result of the termination may be set off against any payments to be made to the principal contractor.

30.2 Construction industry scheme

The purpose of the original legislation was to deal with the problems of tax evasion by subcontractors, particularly the labour-only 'lump' subcontractors. Procedures were established whereby the principal contractor collected tax on behalf of HMRC from those subcontractors who did not hold a tax certificate. Those subcontractors who held the appropriate 714 certificate were eligible for full payment on the basis that they dealt with the Inland Revenue direct. The existing statutory tax deduction scheme clause was replaced with the construction industry scheme (CIS).

The CIS sets out the rules for how payments to subcontractors for construction work must be handled by contractors in the construction industry. The scheme applies mainly to principal contractors and subcontractors in mainstream construction work. However, businesses or organisations whose core activity is not construction, but they have a high annual expenditure on construction works, may also count as contractors and fall under the scheme.

30.2.1 Definitions

The following definitions usually apply:

Act	Income and Corporation Taxes Act 1988, together with any statutory amendment or modification.
authorisation	Form CIS 4, CIS 5 or CIS 6, or a certifying document created on the contractor's headed stationery.
construction operations	Those operations that are defined in section 567 of the Act.
contractor	A person who is a contractor for the purpose of the Act and the Regulations.
direct cost of materials	The direct cost of materials to the contractor.
Regulations	Income Tax (Subcontractors in the Construction Industry) Regulations 1993 SI 743, as amended by the Income Tax (Subcontractors in the Construction Industry) (Amendment) Regulations 1998, SI 2622.
statutory deduction	Referred to in section 559 of the Act.
subcontractor	Any person who is a subcontractor for the purposes of the Act and Regulations.
voucher	A tax payment voucher in the form CIS 25 or a gross payment voucher CIS 24.

30.2.2 Whether employer is a contractor

In the appendix to the various forms of contract, the employer needs to be stated as either *a contractor* or *not a contractor*. Where the former applies, the employer is responsible for making the statutory deduction. Where the employer is *not a contractor*, the appropriate deductions must be made by the main contractor. If the responsible party fails to make the necessary deduction, they may find themselves liable to the authorities for the tax. The status of the employer for tax deduction purposes is also referred to in the fourth recital. An employer may automatically be *a contractor* to cover a wider statutory position; for example, in the case of local authorities. In the case of private employers, it will largely depend upon whether the employer is engaged in *construction operations*. These are defined very broadly in the legislation.

30.2.3 Payment by employer

The employer must not make any payments under the contract until the valid authorisation has been provided.

30.2.4 Validity of authorisation

Where an employer is not satisfied with the validity of the authorisation, the principal contractor must be informed in writing. This will give the reasons why the validity is being questioned. This is to avoid any possibility of fraud taking place. No payments will be made until the authorisation is considered to be valid. The contractor may be able to resubmit the original authorisation if a letter from the contractor's tax office is able to confirm its validity.

30.2.5 Authorisation

Where the authorisation is a CIS 4 registration card, then the contractor must give a statement to the employer, seven days prior to final payment, showing the direct cost of materials. This is to enable the employer to make only a statutory tax deduction in respect of other items that are not defined as materials. When the employer complies, then the contractor must indemnify the employer accordingly. Where the employer feels that the statement provided by the contractor is false, then the employer must make a fair estimate of the direct costs of materials. Where the authorisation is a valid CIS 5, CIS 6 or an appropriate certifying document, then the employer must pay the amount due without making a statutory deduction.

30.2.6 Change of authorisation

When the principal contractor is subsequently issued with a change in CIS certificates, then the employer should be informed immediately. Where the employer considers this to be valid, the appropriate action as described above will be applied. Where the relevant authorisation is withdrawn by HMRC for any reason, the contractor must notify the employer so that appropriate future action can then be taken. Following expiry of a CIS 5 or CIS 6, the employer must make no future payments to the principal contractor until they provide the appropriate authorisation.

30.2.7 Vouchers

Where authorisation CIS 4 applies, and the employer has already made payments to the principal contractor, on the nineteenth day of the month the employer then provides the contractor with a copy of the CIS 25 voucher. This will have been sent to HMRC and will indicate the payments made and the tax deducted. Where authorisation CIS 6 applies, and the employer has made payments to the contractor, then the employer will add their tax reference and send the voucher to HMRC.

30.2.8 Correction of errors in making the statutory deduction

Where an employer makes an error or omission in the calculation of the statutory deduction, the error is usually corrected by repayment or by further deduction from payments due to the principal contractor.

30.2.9 Disputes or differences

The relevant procedures that are outlined in the contract are to be used if there are any disputes or differences in interpretation. This is unless an Act, regulation or statutory instrument overrules such an interpretation.

31 Clauses of a general nature

There are a range of clauses in all forms of contract that, whilst important, are often of a more general nature. These include the following examples:

- Interpretations, definitions, etc.
- Contractor's obligations.
- Architect's instructions.
- Contract documents.
- Antiquities.
- Settlement of disputes – adjudication – arbitration – legal proceedings.

31.1 Interpretations, definitions, etc.

The articles of agreement, the conditions and the appendix are to be read as a whole, and together they are the form of contract. They cannot therefore be used independently, and, in many instances, they have an important bearing upon each other. For example, the details completed in the appendix will generally override the specific requirements indicated in each clause. The appendix will make recommendations for the contract, but the parties have some liberty on the amounts or times that they wish to insert against the appropriate clause.

31.1.1 Definitions

A majority of the definitions listed also refer to particular clauses in the forms of contract:

adjudication agreement	References also to arbitration and for legal proceedings.
all-risks insurance	Referred to especially in the insurance clauses.
appendix	That part of the conditions that are completed by the parties to the contract.
base date	Used for calculating increased costs, for example.
CDM Regulations	The Construction (Design and Management) Regulations 2007, including any amendments at the signing of the contract. The principal contractor will be required to comply with revisions that occur during construction, but will be reimbursed accordingly.
completion date	The date for completion stated in the appendix or any date fixed later for an extension of time.
conditions	The main contract clauses plus any supplemental provisions.

contract bills	The priced bills of quantities signed by or on behalf of the parties to the contract.
contract documents	The contract drawings, contract bills, the articles of agreement, the conditions and the appendix. In some contracts it will also include a specification, and may include schedules and a master programme.
contract drawings	The drawings referred to in the contract conditions which have been signed by or on behalf of the parties to the contract.
principal contractor	The person (or firm) named in the articles of agreement.
date for completion	The date fixed and stated in the appendix.
date for possession	The date stated in the appendix when the principal contractor is given access to the construction site.
defects liability period	The period named in the appendix for making good defects after the completion of the works.
employer	The person named as the employer in the articles of agreement.
excepted risks	Typically, ionising radiations or contamination by radioactivity from any nuclear fuel or waste, but will vary between the different contract conditions.
final certificate	The last certificate to be issued under the contract.
health and safety plan	The plan referred to in the appendix to be provided by the principal contractor.
planning supervisor	The person named in the contract who may have a dual function.
practical completion	When the works are complete.
specified perils	Typically, these will include fire, lightning, explosion, storm, tempest, flood, bursting or overflowing of water tanks, apparatus or pipes, earthquake, aircraft and other aerial devices or articles dropped therefrom, riot and civil commotion, but excluding excepted risks.
works	The project to be constructed as described in the contract documents.

31.1.2 Principal contractor's responsibility

The principal contractor is responsible for carrying out and completing the works in accordance with the contract conditions. This is regardless of any obligations the architect or engineer may have towards the employer or whether or not the employer appoints a clerk of works, or whether the workshops or any other place where work is being prepared are inspected. In some forms of contract, the architect or engineer is not made responsible for the supervision of the works that the principal contractor is to carry out and complete. In essence, the principal contractor's responsibility for work is in no way reduced by the conditions which allow for an inspection of the works. It should be noted that the designer of the project may not be the firm or person who oversees the construction activities on-site.

31.1.3 Employer's representative

The employer may choose to appoint some other person to fulfil this function under the terms of the contract. The principal contractor must be appropriately informed. In order to avoid possible confusion over conflicting roles, neither the architect nor the quantity surveyor should be appointed to this position.

31.1.4 Applicable law

The laws of the country where the project is being constructed are applicable to the contract unless otherwise stated. Of course, if the two parties agree to carry out the works under a different legal system, then this should be stated.

31.2 Contractor's obligations

31.2.1 Contract documents

The principal contractor will carry out and complete the works in accordance with the following combination of documents. These will vary depending upon the contract conditions being used:

- Contract drawings.
- Contract bills.
- Specification.
- Schedule of rates.
- Articles of agreement.
- Conditions of contract.
- Appendix to the conditions of contract.
- Numbered documents (subcontract agreements).

They are collectively known as the contract documents. The principal contractor must use the materials and work standards that have been specified. The approval of the quality of materials or standards of work will be decided by the architect or engineer. These are to be to a reasonable satisfaction. Only the standard described in the contract documents can be enforced and not a higher standard. In the selection of materials and the assessment of work, the documents must be precise. Although objectivity may achieve the appropriate and desired standards, some subjective analysis and interpretation will also be required from time to time.

31.2.2 Contract bills

The contract bills cannot override or modify either the application or the interpretation of these conditions of contract, with the exception of the rules for measurement. The rules used for the preparation of the contract bills must be stated. Any errors subsequently found in the contract bills resulting from their method of preparation must be corrected. The correction of these errors is automatic and usually will not require a variation instruction. The errors referred to cover descriptions, quantities or omission of items. They are errors made by the quantity surveyor during the preparation of the contract bills. Errors in pricing that result from mistakes made by the principal contractor do not come within this definition, and will therefore not be corrected once the contract has been signed. A procedure for their correction prior to the contractor's tender being accepted is described in Chapter 8.

31.2.3 Discrepancies or divergences between documents

If a principal contractor finds any discrepancies or divergences between or within the contract documents, then the architect or engineer should be informed of such differences. Instructions must then be given to the principal contractor, in order to clarify the discrepancy or divergence, so that the works can proceed. The principal contractor cannot assume, for example, that the drawing will automatically take preference over the contract bills. However, it can be reasonably assumed that the conditions of contract will always take preference. Note that the conditions of contract are not capable of discrepancy.

The conditions of contract do not imply that the principal contractor must go looking for differences in the documents. If the compliance with the drawings later revealed that the bills were different but correct, further instructions would need to be issued for the correction of this work, if so desired. The principal contractor should, however, take all reasonable steps to avoid problems arising.

31.3 Instructions

The principal contractor must comply with all instructions issued, unless a reasonable objection in writing is made regarding the non-compliance. A reasonable objection may include the refusal to accept the naming or nomination of a subcontractor where an unsatisfactory relationship existed on a previous project.

The principal contractor may also challenge the architect's or engineer's authority to issue certain instructions. In these circumstances, it may be necessary to request in writing the provision in the contract which empowers the issue of the particular instruction. The architect or engineer must comply with this request. If the two parties cannot agree upon this point, the matter may be referred to arbitration for a decision. If the principal contractor complies with the instruction on the basis of the response, then this shall be assumed to be an instruction under the terms of the contract.

The architect or engineer may write to the contractor requesting the immediate compliance with an instruction. If after, typically, seven days from receipt of this notice the principal contractor does not comply, then the employer may employ other firms to execute the work. The costs involved with this will be deducted from monies due to the principal contractor, or they may be recoverable as a debt by the employer. In more severe cases, particularly where the principal contractor persistently fails to comply with a written instruction, the employer may terminate the principal contractor's employment.

31.3.1 Instructions in writing

All instructions must be in writing to have any contractual effect. However, if oral instructions are given, then the following procedure should be adopted:

- Confirmation should be given in writing within, typically, seven days by the principal contractor.
- Unless this is dissented in writing within a further period, usually seven days, the instruction shall be accepted as a valid instruction within the terms of the contract.
- Alternatively, the oral instruction can be confirmed in writing by the architect or engineer within, typically, seven days.
- If neither party confirms the oral instruction, but the principal contractor executes the work accordingly, then it may be confirmed as a valid instruction at any time prior to the issue of the final certificate.

31.4 Contract documents

The contract documents have been listed in section 31.2.1. They will vary depending on the specific form of contract being used. They will typically comprise some drawn information, a document that specifies the quality of the materials to be used and the standards of workmanship, a document about costs and prices and a form of contract that specifies the conditions under which the work is to be carried out.

The contract drawings and bills or specification are to remain in the custody of the employer, but must be available at all reasonable times for inspection by the principal contractor.

31.4.1 Copies of documents

When the contract has been signed, the principal contractor will normally be provided, free of charge, with:

- one copy of the contract documents certified by the employer
- two further copies of the contract drawings
- two copies of the unpriced bills of quantities.

31.4.2 Descriptive schedules

Within a reasonable period of time after signing the contract, the principal contractor will be supplied with additional information, which might include descriptive schedules and further drawings to amplify the contract drawings. These additional documents are not allowed to impose further obligations beyond those described in the contract documents, but they may provide the detail that is required for construction purposes.

31.4.3 Master programme

Many forms of contract now request the principal contractor to provide a master programme as part of the conditions of contract. If this requires updating, perhaps for an extension of time, then the principal contractor must supply a further copy of the revised master programme. A copy will be retained on the construction site for reference purposes. However, there is often no requirement for the principal contractor to indicate the progress of the works. The master programme remains a tool for checking purposes rather than a method of creating obligations.

31.4.4 Information release schedule

Such a schedule, if this is a part of the contract, informs the principal contractor when additional information will be made available. The schedule is not annexed to the contract. It is important that the information is released to the principal contractor in accordance with that agreed in the information release schedule. In practice, this will be coordinated with the principal contractor's master programme for the works.

31.4.5 Provision of further drawings or details

The principal contractor will be supplied with two copies of any additional drawings or construction details necessary to explain the contract drawings. These may be necessary in order that the principal contractor can properly carry out the works to the appropriate and designated satisfaction of the contract. In some circumstances, the principal contractor will have to request such information, giving reasonable notice of this intent.

31.4.6 Availability of certain documents

The principal contractor must also keep one copy of the contract drawings, one copy of the unpriced bills and other descriptive schedules, documents and drawings on the site for reference by the architect or engineer at all reasonable times. Copies of all the information provided for the construction of works should be retained on the construction site by the principal contractor. A copy of the master programme must also be available, together with a copy of any further drawings and details.

31.4.7 Return of drawings, etc.

Once the principal contractor has received the final payment under the terms of the contract, a request may be made to return all drawings, details schedules and other documents which bear the architect's or engineer's name. The copyright of the design is normally vested in the designer, the architect or the engineer. Where the employer requests the principal contractor to repeat the design, then the architect or the engineer is able to request a further fee.

31.4.8 Limits to use of documents

None of the documents mentioned above must be used for any purpose other than the contract for which they were provided. Also, the principal contractor's rates must not be divulged to others, or be used for any purpose other than this contract. However, where appropriate permission is sought, the contractual information may then be reused for other purposes.

31.4.9 Issue of architect's certificate

Any certificates that are issued must be sent to the principal contractor and the employer concurrently.

31.4.10 Supply of as-built drawings

Before the date of practical completion, the principal contractor, without charge, should supply the employer with drawings and other information that relate to the performance-specified work which the principal contractor will have undertaken. This will show the work as built and also include maintenance and operating schedules, where appropriate.

31.5 Antiquities

Antiquities include fossils and other items of interest or value which may be found on the construction site or during the excavation part of the works. These items become the property of the employer, and if such finds occur, the principal contractor should take the following actions:

- Use best endeavours not to disturb the object and, if necessary, cease construction operations within the vicinity until instructions are received from the architect. Where there is any reasonable doubt about a discovered object, this procedure should be followed as a precaution.
- Take all necessary precautions in order to preserve the condition of the objects from possible damage until they can be dealt with by experts, if necessary.
- Inform either the architect or clerk of works of the discovery and its precise location on-site.

31.5.1 Instructions on antiquities found

The architect or engineer must issue instructions regarding the removal of antiquities that are found on the construction site. This may involve a third party, such as an archaeological society, examining, excavating or removing the object. Such a party will not be described as a sub-contractor but as persons directly employed by the employer.

31.5.2 Direct loss and/or expense

The above instructions will probably involve the principal contractor in direct loss or expense for which reimbursement is not provided elsewhere under the contract. In these circumstances, the amount of such loss or expense should be ascertained and this amount will then be added to the contract sum. Where appropriate, an extension of the contract may be required to allow the removal of the antiquity.

31.6 Settlement of disputes

This clause applies where either party to the contract refers any dispute or difference arising under the contract to adjudication (see Chapter 5).

31.6.1 Mediation

Many contracts now suggest that the parties to the contract should have their differences resolved by mediation. It is often not an obligation that they must do so or even that they consider doing so. This reflects a useful process that has been supported by the courts. Mediation is often seen as a sensible and cost-effective way to resolve a dispute. The courts regularly endorse mediation and are diverting many cases in this direction and penalising in costs those who do not consider it.

31.6.2 Adjudication

The appointed adjudicator is a person to whom both parties agree, and is appointed by the person who will be nominated in the form of contract and who may be named in the appendix to the form of contract. An appointment is normally made within seven days from the date of the notice of the intention to refer. If the adjudicator subsequently becomes ill or is unable to carry out the duties assigned, then the parties may agree to appoint a new adjudicator.

A written statement of the contention or disagreement should be issued to the adjudicator, usually, within seven days. A copy should also be sent to the other party in the dispute. This statement may include other supporting material or information if this is deemed to be appropriate in helping to resolve the dispute. The written statement should be sent by special or recorded delivery or by fax.

The party not making the referral may also send to the adjudicator, within seven days, a written statement of the contentions for the adjudicator to consider.

The adjudicator is acting under the Housing Grants, Construction and Regeneration Act 1996. Adjudicators are not considered as experts or arbitrators. Their decision, in writing, should be sent to the parties, usually, within twenty-eight days, although this may be extended by agreement for up to fourteen days and will vary depending upon the form contract being used. The adjudicator is not obliged to give reasons for the decision reached. In reaching the decision, the adjudicator should set out the procedure in ascertaining the facts and the law. This may include the following:

- Using one's own knowledge and experience.
- Within the terms of the contract, opening up, revising any certificate, opinion, decision, requirement or notice.
- Requesting further information from the parties.
- Requiring the parties to carry out tests, additional tests or opening up the works.
- Visiting the site and workshops.
- Obtaining information and advice from employees or representatives.
- Obtaining information and advice from others on technical or legal matters.
- Having regard to any term of the contract relating to the payment of interest.

If one of the parties chooses not to cooperate with the adjudicator, this will not invalidate the adjudicator's decision. It is assumed that the adjudicator will have the necessary expertise to arrive at a valid decision. Where necessary, the adjudicator can appoint others for advice or assistance. The parties may agree to meet their own costs or allow the adjudicator to direct who should pay the costs involved. The adjudicator will not be liable for anything done or omitted unless the act or omission is done in bad faith.

31.6.3 Arbitration

The second alternative after adjudication is to refer a dispute to arbitration. This process is frequently longer and more costly than adjudication. The provisions of the Arbitration Act 1996, or any amendment, apply to such cases. The arbitration is to be conducted in accordance with the Construction Industry Model Arbitration Rules (CIMAR) that are current at the base date of the contract.

31.6.4 Legal proceedings

Legal proceedings are usually a last resort. They tend to be lengthy and expensive, and decisions usually take some time to be achieved. These form the basis of case law.

Appendix A

Professional bodies in the built environment

There are a large number of professional bodies involved in the built environment. These can be subdivided into a number of groups representing the different professions, although a number of them are multi-disciplinary. A number of these bodies were formed in the nineteenth century, others are more recent and some are the result of mergers and acquisitions. The RICS is the largest of these professional bodies, with approximately 120,000 members. It was formed in 1868 as The Surveyors Institute and received its Royal Charter in 1881. The RIBA is unusual in that it is the only one of these professions that requires registration to allow an individual to use the term of architect. Under the Architects (Registration) Act 1938 it is illegal for anyone to carry on the business as an architect unless they are registered with the Architects Registration Board, which was formed following an Act of Parliament in 1931. The Chartered Institute of Building was formed in 1834 and granted a Royal Charter in 1980.

In all countries around the world, professional associations have been formed to share common interests and goals. Some of the UK bodies now operate on an international basis or have formed links and associations with indigenous professional bodies.

Within the UK, a further distinction has been drawn between the chartered and the non-chartered bodies. A majority of these bodies maintain websites that provide a valuable insight to their role and function together with other information of a more general nature.

Architecture and Surveying Institute	(ASI)
Association for Project Management	(APM)
Association of Building Engineers	(ABE)
Association of Consultant Architects	(ACA)
Association of Consulting Building Surveyors	(ACBS)
Association of Consulting Engineers	(ACE)
Association of Cost Engineers	(ACostE)
Association of Interior Specialists	(AIS)
Association of Planning Supervisors	(APS)
British Institute of Facilities Management	(BIFM)
Chartered Institute of Arbitrators	(CIArb)
Chartered Institute of Architectural Technologists	(CIAT)
Chartered Institute of Building	(CIOB)
Chartered Institute of Housing	(CIH)
Chartered Institute of Purchasing and Supply	(CIPS)
Chartered Institution of Building Services Engineers	(CIBSE)
Chartered Institution of Civil Engineering Surveyors	(ICES)
Chartered Society of Designers	(CSD)
Consultant Quantity Surveyors Association	(CQSA)

Institute of Clerk of Works in Great Britain	(ICWGB)
Institute of Highway Incorporated Engineers	(IHIE)
Institute of Maintenance and Building Management	(IMBM)
Institution of Civil Engineers	(ICE)
Institution of Structural Engineers	(IStructE)
Landscape Institute	(LI)
Royal Incorporation of Architects in Scotland	(RIAS)
Royal Institute of British Architects	(RIBA)
Royal Institution of Chartered Surveyors	(RICS)
Royal Town Planning Institute	(RTPI)

Appendix B

Cases of interest

In some instances where the form of contract is thought to be either lacking or unclear, it is common practice to bring the matter before the courts for learned opinion. This may mean bringing the matter first to arbitration, then to the High Court, the Court of Appeal and finally the Supreme Court, where leave to appeal along the way has been granted. The principles upon which these courts base their decisions are stated, and will then in future affect similar cases. This achieves a measure of consistency on matters on which the various parties to a contract may then rely. Where the higher courts reverse the decisions of lower courts, it is the higher opinions that will count in the future. The resulting body of opinion is then established as case law. It is published in the various law reports of the legal journals and, where relevant to the construction industry, in trade and professional journals. There are now also a number of construction law journals and other sources that record the considerable case law of this industry.

The following cases represent just a sample of some of the disagreements that have reached the courts over a number of years. The complete list would form several books alone. Some of the cases described have become household names for the student of building contracts.

For a comprehensive list of construction case law visit www.atkinson-law.com/cases.

Abbey Developments v. *PP Brickwork* (2003)

Abbey Developments had engaged PP Brickwork as a labour-only subcontractor for the brickwork and blockwork on a housing development. The dispute, which initially went to adjudication, turned on whether Abbey was entitled to take away the remainder of the work which PP had agreed to carry out. The initial development was to be sixty-nine houses, but Abbey relied on provision that allowed it to vary the number of units. It was accepted that if, for market driven reasons, Abbey decided not to proceed with the whole development, it could rely on this clause.

The question was whether this extended to circumstances where Abbey was unhappy with the performance of its subcontractor and wished to have the work carried out by another party. It was found that the variation provision and the termination provision had to be operated with respect for PP's right to carry out the works. If this was to be taken away, there needed to be clear words used in the contract. Although at this stage the court was not looking at PP's right to compensation, it did indicate that a provision that allowed work to be taken away to be given to a third party was not, in principle, unenforceable. If, however, it did not in turn provide for compensation (loss of profit and overheads) to the contractor, there was a risk that the provision would be regarded as 'leonine and unenforceable as unconscionable'.

The basic bargain between the employer and the contractor had to be honoured and the employer could use an omissions provision to get out of a bad bargain. It was also very doubtful that work could be omitted simply because the employer was dissatisfied with the contractor's performance.

Alfred McAlpine Construction Ltd v. *Panatown Ltd* (2000)

Panatown entered into a building contract with McAlpine under which McAlpine undertook to design and construct an office building on a site which was not owned by Panatown but by UIPL, another company in the same group as Panatown. Defects appeared in the building erected by McAlpine and Panatown launched arbitration proceedings claiming substantial damages.

A party to contract with a builder for the construction of a building on land belonging to a third party was not entitled to substantial damages for defects and delays in the performance of the contract if the third party who actually suffered the loss had a direct remedy with the builder.

The House of Lords so held by a majority when allowing an appeal by Alfred McAlpine Construction Ltd from a decision by the High Court. This allowed an appeal from the preliminary ruling of an arbitrator that Panatown Ltd was entitled to claim substantial damages under a building contract even though Panatown, having no proprietary interest in the site, had suffered no loss.

In the case of *St Martin's Property Corporation Ltd* v. *Sir Robert McAlpine Ltd* (1994), it was held that where A entered into a contract with B relating to property and it was envisaged by the parties that ownership of the property might be transferred to a third party, C, so that the consequences of any breach of contract would be suffered by C, A had a cause of action to recover from B the loss suffered by C. However, both cases clearly established that (1) A was accountable to C for any damages recovered by A from B as compensation for C's loss; and (2) the exceptional principle did not apply (because it was not needed) where C had a direct remedy against B. If those were all the relevant facts, the case would be covered by the decision in the St Martin's case and Panatown would be entitled to recover the loss suffered by UIPL. But, critically, there was an additional factor which was absent in the St Martin's case. Under a duty of care deed McAlpine undertook to UIPL that in all matters within the scope of McAlpine's responsibilities under the building contract, McAlpine would exercise all reasonable skill, care and attention and owed a duty of care in respect of such matters to UIPL. The direct cause of action which UIPL had under deed was fatal to any claim to substantial damages made by Panatown against McAlpine.

Amalgamated Building Contractors v. *Waltham Holy Cross UDC* (1952)

A contractor had applied for an extension of time because of labour and material difficulties. The architect acknowledged the request, but did not at this time grant an extension to the contract period. Some time after the completion of the works, the architect granted the contractor an extension in retrospect. It was held that this was valid on the grounds it was a continuing cause of delay. The architect was unable to determine the length of extension until after completion. The parties must therefore have envisaged the retrospective application of the extension clause. The Court of Appeal held that granting of an extension of time after completion would therefore seem to be an explicit possibility.

AMEC Building Contracts Ltd v. *Cadmus Investments Co. Ltd* (1997)

In what circumstances can an employer take away work from one party to give to another? This was one of several issues involved in this case. It concerned the omission by Cadmus from a contract with AMEC of the fitting out of works for a food court. The fit-out was covered by various provisional sums, but Cadmus, through its architect, sought to omit the works and have them executed by a third party. In this case there was a power for the architect to withdraw the work from the contractor if it considered it to be in the best interests of the project and the employer. When AMEC claimed that the omission was a breach in the contract, the court upheld the claim and awarded AMEC damages for loss of profit that it would have earned if it had carried out the work as originally envisaged. The problem here was that the omission and award of the works to a third party appeared to have been exercised arbitrarily rather than in accordance with the contract and that the power within the contract did not allow work to be withdrawn in order to give it to a third party. If the work was to be carried out, then AMEC had the contractual right to do it.

AMF International Ltd v. *Magnet Bowling Ltd and Another* (1968)

Bowling equipment had been installed for Magnet Bowling Ltd by AMF International Ltd. This equipment was damaged after a flood that had been caused by exceptionally heavy rain. The bowling centre was only partially completed at this time. AMF therefore claimed £21,000 in damages against Magnet Bowling Ltd and the main contractor. The point at issue in this case was whether the indemnity clause included in the contract would protect the employer and enable him to claim successfully against the contractor. It was held that:

1 The employer was liable in tort due to their negligence in failing to check that the site had been made safe for AMF.
2 The contractor was liable under the Occupiers Liability Act for failing to take reasonable care.

The damages were apportioned on a 60/40 basis between the main contractor and the employer. The question of whether the employer could recover from the contractor then arose. It was held that:

1 The employer, because they had been found guilty of negligence, could not therefore claim under the indemnity clause. This is based upon a legal principle that the indemnity clause will not protect negligence on the part of the one who relies upon it.
2 The employer was able to recover on contractual grounds since the contract included a clause requesting the contractor to protect the works from damage by water.

Anns v. *London Borough of Merton* (1978)

It was alleged that in 1962 a builder had erected a block of maisonettes with defective foundations. It was further suggested that the defendant local authority had, through its inspector, either failed to inspect the foundations or had inspected them carelessly. The plaintiffs had taken long leases on some of the maisonettes. In February 1972 the plaintiffs issued writs against the builder and against the local authority. This action was brought to decide whether the proceedings against the local authority were barred under the Limitation Act of 1939, which was then in use. Obviously, the plaintiffs' cause of action against the council accrued in 1962. It was therefore too late, under this Act, to start an action in 1972. In fact, although the appeal was nominally on the limitation issue, the speeches in the House of Lords were mainly concerned with the more fundamental question of whether an action would lie against the local authority at all on such facts. In the House of Lords, it was held that such an action would lie, and it would thus not be statute barred.

If the fact that the defects to the maisonettes first appeared in 1970, and the writs were issued in 1972, then the consequences must be that none of these actions were barred by this Act.

Architectural Installation Services Ltd v. *James Gibbons Windows Ltd* (1989)

The plaintiffs were a specialist labour-only subcontractor installing window units to the defendants as main contractors. The subcontract included a clause for termination by notice on specified grounds, including default of the subcontractor for wholly suspending the works or failing to proceed with them expeditiously and then remaining in default for seven days after being given written notice of the default by the main contractor. The plaintiffs sued the defendants for monies alleged due and for wrongful determination of the contract. The defendants counterclaimed stating that they were entitled to determine the contract and to seek damages for defective work. It was ordered that two preliminary issues be tried, each of which assumed that the plaintiffs had wholly suspended or failed to proceed with the works in breach of clause 20 of the subcontract. It was held that:

1. On the assumptions set out in the preliminary issues it was agreed that it was not a rightful termination of the subcontract, because there was no sensible connection between the defendants' notice of default and the notice of determination.
2. However, the defendants' telex was a rightful termination of the subcontract at common law since clause 8 did not exclude common law rights expressly or by implication, but existed side by side with the common law right to terminate.

Bacal Construction v. *Northampton Development Corporation* (1975)

This was a contract for a housing development. The development corporation had provided borehole data on which the contractor had based their substructure design. During the course of the work, tufa was discovered in several areas and this required a redesign of the foundation. It was held that there should be an implied term that the ground conditions would accord with the hypotheses upon which the contractor had been instructed to design the foundation. Because the client had provided the borehole data that was now shown to be inaccurate, the differing costs were to be borne by the employer.

Balfour Beatty Building Ltd v. *Chestermount Properties Ltd* (1993)

In this case it was accepted that a contractor is entitled to an extension of time to the completion date for an employer-caused delay occurring after the date by which the contractor ought to have properly achieved completion. In effect, the contractor is entitled to an extension of time to the completion date for employer-caused delays notwithstanding any contractor-caused (culpable) delay. The completion date is to be the total number of working days in which the contractor ought fairly to have completed the works allowing for the employer-event starting from the date of possession or commencement. The practical effect of this is shown by example. If the work was to be completed in week 52 and a variation is issued in week 59, the extension of time for the variation is to start from week 52 and not from week 59.

***Beaufort Developments (NI) Limited* v. *Gilbert Ash NI Limited* (1998)**

The House of Lords has overruled *Northern Regional Health Authority* v. *Derek Crouch Construction Co. Ltd* (1984) and declared that *Balfour Beatty Civil Engineering Limited* v. *Docklands Light Railway Limited* (1996) was wrong.

The central question in the Beaufort Development case concerned whether a court has the power to 'open up, review and revise' a certificate issued by an architect under JCT 80. The House of Lords decided that the court is entitled to examine the facts and form its own opinion upon them in the light of the evidence that is available. The fact that the architect has formed an opinion on the matter will be part of the evidence. But, as it will not be conclusive evidence, the court can disregard his opinion if it does not agree with it.

The House of Lords went on to overrule the earlier conflicting Court of Appeal decision in the *Crouch* case. In this, the Northern Regional Health Authority appealed against a decision of the Official Referee to allow a stay of court proceedings because of the existence of an arbitration clause in the contract. The arbitration agreement expressly provided that the arbitrator had power to 'open up, review and revise' the certificates and decisions of the architect. In dismissing the appeal, the court considered the question of whether or not the court had a similar power. Despite the fact that the Official Referees had for many years opened up and reviewed certificates, the Court of Appeal decided that the court had no such power.

The point is that the court has no power to open up, review or revise any certificate or opinion of the architect, since the parties have agreed by clause 35(3) of the main contract that that power shall be exercised exclusively by the arbitrator.

***Percy Bilton* v. *Greater London Council* (1982)**

Bilton contracted with the Greater London Council (GLC) to erect a housing estate. The contract was dated 25 October 1976, and was substantially JCT 63. The original completion date was January 1979. The nominated subcontractor for the mechanical services went into liquidation in July 1978. By this time, the contract was already running forty weeks behind the programme. A new subcontractor was nominated, but subsequently withdrew before starting work. A new nominated subcontractor was not appointed until December 1978. The programme for this meant that the work would not now be completed until January 1980. Various extensions of time were granted and the extended completion date became February 1980. However, the contract was not completed by that date and thereafter the GLC deducted liquidated damages. The House of Lords held that the withdrawal of a subcontractor is not a fault or breach of contract on the part of the employer, nor is it covered by JCT 63 clause 23. The architect's clause 22 certificate was valid and the GLC was entitled to deduct liquidated damages.

***Birse Construction* v. *St David Ltd* (1999)**

In a case where the underlying issue was whether the dispute was covered by the arbitration agreement, the court had a choice whether to decide the issue itself or stay the proceedings whilst the matter was referred to arbitration. If the court decided to determine the matter itself and if there was a triable issue, directions should be given for its trial. This was equivalent to the procedure where the court possessed a wider discretion to rule what evidence it needed to decide any particular point. It was for the judge to take account whether the issue was one which could be determined on affidavit evidence without oral evidence in the interest of good case management and cost savings. Additionally, it was worth exploring whether the parties would agree on any factual issues.

Rarely would it be appropriate to adopt the course of resolving issues of fact solely on written evidence unless the parties invited the court to do so. The court would decide to do so in cases where it considered that oral evidence was necessary. If the court decided that the proceedings should be stayed to arbitration, it would be better to view it as acting under its inherent jurisdiction.

Bottoms v. *Lord Mayor of York* (1812)

The contractor in carrying out work for a sewage works had intended to use poling boards to support the excavations. However, the nature of the soil turned out to be different from what was expected and necessitated extra works. The engineer refused to authorise this as a variation. The contractor then abandoned the works and sued the client for the value of the work completed. Neither party had sunk boreholes. Before the signing of the contract, the defendants had been advised that the contractor's price was such that the contractor was bound to lose money on the type of soil conditions that were anticipated. It was held that the plaintiff's claim must fail. The plaintiff was not entitled to abandon the works on discovering the nature of the soil or because the engineer declined to authorise extra payments.

Bovis Land Lease v. *Triangle Developments* (2002)

Bovis entered into a contract with Triangle to refurbish and fit out three existing Victorian school houses into forty-three luxury residential apartments and associated works at Silverthorne Triangle, London. The contract incorporated the JCT Standard Form of Management Contract (1998 edition). Disputes arose as to the contents of the valuations of two interim certificates. The architect had reduced the value of certain works packages from the corresponding sums contained in earlier interim certificates so that each of these two certificates certified a negative value as being due to Bovis. These negative certificates were made the subject of an adjudication with the effect that the two certificates should be amended to reinstate the sums deducted from earlier certificates, plus interest calculated from the last date for payment of each of the original certificates.

Triangle disputed, and still disputes, its suggested obligation to pay this sum on a number of grounds and made no payment towards it. In consequence, Bovis took proceedings to enforce payment.

As is usual in management contracts, the work was divided up into work packages. In each case, the architect, on the advice of the quantity surveyor, decided that Bovis, as management contractor, had been in default in the manner in which it had checked the relevant applications for payment of the three work package contractors and, in consequence, disallowed the entire sum being claimed for each of them. Bovis gave to Triangle an adjudication notice of the dispute as to whether or not the architect was entitled to deduct from interim certificates sums that had been previously certified.

The architect served on Bovis a notice under clause 7.2 of the contract to the effect that Bovis were failing to proceed regularly and diligently with the carrying out of its obligations. In a letter, the architect explained that this suggested failure involved such matters as a severe reduction in labour levels and failures to both replace defective work and administer and manage the works. In a letter to the architect, Bovis responded that the default notice did not comply with the procedural requirements of the contract and that the underlying factual basis for its issue did not exist. Thus, Bovis has always contended that the default notice was invalid and erroneously given and that the determination of its employment was invalid.

Triangle issued a notice under clause 2.10.1 of the contract informing Bovis of its intention to withhold or deduct liquidated damages following certificates of non-completion of two blocks or sections of the works that the architect had previously issued. This notice was followed with the second notice, that clauses 4.3.4 and 4.12.4 of the contract provide for, and which has to be served before liquidated damages may be deducted from sums otherwise payable under interim certificates. Bovis responded with a notice served on Triangle to the effect that Triangle had repudiated the contract by engaging new contractors and that, therefore, Bovis was accepting that repudiation and was treating the contract as being at an end. This amounted to the exercise by Bovis of its common law rights, preserved by clause 7.12 of the contract, to accept what it regarded to be Triangle's repudiatory breach of contract. Triangle immediately responded in a letter to the effect that the contract had not been repudiated by Triangle and that it expected Bovis to improve its performance on-site. On the same day, the architect wrote to Bovis and asked that a programme detailing the steps that Bovis proposed to take to complete the work be delivered within seven days. However, on the following day Triangle sent Bovis a notice under clause 7.2 of the contract as a follow-up to the earlier default notice that determined Bovis' employment under the contract on the grounds that the originally specified default had continued for more than fourteen days.

The principal question is whether the adjudicator's decision that Triangle must pay Bovis £158,020.78 is enforceable or whether, instead, it is superseded by contractual provisions allowing Triangle to withhold payment as a result either of the determination of Bovis' employment or as a result of Bovis' alleged repudiation of the contract or as a result of the service of the withholding notice. These questions involve a consideration of the closely inter-woven contractual provisions concerning payment, the service of withholding notices, termination of the contractor's employment and adjudication.

The decision was that immediate payment of a sum that was the subject of a payment decision of an adjudicator is permissible and judgment could be avoided or stayed on account of a determination where the contract allowed for that or where the cross-claim arose out of a repudiation by the receiving party which had been the subject of an adjudicator's decision favourable to the paying party.

***Bower* v. *Chapel-en-le-Frith Rural District Council* (1911)**

The plaintiffs were the successful tenderers for the erection of a waterworks for the council. The contract was for a lump sum, based on specifications and bills of quantities. The plaintiffs were required by the council to purchase a windmill tower and pump from a named supplier and to fix them. The windmill did not work as expected. It was not argued that the default was in any way due to defective installation by the contractor. The council then requested the plaintiffs to replace it with one that would work efficiently. The plaintiffs argued that they were not responsible, as they had played no part in the choice of the windmill. It was held that the plaintiffs were not liable.

***Bridgeway Construction* v. *Tolent* (2000)**

Some contractors have sought to rewrite the adjudication procedures in their subcontracts to the effect that the party initiating the adjudication process is responsible for the costs that are involved. These include the adjudicator's fees and are to be paid by that party irrespective of whether they win or lose the argument. The provisions to this effect were perhaps surprisingly held to be valid in this case, even though it clearly invalidates the provisions of the Unfair Contracts Terms Act 1977.

Bridgeway and Tolent entered into a construction contract which incorporated the Chartered Institute of Credit Management (CICM) adjudication procedure. This procedure provided that each party should be responsible for their own costs and expenses. Clauses were varied so that each party serving a notice of adjudication should bear the costs and expenses of both parties to the adjudication. Disputes arose under the contract. The adjudicator made an order in favour of Bridgeway but refused to order Bridgeway its costs, declaring that Bridgeway was bound by the provisions of the contract. Tolent paid the sum ordered in the decision, less its own legal costs and other expenses. Bridgeway challenged the validity of the contract provisions that they should pay Tolent's costs of the adjudication on the basis that the provisions inhibited the parties in their pursuit of their lawful remedies.

It was Bridgeway's contention that the effect of the disputed provisions would inhibit subcontractors from pursuing the right which Parliament had conferred upon them under the Housing Grants Construction Regeneration Act 1996. However, the court held that the contract had been freely negotiated between the two parties. The dispute related only to costs, as to which the said Act of Parliament was silent and operated only by way of variation to the procedure, and not to any statutory right of Bridgeway. The clause dealing with costs was therefore upheld.

Brightside Kilpatrick Engineering Services v. *Mitchell Construction (1973) Ltd* (1975)

The plaintiffs were nominated as subcontractors under JCT 63. The subcontract was placed by the defendants, which stipulated that the subcontract documents should consist of a 'standard form of tender, a specification, conditions of contract and a schedule of facilities'. The order continued, 'the form of contract with the employer is the RIBA 1963 edition', and concluded 'the conditions applicable to the subcontract shall be those embodied in the RIBA as the above agreement'. Printed references to the 'green form' of subcontract had been deleted. In the Court of Appeal, it was held that the words should be read as referring to those clauses in the main contract which referred to matters concerning nominated subcontractors. Clause 27 of JCT 63 regulated the nominated subcontract relationship, and the only way to give a sensible meaning to the relationship was to read into it the terms of the 'green form'.

British Crane Hire Corporation Ltd v. *Ipswich Plant Hire Ltd* (1974)

Both of these parties were involved in the plant hire business. A drag line crane was hired over the telephone. The hire charges were agreed, but nothing was discussed about the terms and conditions of hire. After delivery, the plaintiff owners sent to the defendants the then current General Conditions for Hiring Plant. The defendants failed to sign and return the acceptance note. In the Court of Appeal, it was stated that, since both parties were in the plant hire business, and that both used the same standard conditions when hiring plant, it must be assumed that the contract was made on those terms.

Courtney v. *Fairbairn Ltd* v. *Tolaini Brothers (Hotels) Ltd and Another* (1975)

In 1969 Mr Tolaini, a hotel owner, decided to develop a site in Hertfordshire. He contacted Mr Courtney, a property developer and building contractor. Mr Courtney wrote to Mr Tolaini stating that if he (Mr Courtney) found suitable sponsors for the scheme, he would undertake the building work. Mr Courtney found sponsors who were able to reach a satisfactory financial arrangement with Mr Tolaini. Mr Tolaini then instructed his quantity surveyor to negotiate a building contract for the project based upon cost plus 5 per cent. The financial aspects were quickly concluded, but negotiations between the quantity surveyor and the building contractor broke down. Mr Tolaini therefore decided to let the contract to another firm, and was promptly sued for breach of contract by Mr Courtney.

It was held, on appeal, that the exchange of letters did not constitute a contract. Because price was of such fundamental importance to the contract, no contract could be formulated until this was agreed, or there was an agreed method of ascertaining it.

Crowshaw v. *Pritchard and Renwick* (1999)

The plaintiff wrote to the defendant enclosing a drawing and specification, and inviting a tender for some alteration work. The defendant replied, 'Our estimate to carry out the sundry alterations to the above premises, according to the drawings and specification, amounts to £1,230'. The plaintiff wrote next day accepting the defendant's 'offer to execute' the works. Later the defendant refused to go ahead. They contended that by using the word 'estimate', they did not intend it as an offer to do the work, and they contended there was a trade custom that a letter in this form was not to be treated as an offer. It was held that the estimate was an offer which had been accepted by the plaintiff and the defendant was liable in contract. The judge in the Queen's Bench Division stated, 'It has been suggested that there is some custom or well-known understanding that a letter in this form is not to be treated as an offer. There is no such custom, and, if there is, it is contrary to law'.

Davies & Co. Shopfitters Ltd v. *William Old* (1969)

A nominated subcontractor submitted a tender which was accepted by the architect. In placing the order for the work, the main contractor added new terms that had been absent from the earlier documentation. One of those terms stated that no payment would be made to the subcontractor until the main contractor had been paid. In spite of this, the nominated subcontractor started work. It was held that the order from the main contractor constituted a counter-offer which had been accepted by the subcontractor prior to starting work.

Davies Contractors Ltd v. *Fareham UDC* (1956)

A contractor undertook to build seventy-eight houses in eight months for a fixed price. They attached a letter to their tender with the proviso that the price was on the basis of adequate supplies of labour being available. The unexpected shortages of labour on a national scale increased the contract period to twenty-two months. The letter, however, failed to be incorporated into the contract. The courts therefore held that the employer must suffer the delay, but that the contractor must bear the expense.

Dawber Williamson Roofing v. *Humberside County Council* (1979)

A main contractor had, with approval, sublet the roofing part of the project to a domestic subcontractor. The slates were delivered to the site and their value was included in an interim certificate as materials on-site. The certificate was honoured and the sum for these materials was paid to the main contractor. Prior to paying this to the domestic subcontractor, the main contractor went into liquidation. The subcontractor therefore sought permission from the council to remove the slates, but this was refused. The subcontractor sued for the value of the slates on the grounds that they remained the subcontractor's property until fixed. It was held that the title rested with the subcontractor. This resulted in the employer paying twice for these materials.

***Dawneys* v. *F.G. Minter Ltd and Another* (1971)**

This case involved certified payments to a nominated subcontractor being withheld by the main contractor as set-off against damages. Lord Denning summed up this case as follows:

> When the main contractor has received sums due to the subcontractor – as certified or contained in the architect's certificate – the main contractor must pay those sums to the subcontractor. He cannot hold them up so as to satisfy his cross claims. Those must be dealt with separately in appropriate proceedings for the purpose. This is in accord with the needs of business. There must be a cash flow in the building trade. It is the very lifeblood of the enterprise. The subcontractor has to expend money on steelwork and labour. He is out of pocket. He probably has an overdraft at the bank. He cannot go on unless he is paid for what he does as he does it. The main contractor is in a like position. He has to pay his men and buy his materials. He has to pay the subcontractors. He has to have cash from the employers, otherwise he will not be able to carry on. So, once the architect gives his certificates, they must be honoured all down the line. The employer must pay the main contractor; the main contractor must pay the subcontractor; and so forth. Cross claims must be settled later.

***Discain Project Services* v. *Opecprime Development Ltd* (2000)**

A dispute arose between the parties that was referred to adjudication. The adjudicator made an award in favour of Discain, but the respondent, Opecprime, failed to make payment in accordance with the decision within the requisite time period. Discain therefore commenced an action for summary judgment in respect of the award by the adjudicator in its favour. Opecprime sought leave to defend on the grounds that the adjudicator had wrongly held that the dispute as to the payment had come into existence between the parties on the date specified when the obligation to pay had arisen. It also claimed that a letter written on the notepaper of an associated company of the defendant could not amount to good notice under Section 111 of the Housing Grants, Construction and Regeneration Act 1996. It also stated that Discain had breached the rules of natural justice by engaging in secret communications with the adjudicator as to the point in issue in adjudication prior to reaching his decision.

In response to the above, the following decisions were made. The correspondence clearly showed that prior to the stated date, the defence intimated that it would not pay when time for payment arose. The court was satisfied that there was a dispute as to the date of the note of adjudication. The court disagreed with the adjudicator's conclusion as to the validity and the effect of the notices. However, the court was not sitting as a Court of Appeal in respect of the decision and so would not refuse to enforce the award on that ground. The fact that the adjudicator had participated in secret conversations with the claimant meant that it appeared that there was a serious risk of bias. It would therefore be repugnant if the court was obliged to enforce an award which was made in such circumstances. The application for summary judgment was dismissed and the defendant was given unconditional leave to defend.

***Dodd* v. *Churton* (1897)**

A contract provided for the whole of the works to be completed by June 1892. Liquidated damages were included in the contract. There was a provision in the contract that any authority given by the architect for additional work would not vitiate the contract. There were thus no provisions for an extension of time. However, additional works were ordered that delayed the works beyond the completion date. The employer decided on a reasonable amount of time for doing the additional works and then claimed the amount of liquidated damages mentioned in the contract. The Court of Appeal held that by giving the order for additional works, the employer had waived the stipulation for liquidated damages in respect of non-compliance of the works by the stipulated time. It was stated that where one party to a contract is prevented from performing it by the act of another, they are not liable in law for that default.

***Dodd Properties* v. *Canterbury City Council* (1980)**

Canterbury City Council built a multi-storey car park next to a building owned by Dodd Properties. Piling operations caused damage to the building, and Dodd Properties claimed damages for both the cost of necessary repairs and for the loss of business during the carrying out of the repairs. When the case came to court, the work had not been done, since Dodd Properties were awaiting the damages to pay for the repairs. They claimed current rates, whereas the council suggested the rates should be based upon what was relevant at the time the damage was done. It was held that there had been no failure on the part of Dodd Properties to investigate the loss and their claim was therefore upheld.

***John Doyle Construction Limited* v. *John Laing (Scotland) Limited* (2002)**

John Laing, the defendants, were management contractors completing the construction of new headquarters for Scottish Widows in Edinburgh. John Doyle, the claimants, were contracted to carry out a number of work packages. The works were delayed and the claimants brought an action seeking an extension of time of twenty-two weeks and ascertainment for a claim for loss and expense. The defendant made representations regarding the relevancy and particularisation of various aspects of the claimant's pleadings. The defendant said that the claimant had suggested that, despite their best efforts, it was not possible to identify cause or links between each cause of delay and disruption and the cost consequence of these. It was normal for a party making a claim to show a causal link between an event and for each item of the loss and expense. It might be a case that it was impractical to trace the causal nexus between the individual event and the individual item of loss. If it were, a causal link need not be shown if two factors applied.

Occasionally, these events may interact with each other in very complex ways so that it becomes very difficult, if not impossible, to identify what loss and expense each event has caused. However, if all of these events are events for which the defendant is legally responsible, it is unnecessary to insist on proof of which loss has been caused by each event. The logic of the global claim demands, however, that all of the events which contribute to causing the global loss are events for which the defendant is liable.

It was confirmed that the fact that a claimant has chosen to advance a global claim because of the difficulty in relating each causative event to an individual sum of lesser expense does not mean that after evidence has been led it will remain impossible to attribute individual sums of loss or expense to individual causation events. The global claim may fail, but there may be in evidence a sufficient basis to find causal connection between individual losses and individual events or to make a rational apportionment of part of the global claim to the causative events for which the defendant has been held responsible.

The defendant proved that all elements that had caused loss and expense were not caused by them. It was concluded that the global claim was to be allowed to proceed to trial for two main reasons. Firstly, the pleadings made it clear that the claimant relied on concurrent causes of delay and disruption. Upon hearing the evidence, this might overcome the contention that, since one of the causation events was not the responsibility of the defendant, the claim should fail. Secondly, there remained the possibility that the evidence to be heard during the trial would afford a satisfactory basis for an award of a sum less than the full global claim. It was felt that it would be unfair to reject a global claim in its entirety, until the evidence was heard.

Such a ruling could open the floodgates for global claims. It can therefore be expected that this decision will be tested or refined in the near future.

Dutton v. *Bognor Regis UDC* (1971)

The foundations of a house were built upon land that had been used as a refuse disposal tip. The foundations were passed and approved by the local authority building inspector. Some time later, the plaintiff purchased the house from the original owner, and cracks that began to appear in the wall were discovered to be due to settlement. The builder settled the matter for an agreed sum. The local authority, however, was held to be liable for negligence, since it owed a duty of care and this had not been reasonably given. Lord Denning expressed the view that the cause of action arose when the defective foundations were laid.

East Ham Corporation v. *Bernard Sunley and Sons Ltd* (1965)

The corporation claimed damages for defective work in the construction of a school in East Ham. The building contract was in the RIBA form (revision of 1950). The contractor had completed the work and the architect had issued his final certificate three years after practical completion. This final certificate issued by the architect at the end of the job was conclusive evidence of the adequacy of the works. Two years later some cladding panels fell down as a result of faults in the fixings. The courts had to consider two questions:

1. On the true construction of the contract, was the final certificate issued by the architect conclusive evidence as to the sufficiency of the works subject to the exceptions mentioned in clause 24(f)?
2. Did the words ‘reasonable examination’ mean an examination carried out during progress or at the end of the defects liability period?

It was held by the courts that the final certificate was conclusive evidence that the works had been properly carried out, subject to the provisions of the final certificate clause.

The stone panels fell off as a result of the careless and incompetent fashion in which they had been attached. It was extremely fortunate that no one was killed or injured. Many contractors would have been not only too willing but anxious to remedy such defects. Not these contractors. They had relied on an escape clause in the contract and sought to throw the whole of their defective work on to the council. The Court of Appeal concluded that the contractors escaped liability under the contract provision. However, on leave to appeal to the House of Lords, the decision was reversed. The employer was therefore awarded damages based upon the costs at the time of performing it.

***English Industrial Estates Corporation* v. *George Wimpey and Co. Ltd* (1972)**

This case was based upon the JCT conditions relevant at the time. In addition, a special condition allowed for the employer's tenant to install machinery and store materials during the construction of the works. The contractor was, however, to be responsible for insuring the works. During the construction period, an extensive fire damaged the works, with a loss of approximately £250,000. The issue then arose as to whether this loss should be borne by the contractor's insurers or by those of the employer. The contractor claimed that clause 12 of the conditions of contract excluded the possibility of special provisions, particularly where they were in conflict with standard conditions, in this instance clause 16 (clauses 18 and 2 of JCT 98, respectively).

The courts held in favour of the employer, because they were not satisfied that there had been a sufficient taking of possession for clause 16 to apply. It would appear that since, at the outset, both parties envisaged this early occupation by the employer, the insurance risk would remain with the employer.

***A E Farr* v. *The Admiralty* (1953)**

The plaintiffs contracted to construct a jetty at a naval base. Clause 26(2) of the contract stated:

> The works and all materials and things whatsoever including such as may have been provided by the Authority on the site in connection with and for the purpose of the contract shall stand at the risk and be the sole charge of the contractor, and the contractor shall be responsible for, and with all possible speed make good, any loss or damage thereto arising from any cause whatsoever other than the accepted risks.

After the plaintiffs had done a substantial part of the work, the jetty was damaged by a destroyer (owned by the employer) that collided with it. The plaintiffs claimed the costs of repairing the jetty. It was held that, because of the clause, the plaintiffs were liable for the repairs to the jetty.

***A E Farr Ltd* v. *Ministry of Transport* (1965)**

This was an appeal by A E Farr Ltd to the House of Lords which reversed a decision of the Court of Appeal. The case related to the payment of working space on a civil engineering contract using the relevant documents. The method of measurement at the time allowed for separate bill items to be measured for working space which would include their consequent refilling. Although the ICE Conditions of Contract at the time included a clause, 'The contractor shall be deemed to have satisfied himself before tendering as to the correctness and sufficiency of tender for the works', this could not be extended to include errors made by the client's advisers. The court therefore accepted that separate items should have been measured in accordance with the appropriate method of measurement. All bills of quantities are prepared in accordance with a method of measurement and strict compliance with this should be made, or specific amendments stated where necessary.

***Re Fox ex parte Oundle & Thrapston RDC* v. *Trustee* (1948)**

A building contractor became bankrupt and the question the courts had to decide was whether the materials not incorporated into the works were vested in the contractor's trustee. The basis for this was the reputed ownership clause contained in the Bankruptcy Act: 'All goods being at the commencement of the bankruptcy in the possession, order or disposition of the bankrupt in his trade or business by consent or permission of the true owner, he is the reputed owner thereof and his ownership passes to the trustee'. It was held that the trustee had no claim to materials on-site which had been paid for in the interim certificate. He did, however, have a good claim to materials paid for, but retained in the contractor's yard.

Gilbert Ash (Northern) Ltd v. *Modern Engineering (Bristol) Ltd* (1973)

This case involved a contractor who had deducted sums due to a nominated subcontractor for set-off because of a breach of warranty by the subcontractor. The Gilbert Ash form of subcontract had been used, which included a set-off clause. A sentence in this clause allowed for set-off for any breach of the subcontract, however minor and unrelated to the actual damages suffered. This was held to be quite outside any rights which the courts would enforce and declared to be a penalty. Another sentence did, however, allow set-off for bona fide claims of the contractor, and the court held that these sums were capable of being deducted.

Gilbert & Partners v. *Knight* (1968)

A firm of building surveyors undertook to supervise some alterations to a defendant's house for the sum of £30. The alterations were originally estimated at £600, but the final account came to £2,283. The firm of surveyors sent the defendant a bill for £135. This was based on the original £30 that had been agreed plus 100 guineas they considered was a reasonable amount for the additional work that was involved. They received nothing extra. The Court of Appeal held that there had been no fresh agreement and therefore there were no circumstances in which a promise to pay a *quantum meruit* could be applied.

M J Gleeson (Contractors) Ltd v. *London Borough of Hillingdon* (1970)

Gleeson's contract for the erection of 300 houses and associated buildings was based upon drawings and bills of quantities. The bills allowed for the completion of works in various stages, but the appendix to the conditions of contract provided for a single completion date only. It was held that liquidated damages could not be deducted until the date for completion shown in the appendix was exceeded. The decision was based upon the wording of the clause: 'Nothing contained in the Contract Bills shall override, modify or affect in any way whatsoever the application or interpretation of that which is contained in these conditions'. The court expressed the view that it would have been an easy matter to add words to that clause to the effect that it was subject to conditions included in the bills. This, however, is contrary to recommended practice and could lead to confusion and uncertainty.

M J Gleeson v. *Taylor Woodrow Construction Co. Ltd* (1989)

This case concerned a management contract for building works on the Imperial War Museum. Taylor Woodrow were to organise, manage and supervise the works and Gleeson were to carry out their portion of the subcontract works. The subcontract works were delayed. Gleeson were awarded thirty-nine days against a claim for seventy-three days. Taylor Woodrow argued that because of this delay, they had been subjected to claims from other subcontractors and they therefore deducted 'set-off claims' from Gleeson and, in addition, liquidated damages. Gleeson accepted that the dispute concerning the extension of time and the liquidated damages aspect would be dealt with under the arbitration procedures. However, they considered that the set-off sums had been wrongfully deducted. Taylor Woodrow argued that these should also be dealt with under arbitration. The court held that a prerequisite of setting-off was to quantify it precisely. This had not been done. The court also believed that the set-off and the liquidated damages might involve some duplication, since both were concerned with the alleged delay by the subcontractor.

Accordingly, the court found that there was no defence to Gleeson's claim and no dispute that could be referred to arbitration as far as the set-off amount was concerned. Judgment was therefore given for the plaintiffs, with the remaining matters of liquidated damages and the extension of time being referred to arbitration.

Gloucester County Council v. *Richardson* (1968)

This contract was based upon the 1939 form of contract, which is considerably different from that in use now. It illustrates the problems of warranty agreements which the new form seeks to overcome. The main contractor was instructed by the architect to accept a contract from a nominated supplier. This was for the supply of precast concrete columns, on terms and at a price agreed by the employer. After erection, cracks appeared in the units and the question of the contractor's liability arose. It was held that the contractor was not liable since they had been directed to enter into a contract which severely restricted their right of recourse against the supplier in the event of defects. Note that the 1939 form did not give the contractor the right to object to nomination of a supplier or to insist upon indemnity.

Gold v. *Patman & Fotheringham Ltd* (1958)

This contract contained a provision whereby the contractor sought to indemnify the employer against claims in respect of damage to property arising out of the works. The indemnity only occurred where it could be shown that the contractor had been negligent in carrying out the works. Furthermore, the contractor had only insured the works in their own name, and not that of the employer since this was not a requirement of the contract. (The current JCT form of contract requires a joint insurance in the names of the contractor and the employer.) During construction, piling operations damaged adjoining property, but no question of negligence arose since the contractor was carrying this out correctly within the terms of the contract. The owners of the adjoining property therefore sued the employer, and they also attempted to bring an action against the contractor for failure to safeguard the employer's interest. It was held that the contractor's obligation was only to insure themselves and not the employer as well.

Martin Grant & Co. Ltd v. *Sir Lindsay Parkinson & Co. Ltd* (1985)

Martin Grant & Co. were subcontractors for formwork on a large number of local authority housing projects. The subcontract was in non-standard form and contained no provisions for the risk of delay. Substantial delays arose in the performance of the main contracts. Martin Grant & Co. suffered considerable loss, since they had to carry out their work several years later than they anticipated. Martin Grant & Co. contended that the main contractor would make sufficient work available for them in order that they might maintain a reasonable progress of the works in an economic and efficient manner.

The trial judge decided that the express terms of the subcontract left no room for the kind of implied term for which Martin Grant & Co. contended. The case went to the Court of Appeal, where it was eventually dismissed. The express terms of the subcontract left no room for the term contended for to be implied. There is no general rule of law implying such a term in a building subcontract, either because of the relationship of the parties or otherwise.

Greaves & Co. (Contractors) Ltd v. *Baynham Meikle and Partners* (1976)

A firm of consulting engineers were responsible for the design of a warehouse. The two-storey building would be used for the storage of oil drums, and forklift trucks would be used for transporting them around the building. Within a few weeks, cracks occurred in the building. The engineers were sued for the costs of the remedial work that was necessary on the basis that the design was unsuitable for the purpose intended. It was held by the Court of Appeal that, because the engineers knew of the building's proposed use beforehand, they were liable for the costs of the remedial work.

Hadley v. *Baxendale* (1854)

The principles for assessing damages resulting from a breach of contract were stated in this case as that arising naturally from the breach, and that which may reasonably be supposed to have been in the minds of the parties at the time they entered into contract.

Hadley Design Associates v. *The Lord Mayor and Citizens of the City of Westminster* (2003)

Westminster had appointed HDA as consultant surveyors and architects in connection with works to blocks of flats on an estate in Pimlico. When Westminster sought to terminate the engagement, HDA brought a claim for loss of profit. Westminster relied upon a provision in the contract that entitled it to terminate on one month's notice.

One of HDA's contentions was that, during a meeting that took place before the contract was agreed, a representative of Westminster had said that the contract would only be terminated if HDA was in default or if Westminster ran out of money and so was unable to complete the block of flats on the estate. It was not accepted that this was proved as a matter of fact and therefore disregarded this allegation. HDA also sought to argue that there were various implied terms that prevented the contract being terminated in this way, and that the termination clause was unreasonable for the purpose of the Unfair Contract Terms Act 1977 and so could not be relied upon. Both these arguments failed. The reasonableness argument concerned a qualification that had been made to the RIBA standard form, which allowed termination on 'reasonable notice' that Westminster had defined as not less than one month. It was found that this was reasonable and that the parties should not be locked into a contract of this nature.

Hedley Byrne & Co. Ltd v. *Heller and Partners Ltd* (1964)

A firm of advertising agents gave credit to a client in reliance upon a banker's reference, and suffered loss when the client became insolvent. The reference had been given carelessly, but since the bank had expressly disclaimed liability when giving it, the action failed. Nevertheless, the House of Lords stated that, contrary to what had previously been believed, liability for negligence may extend to careless words as well as to careless deeds and that damages may be awarded for financial loss as well as for physical injury to persons and property.

W Higgins Ltd v. *Northampton Corporation* (1927)

Higgins contracted with the local authority to erect fifty-eight houses. However, Higgins had completed their tender incorrectly. They thought they were tendering to erect the houses at £1,670 a pair, when in fact it was £1,613 because of the way in which Higgins had made up the bills. It was held that Higgins were bound by this mistake and could not claim to have the contract set aside or rectified.

Howard Marine v. *Ogden and Sons* (1978)

Ogdens, a firm of contractors, were one of the tenderers for the construction of a sewage works for the Northumbrian Water Authority. Surplus excavation material was to be loaded into barges, shipped downstream and dumped at sea. Howards were a firm who owned barges that were capable of doing this work. A quotation for the hire of the barges specified the volume of material that each barge could carry, and this was 850 cubic metres. On this basis, the contractor could establish the number of trips necessary and their effect upon the programme. But one fact was overlooked. Each vessel travelling on water has a safe loading line which depends upon the weight of the material. The amount carried was therefore 850 cubic metres at the most, but, because the spoil was clay, this turned out to be considerably less in practice. Various telephone conversations took place between Howards and Ogdens, but misunderstandings occurred. To further complicate matters, the payload figure in *Lloyd's Register* was incorrect. This was stated as 1,600 tonnes, but was in fact only 1,055 tonnes. Although Ogdens had prudently based their calculations on a lower figure than the agreed 850 cubic metres, the barges failed to carry even this volume. Ogdens then refused to pay for any more barges, and Howards withdrew their transport. Howards claimed for their outstanding hire, and Ogdens counterclaimed for misrepresentation and damages.

The moral of the story is that, when making enquiries, it is unsafe to rely upon telephone conversations alone. Such answers may be too casual and of little legal consequence. Either independent advice should be sought or, preferably, the salient facts should be in writing.

The principle of *Hedley Byrne* v. *Heller and Partners* (1964) came up for consideration by the Court of Appeal. This stated: 'When an enquirer consults a businessman in the course of his business and makes it clear to him that he is seeking considered advice and intends to act on it in a particular way, a duty to take care in giving such advice arises and the adviser may be liable in negligence if the advice he gives turns out to be bad'.

IBA v. *EMI and BICC* (1980)

EMI were the main contractors on a project for the Independent Broadcasting Authority (IBA). BICC were nominated subcontractors for an aerial mast. The mast collapsed because of BICC's failure to consider the effects of asymmetric ice loading on the struts. Although EMI took no part in the design process, they had accepted the contractual responsibility for the adequacy of the design. It was stated that one who contracts to design and supply an article for a known purpose, must ensure that it is fit for that purpose.

Owing to the complexity of modern buildings, it is often necessary for the architect to employ specialists. Some of these specialists may be independent consultants or subcontractors. One of the functions of the architect is to coordinate the design to ensure it is compatible with the overall scheme and is reasonably fit for the purpose intended.

***Ibmac Ltd* v. *Marshall (Homes) Ltd* (1968)**

The defendants, who were building a council housing estate, employed the plaintiffs to build a roadway to serve it. The plaintiffs quoted a price for the work, supplying a bill of quantities with rates for the work. The plaintiffs found that the work was more difficult than they had envisaged, since the site lay at the bottom of a steep hill. There were also serious difficulties with water. The plaintiffs abandoned the works and claimed payment for the work that they had done. Their claim failed in the Court of Appeal.

***Killby & Gayford Ltd* v. *Selincourt Ltd* (1973)**

In a letter, the architects asked a firm of contractors to price alteration work. The letter concluded: 'Assuming that we can agree a satisfactory contract price between us, the general conditions and terms will be subject to the normal standard form of RIBA contract'. The contractors submitted a written estimate. The architect replied that he was accepting the estimate on behalf of the client and he wanted the contractors to accept his letter as their formal instruction to proceed with the work. No JCT contract was ever signed, but the contractors proceeded with work. In the Court of Appeal, it was held that the exchange of letters incorporated the current JCT form.

***William Lacey (Hounslow)* v. *Davis* (1957)**

The plaintiff tendered for the reconstruction of war-damaged premises belonging to the defendant, who led the plaintiff to believe that they would receive the contract. At the defendant's request, the plaintiff calculated the timber and steel requirements for the building, and prepared various further schedules and estimates which the defendant made use of in negotiations with the War Damage Commission. Eventually, the plaintiff was informed that the defendant intended to employ another builder to do the work. In fact, the defendant sold the premises. The plaintiff claimed damages for breach of contract, and, alternatively, remuneration on a *quantum meruit* basis in respect of the work done by it in connection with the reconstruction scheme. It was held that although no binding contract had been concluded between the parties, a promise should be implied that the defendant would pay a reasonable sum to the plaintiff in respect of the services rendered. Judgment was entered for the plaintiff.

***Lewis* v. *Brass* (1877)**

An architect invited tenders for building works. The defendant's tender read, 'I hereby agree to execute complete, within the space of 26 weeks from the day receiving the instructions to commence, the whole of the work required to be done ... for the sum of £4,193'. The architect replied that he had been instructed to accept the tender and a contract would be prepared, ready for signature within a few days. The defendant had made a mistake in their tender and tried to withdraw from the contract. In the Court of Appeal, it was held that the tender and acceptance formed a contract. A tender and its acceptance may therefore amount to a contract, even though the acceptance refers to a formal contract to be drawn up afterwards.

***Sir Lindsay Parkinson & Co. Ltd* v. *Commissioners for Work* (1949)**

A construction contract was agreed on the basis that the contract price would be the cost of the works, plus a net profit not exceeding £300,000. The project was a typical cost-plus contract carried out at this time. Additional works were ordered, which extensively increased the size of the project. The court held that the contractor was entitled to a *quantum meruit* payment for this additional work, since this work could not have been envisaged at the start of the project.

London Borough of Newham v. *Taylor Woodrow (Anglian) Ltd* (1980)

This infamous case was for a contract to erect a twenty-two-storey block of flats, known as Ronan Point. The building was of a prefabricated concrete construction. One morning, gas caused an explosion in a flat on the eighteenth floor. This resulted in the collapse of the south-east corner. The local authority claimed for the cost of repairing and strengthening a number of similar blocks. The court acquitted the contractor of negligence, largely on the basis that several other local authorities had approved and accepted the design. The court held, however, that the contractor was liable for a breach of contract on the basis that the flats should have been designed and constructed so that they would be safe and fit for their purpose. In addition, the building contract included a provision that the contractor would be responsible for any faults and repair work at their own expense. This case was again reopened in 1984 but the final decision on liability has still to be resolved.

London County Council v. *Wilkins, Valuation Officer* (1956)

This case established that the provision of temporary accommodation on-site by a contractor was rateable if erected for a sufficient period of time.

London School Board v. *Northcroft* (1889)

The quantity surveyor was responsible for measuring buildings up to a value of £12,000. The employers brought an action for negligence resulting from two clerical errors in the quantity surveyor's calculations. This involved overpayments to the builder of £118 and £15, respectively. It was held that the quantity surveyor, who had employed a skilled clerk who had carried out hundreds of intricate calculations correctly, was not liable, since he had done what was reasonably necessary.

Marston v. *Kigrass Ltd* (1989)

The plaintiffs claimed for preparatory works over and above the costs and work of preparing their tender for a design and build contract to provide a factory to replace one that had burnt down. The contract was never placed since the funds from the insurance company were insufficient to cover the costs of rebuilding. The plaintiffs' tender was the best value for money, but because of the tight timescale it needed to be supplemented with further details. The tender was followed by a vital meeting when the plaintiffs were the only tenderers invited to discuss their tender. The chairman of the company said that no contract could be entered into without the insurance money for rebuilding. Nothing was suggested that any preparatory work would be at the contractor's own risk in the event that the insurance money did not materialise. The defendants were well aware that the plaintiffs would need to start the preparatory work before the contract was signed. It was also made clear that, subject to this work being satisfactory, the contract would be given to the plaintiffs.

It was held that there was an express request made by the defendants to the plaintiffs to carry out a small quantity of design work and there was an implied request to carry out preparatory works in general. Both express and implied requests gave rise to a right to payment of a reasonable sum.

Minter v. *Welsh Health Technical Services* (1980)

This case was based upon a claim for loss and expense, the equivalent of clause 26, but on the 1963 form. The amounts paid were challenged on the basis that they had not been certified and paid until long after the contractor had incurred the loss. The courts held that direct loss and/or expense were to be read as conferring a right to recover sums on the same principle as common law damages. In essence, therefore, this might include compensation for the loss of the use of capital.

Molloy v. *Liebe* (1910)

A contract provided that no works beyond those that were included in the contract would be paid for by the employer, unless it was in writing from the employer and the architect. During the progress of the works, the employer required certain work to be carried out. The employer insisted that these were within the terms of the contract. The contractor insisted that they represented extra works and must be paid for separately. It was held that it was open to the arbitrator to decide. If the arbitrator concluded that the works were not part of the original contract, then the employer by implication agreed to pay for them.

Murphy v. *Brentwood District Council* (1991)

A council, in approving the plans for the building of a house, relied on the negligent advice of an independent firm. The plaintiff bought the house on completion, but, owing to the defective design of the foundations, cracks appeared in the walls and service pipes leaked and fractured. The total cost of the repair was £45,000. The plaintiff could not afford to pay these repair costs and sold the house for £30,000, which was £35,000 less than the market price had the house been in a sound condition. He subsequently claimed for damages for negligence against the council since they had approved the plans for the building. Initially, the courts held that the house constituted an imminent danger to health and safety and the plaintiff was awarded £38,777 damages due to the diminution in the value of the property and other losses sustained. The council applied to the Court of Appeal, but this was dismissed. The council then appealed by leave of the Court of Appeal. A careless builder ought to be liable where a latent defect resulted in physical injury to a person or property. The principle in *Donoghue* v. *Stevenson* was not appropriate to extend liability to an occupier who knew the full extent of the defect but continued to occupy the property. The courts had previously characterised the damage as physical damage, but this was mistaken since the damage had been pure economic loss. As such, this had the potential for collision with long-established principles regarding liability in negligence for economic loss.

To permit the plaintiff to recover his economic loss would logically lead to an unacceptably wide category of claims in respect of buildings or chattels which were defective in quality, and would in effect introduce product liability and transmissible warranties of quality into the law of tort by means of judicial legislation.

***Re Newman ex parte Capper* (1876)**

This contract included a provision of a large fixed amount of liquidated damages payable by the contractor in the event of any breach occurring. The court held that this was in effect a penalty since it could not relate with all the varying amounts of loss which might arise. It was not therefore enforceable. The courts did, however, fix an appropriate amount that was upheld as being damages sustained.

***North West Metropolitan Regional Hospital Board* v. *T A Bickerton Ltd* (1976)**

This case placed the responsibility of renomination with the architect rather than the contractor, after a nominated subcontractor had defaulted. A nominated subcontractor went into liquidation before they had completed their work. The main contractor then requested the architect under the terms of the contract (1963 edition) to appoint a new firm as a successor to this nominated subcontractor. Because a new price was required, and that this was likely to be a higher price, the employer suggested that the contractor was responsible for completing this work in any way, but to the approval of the architect. This may result in the contractor doing the work themselves or subletting with approval, but that any extra cost must be borne by the main contractor.

The 1963 edition of the form of contract is unclear on this point, but the case was decided in the contractor's favour. It is the architect's responsibility to renominate in the event of default, and the employer is therefore obliged to pay the additional cost of the new firm employed. JCT 80 clarified this point along the lines of this decision.

***Ocean Leisure Ltd* v. *Westminster City Council* (2004)**

Statutory power given to a local authority to carry out work on the highway did not relieve it of the duty not to harm the occupiers of property adjoining the highway.

The current state of statutory law governing compensation for such damage was unsatisfactory. The Court of Appeal so held in dismissing the appeal of the defendant, Westminster City Council, against the decision of the Lands Tribunal, which held that the site hoarding erected and left by the council along Northumberland Avenue outside the shops of the claimant, Ocean Leisure Ltd, along Victoria Embankment during construction of two footbridges across the Thames on either side of Hungerford railway bridge caused injurious affection to the claimant's property. The bridges, known as the Golden Jubilee Bridges, were a project promoted by a consortium comprising Westminster City Council, Lambeth London Bridge Council, Railtrack plc, London Underground Ltd and the Port of London Authority. Westminster was to carry out the work.

Ocean Leisure brought proceedings for compensation against Westminster relying on Section 10 of the Compulsory Purchase Act 1965, which provided:

> if any person claims compensation in respect of any land, or any interest in land, which has been taken for or injuriously affected by the execution of the works, and for which the acquiring authority have not made satisfaction under the provisions of this Act, any dispute arising in relation to the compensation shall be referred to and determined by the Lands Tribunal.

***Oram Builders Ltd* v. *M J Pemberton and C Pemberton* (1985)**

This project was on the minor works agreement (JCT 1980 edition). Article 4 is an arbitration agreement in general terms. The supervising officer issued instructions to the plaintiffs which constituted a substantial variation to the contract. The plaintiffs claimed to have done the extra work and so required a certificate for its payment. This was not forthcoming and the plaintiffs claimed extra costs for the work or, alternatively, damages. The defendants disputed the plaintiffs' claim and counterclaimed for alleged breaches of contract. The preliminary issue, with which this judgment is concerned, was whether the courts had powers to review the exercise of the supervising officer's discretionary powers under the contract and open up certificates. It was held that the courts had no jurisdiction to go behind a certificate of the architect or supervising officer. Where there is an arbitration clause in general terms referring any dispute or difference between the parties concerning the contract to the arbitrator, then even on a narrow interpretation of the reasoning of the Court of Appeal in *Northern Health Authority* v. *Derek Crouch Construction Co. Ltd* (1984), the High Court has no jurisdiction to go behind an architect's certificate.

Peak Construction Ltd v. *McKinney Foundations Ltd* (1970)

This case considered two separate matters concerning a project that overran its contract period. The first problem was concerned with an extension of time. McKinney Foundations Ltd were the nominated subcontractors for piling work. Defects were found upon completion of this work, but there was a prolonged delay on the part of the employer in both obtaining an engineer's advice and deciding upon the remedial work to be carried out. When the project overran, the employer levied liquidated damages on the main contractor. The main contractor then attempted to recover these damages from the subcontractor. Furthermore, the above was not a reasonable cause for granting an extension of time within the terms of the contract. It was held, however, that because the delay was due largely to the employer's default, liquidated damages could not be deducted. This was the case if the extension of time clause did not make provision for such a delay, or even if there was failure to extend the time.

The second matter was concerned with the price fluctuation clause. The court held that this clause continued to operate after the time for completion was past. This matter was clarified in JCT 80.

Peter Lind & Co. Ltd v. *Mersey Docks and Harbour Board* (1972)

The contractor submitted to the Board alternative tenders for the construction of a freight terminal. One was on a fixed-price basis and the other using cost reimbursement. The Board accepted the tender but failed to specify which alternative would be used for payment purposes. The contractor carried out the work and claimed payment on a *quantum meruit* basis. It was held that there was no concluded contract because the defendant's acceptance did not specify which tender was to be accepted and that the plaintiffs were entitled to payment on this basis.

Porter v. *Tottenham Urban District Council* (1915)

The plaintiff contracted to build a school for the council upon land belonging to them. The contract provided that the plaintiff should be entitled to enter the site immediately, and that the works should be completed by a specified date. The only access to the site was from an adjoining road and to lay a temporary sleeper road and then subsequently a permanent pathway. The plaintiff began work, but was forced to abandon it because of a threatened injunction from an adjoining owner, who claimed that the road was their property. The third party's claims were held to be unfounded. The plaintiff completed the works then claimed damages against the council in respect of the delay caused by the third party's action. In the Court of Appeal, the plaintiff's claim failed. It was stated that there was no implied warranty by the council against wrongful interference by third parties with free access to the site.

Raffles **v. *Wichelhaus* (1864)**

Raffles made a contract to sell 125 bales of cotton to Wichelhaus. The goods would be shipped from Bombay in India to Liverpool in England on a ship named *Peerless*. Unbeknown to either party, there were two ships with this identical name that each carried cargo from Bombay to Liverpool. One of the ships would arrive in October and the other ship in December. Wichelhaus believed that he would receive his cotton arriving on the October ship, but Raffles sent his cotton on the ship arriving in December. Wichelhaus refused to accept delivery of the cotton arriving from the December ship and Raffles then tried to sue for a breach of contract. The case established that when both parties are mistaken as to an essential element of the contract, the court will first attempt to find a reasonable interpretation from the context of the agreement before making the contract void. Raffles argued that the only purpose of naming the ship was that in the event of the ship sinking the contract would then become void. The issue then before the court was whether Wichelhaus should be bound to buy the cotton from the other *Peerless* ship which arrived in December.

The court ruled that no binding contract existed since the parties did not agree on the same thing. There was no meeting of minds and no mutual understanding of the contract. The main defence was that the contract was too ambiguous and this was accepted by the court. If a misunderstanding concerns a material fact and neither party knows or has reason to know of the misunderstanding, there is no contract.

Regalian Properties plc **v. *London Dockland Development Corporation* (1994)**

Regalian Properties plc offered to purchase a licence for the residential development of land from the defendant on terms that the defendant would grant them a building lease. This offer was accepted in a letter that included three conditions. Firstly, it would be subject to contract; secondly, it would be subject to the District Valuer's market value; and, thirdly, the scheme must achieve the desired design quality and planning consent obtained.

The building lease was never granted because of delays caused by the defendant in requesting further designs together with delays in them becoming owners of all the land. By the time these difficulties had been overcome, the value of the land had fallen dramatically. The plaintiffs sought to recover from the defendant nearly £3 million in professional fees which they had incurred in connection with the proposed development. It was held, however, that by using the words 'subject to contract' each party had taken the risk that the transaction would not come to fruition.

Roberts & Co. Ltd **v. *Leicestershire County Council* (1961)**

A contractor submitted a tender which specified a completion of eighteen months. The county architect, unknown to the contractor, decided that thirty months was a more appropriate contract period. Prior to the signing of the contract, there were two meetings at which the contractor referred to their plans to complete the work according to a progress schedule in eighteen months. It was held that the contractor was entitled to rectification on the grounds that the defendants were estopped by their conduct from saying that there was no mistake.

***Rotherham MBC* v. *Frank Haslam Milan and Co. Ltd* (1996)**

What amounts to a reasonable degree of skill and care for a product fit for its purpose will depend upon the circumstances of a particular case. Fitness for purpose is an absolute obligation, and the purpose may be express or implied. Where liability is strict, if something is required to be fit for purpose but is not, a contractor will not escape liability for demonstrating that reasonable skill and care were exercised.

But the situation may not be straightforward – few things are. In this case, the employer specified in some detail the hardcore to be used as fill around foundations and provided testing and approval by the architect. The contractor used the hardcore described. It failed. The issue was the extent of the contractor's obligations and its liability for defects.

The Court of Appeal refused to place on the contractor an obligation of ensuring fitness for purpose of the fill material where the contractor had used a material that complied with the architect's specification, even though it was unsuitable. The architect had described several materials which the contractor could use. It was not a good specification because, by being too broad, it allowed the contractor to use an unsuitable material. The court held that the employer was relying on the expertise of the architect and not the contractor.

***Scott* v. *Avery* (1856)**

This case concerned an insurance policy which indicated that only arbitration proceedings were permissible in the event of a dispute occurring. It therefore sought to prohibit direct legal actions in the courts. It was held that this provision was valid. The majority of the standard forms of building contract contain arbitration provisions. This course of action is agreed to between the parties, who may consider arbitration more appropriate. If one of the parties wishes to hold to arbitration, they may apply for a stay of proceedings on the basis that the contract follows the *Scott* v. *Avery* principle. In practice the courts, in deciding whether to grant this, are likely to take into consideration any hardship that may be caused.

***Sport* v. *Zittrer* (1988)**

The receiver of a bankrupt company sold a large part of its assets to a company which, on the same day, resold the assets to the appellant. Clause 2.01 of the agreement with regard to this latter sale provided that the valuation of the inventory would be reviewed by the bankrupt's auditors, the respondents, who were to take into consideration the representations of the appellant. The respondents had then to deliver a written opinion to all the parties to the effect that such inventory count and valuation was fairly presented, the whole at the cost of the bankrupt. Upon delivery of such opinion, the inventory count and valuation was to be deemed to be 'definitively determined'. In a letter sent to the appellant, the bankrupt agreed that the appellant attend the valuation of the inventory and make any representations, adding that the respondents' valuation of the inventory would be 'final and binding'. The respondents confirmed the bankrupt's valuation of the inventory and the appellant paid the amount so established. A year later, the appellant brought an action for damages in the Superior Court against the respondents in the amount representing the difference between the price it had paid and the price it would have paid if the respondents had not been negligent in performing their task. The respondents made an exception alleging that they were acting in this matter as arbitrators and that, as such, they were covered by immunity. The Superior Court dismissed their motion. The Court of Appeal set aside that judgment. In its opinion, the respondents were acting as arbitrators and, in the absence of fraud or bad faith, they enjoyed the immunity from civil liability.

This appeal is to determine whether the parties had agreed to submit a dispute to arbitration by a third party. This appeal must be disposed of according to the provisions then applicable, namely parts 940 to 951 of the 1965 Code of Civil Procedure in force prior to the 1986 amendments. It was held that the appeal should be allowed.

Sutcliffe **v.** ***Thackrah and Others*** **(1974)**

The facts of this case were that Mr Sutcliffe employed Thackrah and Others, a firm of architects, to design and supervise the construction of a house. During construction, two certificates (Nos 9 and 10) were issued and duly paid by Mr Sutcliffe. Shortly afterwards, the contract with the builders ended when the builders became insolvent and another firm completed the work.

It was then discovered that the two certificates had included defective work. Because it could not be recovered from the builders, Mr Sutcliffe sued the architects for his loss because of their negligence in issuing the certificates. The negligence was not due to a failure to detect the defective work, since one of the architects was aware of this; it was due to a failure to pass the information to the quantity surveyor, who had valued it assuming that it was in accordance with the contract.

The official referee who tried the case in arbitration, held that the architect was negligent and awarded damages to Mr Sutcliffe. The Court of Appeal reversed this decision, but this was finally overruled by the House of Lords. It was held that the architect was discharging a duty under the contract between the parties, but that there was no dispute over the subject matter and so it could not be contended that he was acting in arbitration.

In order for quantity surveyors to safeguard themselves from possible repercussions in the future, they now include on their valuations words to the following effect: the valuation assumes that the work is in accordance with the specification of materials and work standards. A quantity surveyor would clearly not include in a valuation work that was obviously defective, but the matter of quality control is more rightly the responsibility of the architect. The case of *Chambers* v. *Goldthorpe* was cited, but this was overruled by the House of Lords.

George E. Taylor & Co. Ltd **v.** ***G. Percy Trentham*** **(1980)**

Taylor & Co. were nominated subcontractors to Trentham. They also had a contract with the employer whereby they warranted due performance of the subcontract works so that the main contractors should not become entitled to an extension of time. The employer paid Trentham only £7,526 against an interim certificate of £22,101. The amount withheld was the balance payable to the subcontractors after deduction of the main contractor's claim against them for delay. It was held that the employer was not entitled to withhold the money, the contract between the employer and the subcontractor being *res inter alios acta*.

Trollope & Colls and Holland and Hannan and Cubitts Ltd **v.** ***Atomic Power Construction*** **(1962)**

The contractor tendered for work under a contract which included provisions for variations and fluctuations. After a delay of four months, the contractor was requested to commence work upon the site. This request included a letter of intent to enter into a formal contract when the terms of the contract were settled. Substantial changes to the scheme had been notified to the contractor and this continued to occur until the terms of the contract were agreed ten months later. The contractor therefore claimed that they should be paid on a *quantum meruit* basis for the work that had already been completed prior to the signing of the formal agreement. This, they claimed, was more equitable than on the basis of their original tender adjusted by variations, since at that time no contract was in existence.

It was held, however, that both parties had contemplated entering into a contract which would affect all the work associated with the contract. The contract was formulated in the light of the work going on at the time and it therefore had a retrospective effect.

Trollope & Colls Ltd v. *North West Metropolitan Regional Hospital Board* (1973)

This was a very large building contract where the Court of Appeal reversed the decision of the trial judge, but the House of Lords confirmed the trial judge's opinion. The Hospital Board wanted the project completed in three phases, using three largely similar sets of conditions. Phase 1 was delayed by fifty-nine weeks and the architect granted an extension of time by forty-seven weeks. Phase 3 was to start contractually six months after phase 1 had received its certificate of practical completion. Phase 3 did, however, still retain its original completion date, which then envisaged a sixteen-month contract period rather than the previously accepted thirty-month contract period.

In these circumstances, could the time for completion of phase 3 be extended by the period of forty-seven weeks granted on phase 1? Unexpectedly, it was the contractor who suggested that it could not, but for their own advantage. The contractor professed to being able to complete in time, and they therefore called upon the Hospital Board to nominate their appropriate sub-contractors. This they were unable to do. In this situation the contractors would require new prices to be agreed relevant to the prevailing date. The contractors therefore sought a declaration in the High Court that the date for completion of phase 3 was unaffected by phase 1 being behind schedule.

The decision was made in favour of the contractor on the basis that the business efficacy of this large contract did not necessitate implying the kind of term the Board required. The House of Lords was not satisfied that the parties had overlooked the effect of delays on earlier phases, in fixing the completion date for phase 3.

Tyrer v. *District Auditor for Monmouthshire* (1974)

A firm of contractors who went into liquidation had been employed upon several contracts by a local authority. The local authority found that this firm had been overpaid on interim certificates. The district auditor established that overpayment was due to the negligence of the quantity surveyor, in accepting excessively high rates for work carried out and failing to check simple arithmetic. The auditor then surcharged the quantity surveyor for the sums that he was unable to recover from the liquidator. The quantity surveyor appealed on the grounds that he had undertaken his duties in the capacity of an arbitrator. The court, however, rejected this appeal stating that 'there was nothing to show that the appellant was in quasi-judicial position when carrying out his duties here'.

Victoria Laundry Ltd v. *Newman Ltd* (1948)

This case provides an interesting contrast with the case of *Hadley* v. *Baxendale*. The plaintiffs were launderers and dyers and required a large boiler to extend their plant and to help them win some lucrative contracts. A firm of engineers contracted to sell them such a boiler, but certain faults arising meant its delivery was seriously delayed. The plaintiffs claimed damages: firstly, equivalent to the estimated loss of the increased profits the use of the boiler would have acquired for them; and, secondly, the amount that they would have earned from dyeing contracts during the same period. The court held that the engineers were liable for the ordinary losses which they must have known from the particular circumstances. They were not liable in the second instance since this would have required special knowledge which they did not have.

University of Warwick v. *Sir Robert McAlpine* (1989)

The buildings at University of Warwick had ceramic tile cladding on a concrete frame. The cladding, which had been carried out by subcontractors, began to fail. The architects blamed bad work and McAlpine, the main contractors, blamed the design. The university decided to remedy the defects by an innovative resin injection process. The sole licensees of the process were a firm known as CCL. CCL were recommended by the architect. The main contract conditions were varied and, though CCL were technically not nominated subcontractors, they were employed by McAlpine to carry out the remedial works.

McAlpine had considerable reservations about the use of this new process. The remedial works were not successful and the university alleged that McAlpine were in breach of an implied fitness for purpose. It was held that a term that resin be fit for its purpose could only be implied in the main contract if the university had relied on McAlpine. As they had not done so, no such term was to be implied.

Wells v. *Army & Navy Co-operative Society Ltd* (1902)

A building contract provided that certain matters causing delays and other causes beyond the contractor's control were to be submitted to the decision of the directors of the employer. The directors were to adjudicate thereon and make due allowance therefore if necessary, and their decision would be final. In the contract there was a liquidated damages clause. There was a year's delay in completion, and the directors allowed a three-month extension of time for delays caused by subcontractors. Other breaches were established, including failure by the employer to give possession of the site and failing to provide plans and drawings in due time. It was held that the words 'beyond the contactor's control' did not extend to the delay caused by the interference of the employers or their architect. Liquidated damages could not therefore be deducted and the contractor's obligation was to complete within a reasonable period of time.

City of Westminster v. *J. Jarvis & Sons Ltd and Peter Lind* (1970)

Jarvis contracted with Westminster for the erection of a multi-storey car park, with flats, offices, showrooms and ancillary works. The contract was in JCT 63 form. Lind were nominated subcontractors for the piling work. Lind carried out the work and purported to complete by the subcontract completion date. Lind left the site. Some weeks later it was discovered that many of the piles were defective, either as a result of bad work standards or poor materials. Lind carried out the remedial works. The main contract works were delayed for some 21.5 weeks. Jarvis, the main contractor, claimed an extension of time under clause 23(g). In the House of Lords, it was held that Jarvis was not entitled to an extension of time since the works had apparently been completed by the due date. A 'delay' within the meaning of the clause occurs only if, by the subcontract completion date, the subcontractor has failed to achieve such completion of their work that they cannot hand it over to the main contractor.

Re Wilkinson ex parte Fowler (1905)

This contract allowed for the right to make direct payment to certain firms, where the main contractor defaulted or delayed in proper payment to them. The contractor became bankrupt and the employer decided to make payments out of the retentions owing to the contractor. It was held by the courts that such payments were valid against the trustee.

William Sindall v. *North West Thames Regional Health Authority* (1977)

The main contractor introduced a bonus incentive scheme on a site in accordance with the principles recommended by the National Joint Council for the Building Industry (NJCBI). The contract was on a fluctuations basis incorporating clause 31A. An increase in the basic rates of wages occurred and this, in a bonus scheme, was a voluntary decision and not an unavoidable consequence of following the obligatory rules or decisions, and the cost of its operation was therefore outside the scope of clause 31A. This was in spite of the fact that once the contractor had introduced their bonus scheme on the basis of the NJCBI rules, they had also to pay the eventual increase.

Williams v. *Fitzmaurice* (1859)

This involved the contractor in building a house in accordance with the drawings and specification supplied by the architect. The specification included a clause that the contractor undertook 'to provide the whole of the materials necessary for the completion of the works and to perform all the works of every kind mentioned'. Floorboarding, although shown on the drawings, was not included in the specification. The contractor therefore refused to carry out this work unless it was paid for as extra works. It was held that the boarding was necessary and was therefore included in the contract price, even though it had been omitted from the specification.

Appendix C

Contract and procurement websites

ABE	www.abe.org.uk
Achieving Excellence in Construction	www.ogc.gov.uk
Adjudication	www.adjudication.co.uk
Association of Researchers in Construction Management	www.arcom.ac.uk
Bar Council	www.barcouncil.org.uk
Building bookshop	www.constructionbooks.net
Building magazine	www.building.co.uk
Building regulations	www.communities.gov.uk
Building Research Establishment	www.bre.co.uk
Centre for Education in the Built Environment	www.heacademy.ac.uk/cebe
Chartered Institute of Building (CIOB)	www.ciob.org.uk
CIB research and innovation	www.cib.nl
CITB-Construction Skills	www.citb.co.uk
Constructing Excellence	www.constructingexcellence.org.uk
Construction Index	www.theconstructionindex.co.uk
Construction Industry Council	www.cic.org.uk
Construction Industry Research and Information Association	www.ciria.org.uk
Construction law	www.building.co.uk/legal/case-law
Construction law cases	www.i.law.com/construction
Construction Law Directory	www.constructionlawdirectory.com
Construction News	www.cnplus.co.uk
Construction procurement	www.constructingexcellence.org.uk
Construction Procurement Manual	www.scotland.gov.uk
Construction Skills Certification Scheme	www.cscs.uk.com
Construction Web Links	www.constructionweblinks.com
Department of Business, Innovation and Skills	www.bis.gov.uk
Health and Safety Executive	www.hse.gov.uk
Institution of Civil Engineers (ICE)	www.ice.org.uk
Joint Contracts Tribunal	www.jctcontracts.com
Law on the Web	www.lawontheweb.co.uk/construction
Law Society	www.lawsociety.org.uk
Office for National Statistics	www.ons.gov.uk
Pearson Education	www.pearsoned.com
Pinsent Masons	www.pinsentmasons.com
Rethinking Construction	www.dti.gov/construction
RIBA	www.architecture.com
RICS	www.rics.org.uk
Society of Construction Law	www.scl.org.uk

Strategic Forum for Construction	www.strategicforum.org.uk
Sweet and Maxwell	www.sweetandmaxwell.co.uk
UK Construction	www.ukconstruction.com
UK Legislation	www.legislation.gov.uk
UK Register of Expert Witnesses	www.jspubs.com
Workplace Law Network	www.workplacelaw.net

Bibliography

Abeyratne, S. A. and Monfared, R. P. (2016) 'Blockchain ready manufacturing supply chain using distributed ledger', *International Journal of Research in Engineering and Technology*, Vol. 5, Iss 9, pp. 1–10, http://esatjournals.net/ijret/2016v05/i09/IJRET20160509001.pdf (accessed 6 October 2017).

Addis, B. and Talbot, R. (2001) *Sustainable Construction Procurement: A Guide to Delivering Environmentally Responsible Projects*. Construction Industry Research and Information Association.

Arrowsmith, S. (ed.) (2000) *Public Private Partnerships and PFI*. Sweet and Maxwell.

Ashworth, A. (2008) *Pre-Contract Studies: Development Economics, Tendering and Estimating*, 3rd edn. Blackwell.

Ashworth, A. (2010) *Cost Studies of Buildings*, 5th edn. Prentice Hall.

Ashworth, A. and Hogg, K. I. (2000) *Added Value in Design and Construction*. Pearson Education.

Ashworth, A. and Hogg, K. I. (2007) *Willis's Practice and Procedure for the Quantity Surveyor*, 12th edn. Blackwell Science.

Ashworth, A. and Perera, S. (2015) *Cost Studies of Buildings*, 6th edn, Routledge.

Atkin, B., Flanagan, R. and Marsh, L. (1995) *Improving Value for Money in Construction*. Royal Institution of Chartered Surveyors.

Babatunde, S. O. and Perera, S. (2017a) 'Analysis of traffic revenue risk factors in BOT road projects in developing countries', *Transport Policy*, Vol. 56, pp. 41–9.

Babatunde, S. O. and Perera, S. (2017b) 'Analysis of financial closure delay in PPP projects in transition countries', *Benchmarking*, Vol. 24, Iss 7, pp. 1690–708.

Babatunde, S. O. and Perera, S. (2017c) 'Barriers to bond financing for public–private partnership infrastructure projects in emerging markets: a case of Nigeria', *Journal of Financial Management of Property and Construction*, Vol. 22, Iss 1, pp. 2–19.

Babatunde, S. O., Perera, S. and Zhou, L. (2016) 'Methodology for developing capability maturity levels for PPP stakeholder organisations using critical success factors', *Construction Innovation*, Vol. 16, Iss 1, pp. 81–110, www.emeraldinsight.com/doi/full/10.1108/CI-06-2015-0035 (accessed 11 October 2017).

Banwell, P. (1964) *The Placing and Management of Contracts for Building and Civil Engineering* [known as The Banwell Report]. HMSO.

Banwell, P. (1967) *Action on The Banwell Report*. HMSO.

Barlow, J. (1996) *A Statement on the Construction Industry* [known as the Barlow Report]. Royal Academy of Engineering.

Bennett, J. and Jayes, S. (1995) *Trusting the Team: The Best Practice Guide to Partnering in Construction*. Centre for Strategic Studies in Construction, University of Reading.

Bennett, J. and Jayes, S. (1998) *The Seven Pillars of Partnering*. Reading Construction Forum and Thomas Telford.

Bowmaster, J., Rankin, J. and Perera, S. (2016) 'E-business in the Atlantic Canadian architecture, engineering and construction (AEC) industry', Report published by CIB TG83: e-Business in Construction, University of New Brunswick, Canada, and Western Sydney University, Australia, doi: 10.13140/RG.2.2.15753.44647.

BPF (1983) *Manual of the British Property Federation System*. British Property Federation.

Bruges Group (1983) *Construction Contract Arrangements in the European Union (EU) Countries*. European Construction Institute.

BSRIA (2017) US Smart Connected HVAC in commercial buildings soars according to 2017 study, BSRIA, www.bsria.co.uk/news/article/us-smart-connected-hvac-in-commercial-buildings-soars-according-to-2017-study/ (accessed 30 October 2017).

Burgess, R. A. (ed.) (1982) *Construction Projects, Their Financial Policy and Control*. Longman.

Cabinet Office (2011) *Government Construction Strategy (2011–2015)*, www.gov.uk/government/publications/government-construction-strategy (accessed 29 September 2017).

Cabinet Office (2012) *Government Construction Strategy: Final Report to Government by the Procurement/Lean Client Task Group*, www.gov.uk/government/publications/government-construction-task-groups (accessed 29 September 2017).

Cabinet Office and Infrastructure and Projects Authority (2012) *Project Bank Accounts: Briefing Document and A Guide to the Implementation of Project Bank Accounts (PBAs) in Construction for Government Clients* [two documents], www.gov.uk/government/publications/project-bank-accounts (accessed 29 September 2017).

Cabinet Office and Infrastructure and Projects Authority (2016) *Government Construction Strategy 2016–2020*, www.gov.uk/government/publications/government-construction-strategy-2016-2020 (accessed 29 September 2017).

Cabinet Office and Office of Government Commerce (2003) *Building on Success: The Future Strategy for Achieving Excellence in Construction*, Constructing Excellence in the Built Environment, http://webarchive.nationalarchives.gov.uk/20110802160126/http://www.ogc.gov.uk/documents/BuildingOnSuccess.pdf (accessed 11 October 2017).

Cain, C. T. (2004) *Profitable Partnering for Lean Construction*. Blackwell.

Carroll, B. and Turpin, T. (2009) *Environmental Impact Assessment: A Practical Guide for Planners, Developers and Communities*. Thomas Telford.

CBI (2008) *Property for Business: An Essential Guide for Senior Executives*. Confederation of British Industry ebooks.

Centre for Strategic Studies in Construction (1996) *Design and Building: A World-Class Industry*. University of Reading.

CIOB (2010) *Code of Practice for Project Management for Construction and Development*. Wiley-Blackwell.

CITB (2003) *Construction Skills Foresight Report 2003*. Construction Industry Training Board.

CITB (2005) *Construction Skills Network Forecast Report 2005*. Construction Industry Training Board, www.citb.co.uk/documents/research/csn%202017-2021/csn-national-2017.pdf (accessed 4 October 2017).

Clients Construction Forum (1998) *Survey of Construction Clients' Satisfaction 1998–99*. Clients Construction Forum.

Colverson, S. and Perera, O. (2012) *Harnessing the Power of Public-Private Partnerships: The Role of Hybrid Financing Strategies in Sustainable Development*. IISD, www.iisd.org/pdf/2012/harnessing_ppp.pdf (accessed 1 October 2017).

Construction Industry Board (1996) *Towards a 30% Productivity Improvement in Construction*. Thomas Telford.

Construction Industry Board (1997a) *Code of Practice for the Selection of Main Contractors*. Thomas Telford.

Construction Industry Board (1997b) *Constructing Success: Code of Practice for Clients of the Construction Industry*. Thomas Telford.

Construction Industry Board (2000) *Key Performance Indicators*. Construction Industry Board.

Construction Industry Council (1999) *How to Rethink Construction: Practical Steps in Improving Quality and Efficiency*. Construction Industry Council.

Construction Industry Council (2000) *Project Team Partnering*. Construction Industry Council.

Construction Industry Training Board (2003) *Construction Skills Foresight Report 2003*. Construction Industry Training Board.

Construction Sponsorship Directorate (1995) *Fair Construction Contracts*. Department of the Environment.
Construction Sponsorship Directorate (2002) *Construction Clients' Forum*. Department of the Environment and the Central Unit on Procurement of HM Treasury.
Dallas, M. (2006) *Value and Risk Management: A Guide to Best Practice*. Blackwell Publishing.
Davies, W. H. (1982) *Construction Site Production: Level 4 Checkbook*. Butterworth.
Davis, L. (2001) *Buildings That Work for Your Business*. Institute of Directors/Kogan Page.
Dearle and Henderson (1988) *Management Contracting: A Practice Manual*. E & F N Spon.
Department for Communities and Local Government (2017) *Fixing our Broken Housing Market*, Housing White Paper, www.gov.uk/government/publications/fixing-our-broken-housing-market (accessed 29 September 2017).
Department of Business, Innovation and Skills (2010) *Low Carbon Construction*. HMSO.
Department of Business, Innovation and Skills (2013a) *Construction 2025: Industrial Strategy for Construction – Government and Industry in Partnership*, www.gov.uk/government/publications/construction-2025-strategy (accessed 29 September 2017).
Department of Business, Innovation and Skills (2013b) *Construction Sector Infographic: Why the Sector is Important to the UK economy*, www.gov.uk/government/uploads/system/uploads/attachment_data/file/229339/construction-sector-infographic.pdf (accessed 4 October 2017).
Department of the Environment (2008) *Environmental Assessment: A Guide to Procedures*. HMSO.
Digital Built Britain (2015) http://digital-built-britain.com/ (accessed 30 October 2017).
Eadie, R. and Perera, S. (2016) 'The state of construction e-business in the UK', Report published by CIB TG83: e-Business in Construction with Ulster University, UK, Northumbria University, UK, and Construct IT for Business, UK, www.researchgate.net/profile/Robert_Eadie/publication/305430113_The_State_of_Construction_e-Business_in_the_UK/links/578e7ecf08aecbca4caad1b5/The-State-of-Construction-e-Business-in-the-UK.pdf (accessed 11 October 2017).
Eadie, R., Perera, S. and Heaney, G. (2011a) 'Key process area mapping in the production of an e-capability maturity model for UK construction organisations', *Journal of Financial Management of Property and Construction*, Vol. 16, Iss 3, pp. 197–210, doi: 10.1108/13664381111179198.
Eadie, R., Perera, S. and Heaney, G. (2011b) 'Analysis of the use of e-procurement in the public and private sectors of the UK construction industry', *Journal of Information Technology in Construction*, Vol. 16, Special Issue Innovation in Construction e-Business, pp. 668–86.
Eadie, R., Perera, S. and Heaney, G. (2012a) 'Capturing maturity of ICT applications in construction processes', *Journal of Financial Management of Property and Construction*, Vol. 17, Iss 2, pp. 176–94, doi: 10.1108/13664381211246624.
Eadie, R., Perera, S. and Heaney, G. (2012b) 'Electronic procurement in the construction industry', Chapter 7. In: *Public Sector Transformation Processes and Internet Public Procurement: Decision Support Systems*, Pomazalová, N. (Ed), IGI Global.
Eadie, R., Millar, P., Perera, S., Heaney, G. and Barton, G. (2012) 'E-readiness of construction contract forms and e-tendering software', *International Journal of Procurement Management*, Vol. 5, Iss 1, pp. 1–26, doi: 10.1504/IJPM.2012.044151.
ECI (1997) *Partnering in the Public Sector: A Toolkit for the Implementation of Post Award, Project Specific Partnering on Construction Projects*. European Construction Institute.
Egan, J. (1998) *Rethinking Construction: The Report of the Construction Task Force* [known as the Egan Report]. Department of Environment, Transport and the Regions.
Elliott, C. and Quinn, F. (2012) *The English Legal System*, 3rd edn. Pearson Education.
Emmerson, R. (1962) *Survey of Problems before the Construction Industries* [known as the Emmerson Report]. HMSO.
Eyers, J. (2017) 'Blockchain "smart contracts" to disrupt lawyers', *Financial Review*, www.afr.com/technology/blockchain-smart-contracts-to-disrupt-lawyers-20160529-gp6f5e#ixzz4qCPlamYi (accessed 27 August 2017).
Fairclough, J. (2002) *Rethinking Construction Innovation and Research: A Review of Government R&D Policies and Practices* [known as the Fairclough Report]. Department of Trade and Industry.

Farmer, M. (2016) *The Farmer Review of the UK Construction Labour Model: Modernise or Die, Time to Decide the Industry's Future*. Construction Leadership Council.

Feng, Y., Teo, E. A. L., Ling, F. Y. Y. and Low, S. P. (2014) 'Exploring the interactive effects of safety investments, safety culture and project hazard on safety performance: an empirical analysis', *International Journal of Project Management*, Vol. 32, Iss 6, pp. 932–43.

Flanagan, R. (1999) *Linking Construction Research and Innovation to Research Innovation in Other Sectors*. CRISP.

Gajendran, T. and Perera, S. (in press), 'The Australian Construction e-business review', Report published by CIB TG83: e-Business in Construction with University of Newcastle, Australia, Western Sydney University, Australia, and the Australian Institute of Building (AIB).

Graves, A. (1998) *Constructing the Best Government Client: The Government Client Improvement Study*. Agile Construction Initiative, Bath University/Government Client Construction Panel.

Gray, C. (1996) *Value for Money*. Reading Construction Forum.

Halsbury, Earl of [Davies, A. and Roydhouse, E. (Eds)] (1991) *Halsbury's Statutes of England and Wales*, 4th edn. LexisNexis.

Harvey, R. C. and Ashworth, A. (1997) *The Construction Industry of Great Britain*, 2nd edn. Butterworth-Heinemann.

HM Treasury (2008) The Private Finance Initiative (PFI). *Public Private Partnerships* (archived), http://webarchive.nationalarchives.gov.uk/+/http://www.hm-treasury.gov.uk/documents/public_private_partnerships/ppp_index.cfm (accessed 30 October 2017).

HSE (2010) *Self-reported Work-related Illness and Workplace Injuries in 2008/09: Results from the Labour Force Survey*. HSE Information Services.

HSE (2014) *Risk Assessment: A Brief Guide to Controlling Risks in the Workplace*, www.hse.gov.uk/risk/controlling-risks.htm (accessed 5 October 2017).

HSE (2017) *Fatal Injuries*, www.hse.gov.uk/statistics/pdf/fatalinjuries.pdf (accessed 24 August 2017).

Iansiti, M and Lakhani, K. (2017) 'The truth about blockchain'. *Harvard Business Review*, https://hbr.org/2017/01/the-truth-about-blockchain (accessed 6 October 2017).

ICE (2015) *Civil Engineering Procedure*, 7th edn. Institution of Civil Engineers.

Ingirige, B., Perera, S., Ruikar, R. and Obonyo, E. (2017) 'Conclusions-summary, the status core and future trends', Chapter 20, In: *Advances in Construction ICT and e-Business*, Perera, S., Ingirige, B., Ruikar, R. and Obonyo, E. (Eds), Routledge.

Joint Contracts Tribunal (2016) *The Standard Form of Building Contract*, www.jctltd.co.uk/ (accessed 27 September 2017).

Kelly J., Morledge, R. and Wilkinson, S. (Eds) (2002) *Best Value in Construction*. Blackwell Science.

Kwayke, A. A. (1993) *Alternative Dispute Resolution (ADR) in Construction*. CIOB Construction Papers No. 21.

Latham, S. M. (1993) *Trust and Money: Interim Report of Latham Report*. Department of the Environment, Transport and Regions/HMSO.

Latham, S. M. (1994) *Constructing the Team: Joint Review of Procurement and Contractual Arrangements in the United Kingdom Construction Industry. Final Report* [known as the Latham Report]. HMSO.

Legislation.gov.uk (2015) 'Public Contracts Regulations 2015', www.legislation.gov.uk/uksi/2015/102/contents/made (accessed 27 September 2017).

Lock, D. (2007) *Project Management*, 9th edn. Gower.

McCabe, S. (2001) *Benchmarking in Construction*. Wiley-Blackwell.

Merlicek, R. (2017) 'Beyond bitcoin: what emerging blockchain technology means for your organisation', *The Weekend Australian*, www.theaustralian.com.au/business/technology/opinion/beyond-bitcoin-what-emerging-blockchain-technology-means-for-your-organisation/news-story/80a9b7ae83e446885271e7a72bfcbdea (accessed 6 October 2017).

Morledge, R., Smith, A. and Kashiwagi, D. (2006) *Building Procurement*. Wiley-Blackwell.

Murphy, M., Perera, R. S. and Heaney, S. G. (2011) 'A methodology for evaluating construction innovation constraints through project stakeholder competencies and FMEA', *Journal of*

Construction Innovation: Information, Process, Management, Vol. 11, Iss 4, pp. 416–40, doi: 10.1108/14714171111175891.
Murphy, M., Perera, R. S. and Heaney, S. G. (2015) 'Innovation management model: a tool for sustained implementation of product innovation into construction projects', *Construction Management and Economics*, Vol. 33, Iss 3, pp. 209–32, doi: 10.1080/01446193.2015.1031684.
National Audit Office (2001) *Modernising Construction*. HMSO.
National Building Specification (NBS) (2015) *National Construction Contracts and Law Survey*. NBS.
National Economic Development Corporation (NEDC) (1983) *Faster Building for Industry*. HMSO.
National Economic Development Corporation (NEDC) (1985) *Thinking About Building*. HMSO.
National Economic Development Corporation (NEDC) (1998) *Faster Building for Commerce*. HMSO.
National Joint Consultative Committee for Building (NJCC) (1959) *A Code of Procedure for Selective Tendering*. NJCC.
National Joint Consultative Committee for Building (NJCC) (1979) *Code of Procedure for Two Stage Selective Tendering*. RIBA Publications.
National Joint Consultative Committee for Building (NJCC) (1996) *Code of Procedure for Single Stage Selective Tendering*. RIBA Publications.
Ndekugri, I. and Rycroft, M. (2009) *The JCT 05 Standard Building Contract Law and Administration*. Elsevier Science.
Office for National Statistics (2016) *Construction Statistics Annual 2016*. Office of Public Sector Information, www.statistics.gov.uk (accessed 4 October 2017).
Office for National Statistics (2017) *Construction Statistics Annual 2015*, www.ons.gov.uk/business-industryandtrade/constructionindustry/datasets/constructionstatisticsannualtables (accessed 5 October 2017).
Office of Government and Commerce (2008) *The OGC Gateway Process: A Manager's Checklist*. Office of Government and Commerce [now handled by the Cabinet Office, www.cabinetoffice.gov.uk].
Panayides, P. M. and Venus Lun, Y. H. (2009) 'The impact of trust on innovativeness and supply chain performance', *International Journal of Production Economics*, Vol. 122, Iss 1, pp. 35–46, www.sciencedirect.com/science/article/pii/S092552730900142X (accessed 6 October 2017).
Perera, S. and Pearson, J. (2011), 'Alignment of professional, academic and industrial development needs for quantity surveyors: post recession dynamics', RICS Education Trust Funded Research Full Report, doi:10.13140/RG.2.2.26083.14886.
Perera, S. and Pearson, J. (2013), 'RICS professional competency mapping framework for programme appraisal and benchmarking', RICS Research Trust Funded Research Main Report, RICS, October 2013, doi:10.13140/2.1.2109.3444.
Perera, S. and Rodrigo, A. (2017), 'Capability maturity of construction e-business processes', Chapter 12. In: *Advances in Construction ICT and e-Business*, Perera, S., Ingirige, B., Ruikar, R. and Obonyo, E. (Eds), Routledge.
Perera, S. and Victoria, M. (2017) 'The role of carbon in sustainable development', Chapter 8. In: *Future Challenges for Sustainable Development Within the Built Environment*, Lombardi, P., Shen, G. Q. and Brandon, P. S. (Eds), John Wiley & Sons.
Perera, S., Hayles, C. S. and Kerlin, S. (2011) 'An analysis of value management in practice: the case of Northern Ireland's construction industry', *Financial Management of Property & Construction*, Vol. 16, Iss 2, pp 94–110.
Perera, S., Victoria, M. and Brand, S. (2017), 'Social media in construction: an exploratory case study', Chapter 16. In: *Advances in Construction ICT and e-Business*, Perera, S., Ingirige, B., Ruikar, R. and Obonyo, E. (Eds), Routledge.
Perera, S., Babatunde, S. O., Pearson, J. and Ekundayo, D. (2017) 'Professional competency-based analysis of continuing tensions between education and training in higher education', *Higher Education, Skills and Work-Based Learning*, Vol. 7, Iss 1, pp. 92–111, doi: 10.1108/HESWBL-04-2016-0022.

Perera, S., Ingirige, B., Ruikar, K. and Obonyo, E. (Eds) (2017a) *Advances in Construction ICT and e-Business*. Routledge.

Perera, S., Ingirige, B., Ruikar, R. and Obonyo, E. (Eds) (2017b) 'Introduction', Chapter 1. In: *Advances in Construction ICT and e-Business*, Perera, S., Ingirige, B., Ruikar, R. and Obonyo, E. (Eds), Routledge.

Perera, S., Babatunde, S. O., Pearson, J., Zhou, L. and Ekundayo, D. (2016) 'Competency mapping framework for regulating professionally oriented degree programmes in higher education', *Studies in Higher Education*, Vol. 41, Iss 12, pp. 2316–42, doi: 10.1080/03075079.2016.1143926.

Perera, S., Zhou, L., Udeaja, C., Victoria, M. and Chen, Q. (2016) *A Comparative Study of Construction Cost and Commercial Management Services in the UK and China*, RICS, www.rics.org/uk/knowledge/research/research-reports/construction-cost-and-commercial-management-services-in-the-uk-and-china-/ (accessed 11 October 2017).

Photis, M., Panayides, Y. H. and Venus L. (2009) 'The impact of trust on innovativeness and supply chain performance', *International Journal of Production Economics*, Vol. 122, Iss 1, pp. 35–46, doi: 10.1016/j.ijpe.2008.12.025.

Porter, M. (1980) *Competitive Strategy: Techniques for Analyzing Industries and Competitors*. John Wiley & Sons.

Raftery, J. (1993) *Risk Analysis in Project Management*. E & F N Spon.

Rahman, S., Odeyinka, H., Perera, S. and Bi, Y. (2012) 'Product-cost modelling approach for the development of a decision support system for optimal roofing material selection', *Expert Systems with Applications*, Vol. 39, Iss 8, pp. 6857–71, doi: 10.1016/j.eswa.2012.01.010.

Rethinking Construction (1998) *Rethinking Construction: The Report of the Construction Task Force*. HMSO.

Rethinking Construction (2000) *A Commitment to People, Our Biggest Asset*. Rethinking Construction.

Rethinking Construction (2002) *Rethinking Construction Achievements*. Rethinking Construction.

RIBA (2007) *The RIBA Plan of Work Stages*. Royal Institute of British Architects.

RIBA (2008) *RIBA Outline Plan of Work*. Royal Institute of British Architects.

RICS (1984) *A Study of Quantity Surveying Practice and Client Demand*. Royal Institution of Chartered Surveyors.

RICS (1990) *A Client's Guide to Successful Building: The Role of the Chartered Quantity Surveyor*. Royal Institution of Chartered Surveyors.

RICS (1996) *The Procurement Guide*. Royal Institution of Chartered Surveyors.

RICS (2007) *Definition of Prime Cost of Daywork Carried out Under a Building Contract*. Royal Institution of Chartered Surveyors.

RICS (2009) *RICS New Rules of Measurement: Order of Cost Estimating and Elemental Cost Planning*. Royal Institution of Chartered Surveyors.

RICS (2010) *Contracts in Use Survey*, 11th edn. Royal Institution of Chartered Surveyors.

RICS (2012) *RICS New Rules of Measurement 2: Detailed Measurement for Building Works*. Royal Institution of Chartered Surveyors.

RICS and Davis Langdon (2012) *Contracts in Use Survey*, 12th edn. Royal Institution of Chartered Surveyors Construction Faculty.

Sarmento, Joaquim Miranda (2010) 'Do public–private partnerships create value for money for the public sector? The Portuguese experience', *OECD Journal on Budgeting*, Vol. 10, Iss 1, pp 1–27, doi: 10.1787/budget-10-5km8xx3fgws5.

Simon, E. (1944) *The Placing and Management of Building Contracts* [known as The Simon Report]. HMSO.

Smith, N., Merna, T. and Jobling, P. (2006) *Managing Risk in Construction Projects*, 2nd edn. Wiley-Blackwell.

Strategic Forum for Construction (2002) *Accelerating Change*. Rethinking Construction.

Szabo, N. (1996) *Smart Contracts: Building Blocks for Digital Free Markets*. www.fon.hum.uva.nl/rob/Courses/InformationInSpeech/CDROM/Literature/LOTwinterschool2006/szabo.best.vwh.net/smart_contracts_2.html (accessed 30 October 2017).

The Tavistock Institute (2000) *The Handbook of Supply Chain Management*. The Tavistock Institute.

UNISDR (2009) *UNISDR Terminology on Disaster Risk Reduction*. United Nations.

University of Reading (1991) *Building Towards 2001*. University of Reading.

Walker, A. (2007) *Project Management in Construction*, 5th edn. Wiley-Blackwell.

Womack, J. and Jones, D. (2003) *Lean Thinking: Banish Waste and Create Wealth in Your Corporation*. Free Press.

Womack, J. P., Jones, D. T. and Roos, D. (2007) *The Machine That Changed the World*. Simon & Schuster.

Zhao, X., Feng, Y., Pienaar, J. and O'Brien, D. (2017) 'Modeling paths of risks associated with BIM implementation in architectural, engineering and construction projects', *Architectural Science Review*, http://dx.doi.org/10.1080/00038628.2017.1373628 (accessed 11 October 2017).

Index